EX FIGLINIS

EX FIGLINIS

The Network Dynamics of the Tiber Valley Brick Industry in the Hinterland of Rome

Shawn Graham

BAR International Series 1486
2006

Published in 2016 by
BAR Publishing, Oxford

BAR International Series 1486

EX FIGLINIS

ISBN 978 1 84171 738 8

Volume Editor: John W. Hedges

BAR Publishing is the trading name of British Archaeological Reports (Oxford) Ltd. British Archaeological Reports was first incorporated in 1974 to publish the BAR Series, International and British. In 1992 Hadrian Books Ltd became part of the BAR group. This volume was originally published by John and Erica Hedges Ltd. in conjunction with British Archaeological Reports (Oxford) Ltd / Hadrian Books Ltd, the Series principal publisher, in 2006. This present volume is published by BAR Publishing, 2016.

Printed in England

BAR
PUBLISHING

BAR titles are available from:

BAR Publishing
122 Banbury Rd, Oxford, OX2 7BP, UK
EMAIL info@barpublishing.com
PHONE +44 (0)1865 310431
FAX +44 (0)1865 316916
www.barpublishing.com

Abstract

The growth of the city of Rome was dependent on its ability to exploit successfully the human and natural resources of its hinterland. Although this hinterland eventually extended to incorporate the entire Mediterranean seaboard, the resources of the Tiber valley originally nourished the city and continued to do so despite the growth in imports from elsewhere in the Roman world. One of the most important industries to exploit the valley was the building industry, relying on (amongst other resources) extensive clay deposits to provide bricks. The study examines the way the Tiber valley (the immediate hinterland of Rome) functioned in terms of its economic and social geography, as evidenced by the organisation and dynamics of the brick industry. It concentrates on assemblages of stamped bricks from a number of sites in the Valley. Through an archaeometric approach to the fabrics of these bricks, coupled with a social networks analysis approach to the patterning of social and physical connections represented by the bricks and their associated stamps, the study arrives at an understanding of the social and economic relationships which characterised the city-hinterland relationship. Patterns of land exploitation are studied by locating the clay sources for bricks carrying the stamps of various *figlinae* and *praedia*. These different patterns suggest particular methods of land-tenure, which in turn allows the exploration of the sources of social power. The complex dynamics of how these sources of power change over time point to the conscious manipulation of social and physical networks in the industry. The importance of landed wealth for political and social power in Rome is a commonplace; the relationships which can be discerned in brick therefore mirror the political and social life of not only the élite, but also of their clients and tenants as well.

Table of Contents

List of Tables

List of Figures

To the memory of

Rena Miller Graham

who gave so much to everyone

Acknowledgements

There are many people to whom I owe a debt of gratitude and thanks, for technical help, for inspiration, and for moral support, not least of whom are Janet DeLaine, who had faith in me from the beginning, Mike Fulford who asked the tough questions, Andrew Parker, who had unwavering enthusiasm and guidance in the ways of geology, and Ray Laurence, whose door was always open. Mike Andrews and Franz Street performed the actual X-ray analysis, but left the interpretation up to me. Permission to study the SES collection and ecouragement to do so came from the BSR's director, Andrew Wallace-Hadrill. Helen Patterson and dott.ssa Helga di Giuseppe provided invaluable support in working with the SES collection, laughs, and days out in the Ducato. Paul Roberts, Kris Strutt, Helen Goodchild, Rob Witcher, Stephen Kay, Letizia Ceccarelli, Gord Graham, the Bradleys, the Crew (you know who you are) and Jonathan Murphy all proved to be excellent sounding boards (and sometimes provided much appreciated logistical support!) Dott. Giorgio Filippi and Dott. Enrico Stanco were unfailingly supportive and generously shared their own data with me. Barbara Haughton opened her private collection of materials related to the Ottawa Valley timber industry to me. Mark Wakefield's enthusiasm for the problems of intersite relations turned a model using a system of differential equations into a functioning computer programme; T.E. Rihll and A.G. Wilson gave permission for that model to be used. C. Bruun and A. Barabási responded helpfully to my sometimes naïve inquiries. Garry and Norma Graham helped me through the rough bits. I would also like to thank Paola Moscati and *Archeologia e Calcolatori* for permission to reproduce figures 2.6 and 2.7.

This study has its genesis in my PhD thesis, in that it represents a modified, edited, and sometimes expanded version of that thesis. '*Ex Figlinis*: The Complex Dynamics of the Roman Brick Industry in the Tiber Valley During the 1st to 3rd Centuries AD' was completed at the University of Reading in 2002. The thesis was made possible through the generous financial support provided by the Social Sciences and Humanities Research Council of Canada and the Overseas Research Studentship Scheme administered by the Committee of Vice-Chancellors and Principals of the Universities of the UK. That I had the opportunity to ready this study for publication is thanks to the generous support of the Canada Research Chair in Roman Archaeology Dr. Lea Stirling and the Department of Classics at the University of Manitoba. Any mistakes are of course my own.

And Tamara, for your love and support, and for sharing my everyday: thank you.

Shawn Graham, RPA, MIFA
Postdoctoral Research Fellow in Roman Archaeology
Department of Classics
University of Manitoba

...Although I have always been deeply interested in the works of the Roman poets and the Roman historians, I have never been able to summon up much enthusiasm for Roman architecture. In fact, the contemplation of a Roman brick seems to leave me cold – quite cold. So I would dearly like to know why it is that you find yourself so enthusiastic...

-Chief Inspector Morse in *The Jewel That Was Ours*, Colin Dexter, 1991

Chapter 1: Introduction

Networks are present everywhere. All we need is an eye for them.

(Barabási, 2002: 7)

1.1 Introduction

This study is an investigation into the organisation of the brick industry in Central Italy from archaeometric and social networks viewpoints. The study has a number of objectives, but chief of these is to locate the centres of production and to draw out the social and physical networks which enabled the exploitation of clay for brick. However, this focus on one particular industry has a greater purpose. The aim of the study is to understand and explain the dynamics of the socio-economic relationships between the city of Rome and its hinterland in the Tiber Valley over the first three centuries AD, using brick and tile as an indicator of these relationships. Brick and tile are suitable for this purpose because the archaeometric relationships in the fabrics can be used to interpret the changing patterns of clay exploitation. Also, the stamps record the names of the landed estates where the bricks were made. From the appearance of named individuals in the stamps, something of the social relationships between manufacturers and landlords, and their social equals, can be deduced. The *shape* of the networks of relationships evident in the fabric of brick and tile itself or in the stamps (networks of manufacturers using the same clay sources; networks of family relationships; networks of patrons and clients) have ramifications for our understanding of Roman society as a whole.

1.2 Aims

At a very basic level, the growth of the City of Rome was dependent on Rome's ability to exploit successfully the resources of its immediate hinterland, the Tiber Valley. The principal aim of the study is to explore and explain the dynamics of this city – hinterland relationship. One of the most important industries to utilise the valley and so encapsulating that relationship was the building industry, relying on (amongst other resources) extensive clay deposits to provide bricks. Consider a city of over a million people, a city whose fabric was made of brick-faced concrete. Calculate what it would take in manpower and in natural resources, to make, to fire, and to transport those bricks to the building sites. How does it work? Is the industry vertically integrated, with the landlord dictating every step in the process from manufacture to consumption? Is it under the bureaucratic control of the Urban Prefect perhaps? What sort of impact does it have on the human landscape? How do the answers to these feed back into the functioning of this industry? These are the sorts of dynamics that the study seeks to explore.

The current interpretation of stamped bricks leaves certain questions in my mind (described below, 2.2). I spend a certain amount of effort in the first instance carefully examining that interpretation, to arrive at a more satisfactory picture of the industry. I am not concerned *primarily* with disproving or proving that interpretation; rather my aim in this study is to *transcend* that discussion altogether. I take the brick industry, since it is so obviously connected with the landed holdings of the élite (and therefore the sources of their political and social legitimacy), as an indicator for larger patterns in Roman society. My study aims to examine the way, as evidenced from brick and tile, individual interactions in their aggregate gave rise to the social power enjoyed by the biggest players in this industry, including the Emperor himself. How this power is exercised characterises the socio-economic relationship between the city and the hinterland. This power does not exist, however, in any one interaction, nor will it be found in any one piece of archaeological evidence: "there is more going on in the dynamics of the system than simply aggregating little pieces into larger units" (Torrens 2000: 16). Social power is an emergent property, that is, a property evident only of the whole but not of the parts of the system represented by the ancient brick and tile industry.

Drinkwater (2001: 306) writes, in discussing the differences between ancient and medieval manufacturers and traders, and the failure of ancient traders to gain political power:

> "Modern historians can characterize Edward III as 'woolmonger extraordinary' and not offend the ear; but Antoninus Pius 'merchant emperor' rings wholly false [....] this difference came about because Roman emperors had a cultural blindspot. They remained unaware of the potential, and hence the advantage to themselves, of the woollen or any other industry".

In its strong version, my argument in this study is explicitly to the contrary: that the major players were indeed alert to the advantage industry could present, and took active steps to ensure their predominance.

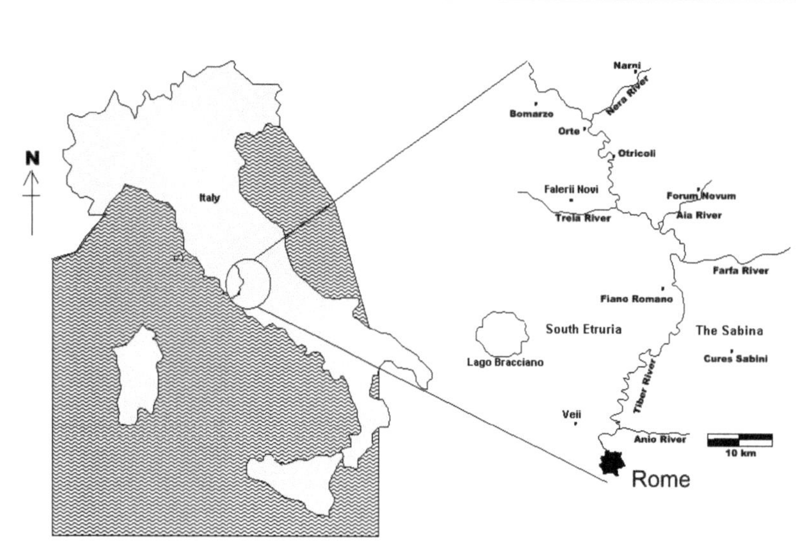

Figure 1.1 Study area

1.3 Objectives

To achieve these aims, I have to answer a series of questions of increasing complexity. The objectives of the study are to answer the following:

Questions of Production:

1. Where are the locations of production, or at least, the sources of clay? (cf. 3.2, 3.6, 4.2.1)

2. What is implied by this pattern of land exploitation? (cf. 4.2.2)

3. What are the modes of production employed? (cf. 4.3)

4. What is the economic value of brick? (cf. 4.5)

Answering these basic questions, through the archaeometric study of brick, provides the foundations for the next group of questions. For if I know where the sources of exploited clay were, and the modes of production employed, then I should be able to answer:

Questions of Meaning:

5. What is a brick stamp? (cf. 4.4)

6. What is the purpose of a brick stamp? (cf. 5.3)

7. What is the relationship between stamped and unstamped bricks? (cf. 5.3, 5.4)

There are many possibile explanations regarding the practice of brick stamping, but if I know how and where the bricks were produced, the number of possibilities is reduced. Having settled on the few most likely possibilities (one of which is related to distribution), I can begin to place the industry in wider Roman society.

Questions about the place of this industry in society:

8. How does the brick distribution infrastructure work? (cf. 5.3.2)

9. Does the *dominus* (landlord) have an active role in the industry? (cf. 4.5, 5.2)

10. Does the state have a role in this industry? (cf. 5.3.4, 6.4)

11. What is the impact of the industry on the human landscape? (cf. 6.3)

The answers to these questions contribute towards the principal aim of exploring the socio-economic relationships between the city of Rome and its hinterland in the Tiber valley. But these questions are still pitched at too low a level. To achieve my aim, I have to consider ways in which the brick industry can serve as an indicator for wider social and economic patterns.

Questions about the Implications

12. What is the relationship between the city and its hinterland, and why is it like this? (cf. 6.3, cf. 6.4)

13. What are the implications of these results for Roman social and economic history? (cf. 6.2, 6.3, 6.4, 6.5, 6.6)

In answering these questions, I consider the pattern of social relationships between individuals uncovered in the brick industry to be indicative of the deeper structure of Roman society. This structure can be mapped over time and the kinds of higher-scale phenomena which emerge, from such a structure, will characterise for us the city-hinterland dynamic.

1.4 Scope of the Study

1.4.1 The Tiber Valley Project: Rome and its Hinterland

This study contributes to the British School at Rome (BSR) Tiber Valley Project, and so takes as its study area the same geographic region as the larger project (Figure 1.1, the study area). The Tiber Valley has long been studied by archaeologists and historians, but since the Second World War there has been an increasing emphasis on the significance of Rome and its fluctuating influence on the Valley. J.B. Ward-Perkins' realization that land reform was in danger of completely obliterating the archaeological record (Patterson and Millett 1998: 7-8) was the impetus for the British School at Rome's celebrated South Etruria Survey (SES), the first systematic field survey in Italy. Those fears have been borne out, leaving the collection of material from the Survey an irreplaceable resource. In the twenty-odd years since, refinements in pottery chronologies and other materials studies have made the only published synthesis of the SES, Potter's *The Changing Face of South Etruria*, now somewhat out of date (1979). The Tiber Valley Project was initiated at the BSR in 1997 to redress the balance. A major component of the project is the re-evaluation of the South Etruria corpus, and the initiation of new projects to 'fill in the gaps', especially on the other side of the river in the Sabina. The Project's ultimate aim is to

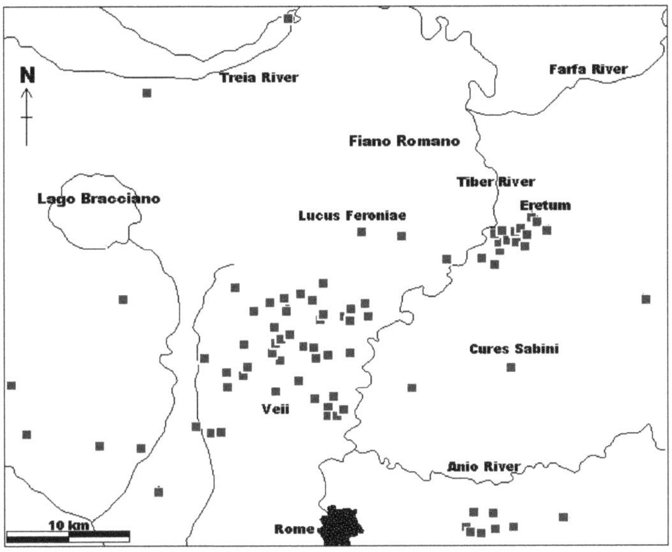

Figure 1.2 Sites where the South Etruria Survey found stamped bricks

Figure 1.3 SE 26 with stamp *CIL* XV.1 189

write a new materials-based history of the Tiber Valley from pre-history to the early medieval period. The Project is set up in such a way as to act as an 'umbrella' for different individuals and groups to pursue different aspects of the city-countryside relationship. Each sub-project's findings are collated in a central Geographic Information System (GIS), creating a vast spatially-referenced database. At the conclusion of the Project, this database (the core of which is the material collected by the SES) will be made available to the wider academic community (Patterson and Millett 1998: 17).

1.4.2 The SES Collection of Stamped Brick

The SES collection of stamped bricks (nearly two hundred examples) covers a temporal span from the first to fourth centuries AD and a geographical area of the middle Tiber Valley from Veii in the west to Cures Sabini in the east, the Treia River in the North and Rome in the South (Figure 1.2). However, this was only roughly half of the SES's study area (Patterson and Millett 1998: 3, 4; Figures 1.3 and 1.4, photographs of examples from the collection). Only a portion of these were ever published, and only two of the stamp types ever received any in-depth analysis. Peña (1987:312-319) analysed the fabric of one of these stamped bricks, which until the present study was the only scientific analysis carried out on any portion of the SES collection. This study presents the first systematic study of the SES collection of stamped bricks.

The SES as a whole was governed by various factors of visibility and accessibility, and by the fact that approaches to systematic field survey were in their infancy (Witcher 2000). That stamps were only recovered from such a limited area suggests that the investigators, over the twenty years of the survey, perhaps lost interest in collecting stamps after a while: the *Forma Italiae* series and the work of Filippi and Stanco (unpublished) demonstrate a much wider distribution of nearly four hundred more stamps. Another possibility is that the individual most concerned with brick stamps during the SES, Anne Kahane (Kahane *et al.* 1968; Kahane 1972, Kahane and Ward-Perkins 1977), stopped working in the field in the early 1980s. Moreover, strange outliers from this distribution (some notably on the banks of the Anio and not in the South Etruria Survey area at all) are present in the collection, which suggests that various British School at Rome outings over the years have haphazardly taken examples back and placed them with the others in the South Etruria stores. This is a practice which continues: recent field survey concerned with the pre- and proto-historic periods in the Galatina area (south west of Forum Novum) in the Sabina recovered the occasional Roman brick stamp, and these were added to the SES collection. Brick stamps present in the collection but not recorded in the SES record cards might be ones recovered in a non-SES expedition; other such stamps might be those for which the provenance (in the form of eastings and northings) is not marked on the brick. Finally, the lack of anepigraphic stamps in the collection, suggests that the participants in the SES did not recognize or recover all the stamps when visiting sites.

Chronologically, the entire collection of stamped bricks held at the British School at Rome covers the entire range of known dated examples[1], from the middle of the 1st century AD to post-Diocletian. All the stamps, where they can be read with certainty, have been identified in other collections of stamped bricks. There is, in fact, very little to distinguish the collection on epigraphic grounds alone, other than the periodisation of the stamped bricks: 51% of stamps which can be

[1] There is, as it happens, one fascist-era stamped brick dating to Year 12 of Mussolini's reign conserved with the collection.

Figure 1.4 SE 33 with stamp *CIL* XV.1 2194

dated (either on the basis of consular date, or stamp shape) fall within the first century.

Julio-Claudian ~ 80 examples
Flavian ~ 14 examples
First half of the 2nd century (to end of reign of Hadrian) ~ 27 examples
Second half of the 2nd century (to end of reign of Commodus) ~ 15 examples
Severans ~ 17 examples
Diocletianic or later ~ 5 examples
Impossible to date ~ 22 examples

Of the 177 stamps in the collection, 15 are of unknown provenance. The remainder come from 102 sites. Of those 102 sites, two or more stamps were recovered from only 22 sites (in total, 75 stamped bricks). This study will focus on brick from these 22 rural sites.

There is a problem in knowing what one stamp in isolation actually means. Does it represent one shipment of bricks, or does its presence on a site represent a chance inclusion with shipments or assemblages of building materials to which it bears no relation? Concentrating on sites from which at least two stamps were recovered reduces the likelihood that a stamped brick has been deposited by chance. Figure 1.5 indicates how the relationships between sites in the Tiber valley can be approached using brick assemblages. Information in the stamps can connect otherwise disparate sites at different levels (access to work by the same manufacturers, or from the same landlord).

1.4.3 Unstamped Bricks

Stamped bricks are comparatively rare finds, while unstamped brick is virtually ubiquitous at Roman sites throughout the Tiber valley. For this reason some unstamped material was included in this study, to connect the historical framework for stamped bricks to the archaeometry of unstamped brick. Unstamped material was collected from Falerii Novi and Forum Novum (see Figure 1.1 for location). These two towns

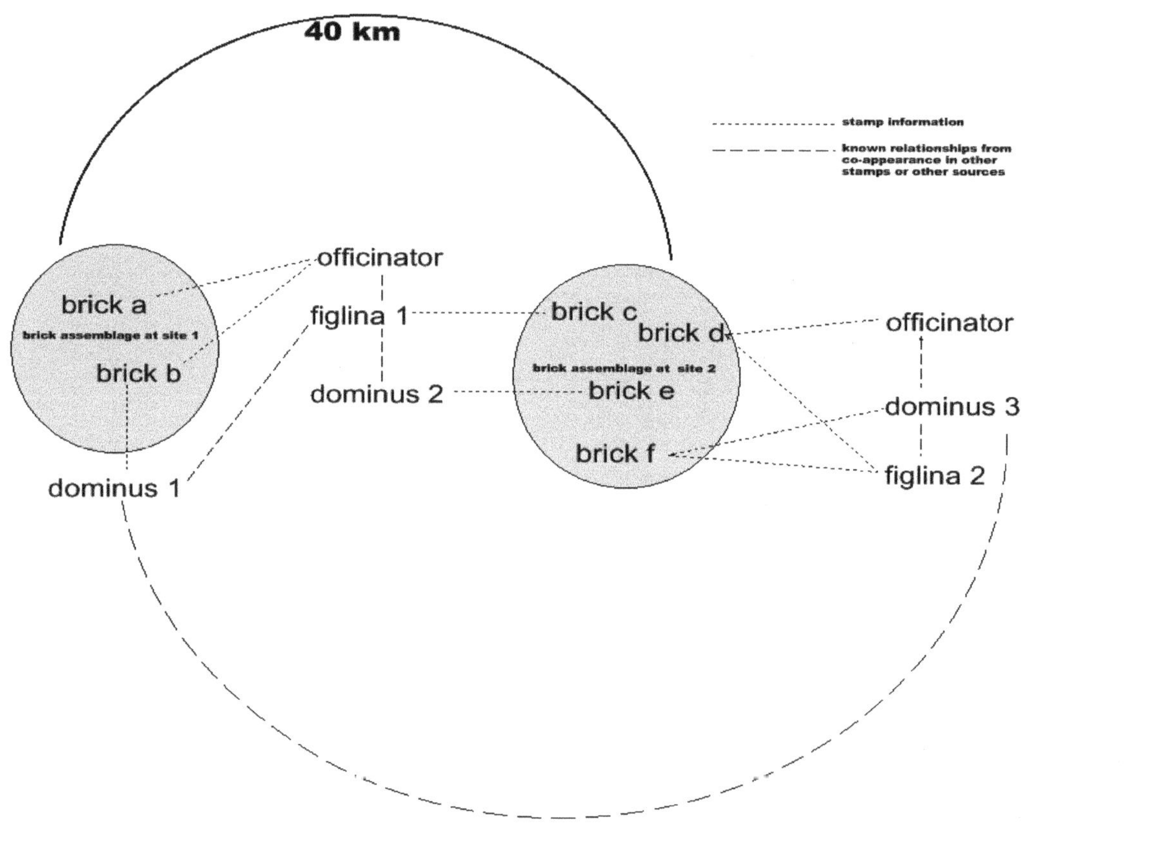

Figure 1.5 Intersite relationships as evidenced in brick assemblages. The diagram depicts relationships between two hypothetical sites, located 40 km distant from each other; the relationships are based on co-appearance of names in different stamps and so on.

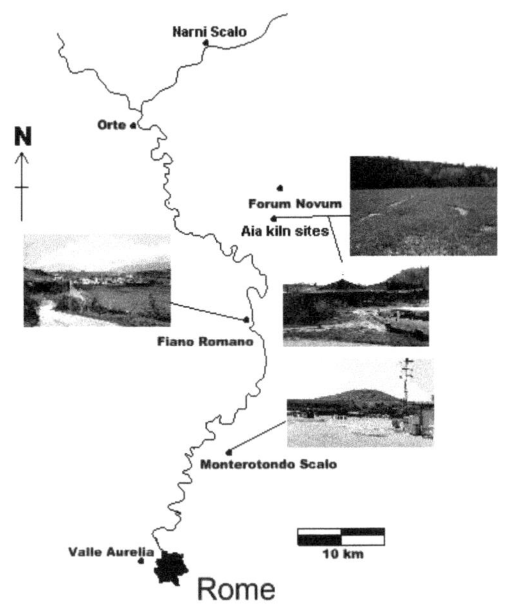

Figure 1.6 Locations of modern brick yards in the Tiber Valley; permission was not granted to take photographs at the Narni, Orte, and Velle Aurelia yards.

sit on opposite sides of the Tiber, each at roughly the same distance from Rome. A monumental villa at Forum Novum has been the subject of a recent excavation by the BSR (Gaffney *et al.* 2001). No stamped bricks have yet been discovered at this site. Falerii Novi is a large walled town, created when the Romans forcibly removed the Faliscan occupants of Falerii Veteres (modern Cività Castellana) to the less defensible plain in the second century BC. It could not be more different from the *municipium* of Forum Novum. There are no stamped bricks from Falerii Novi in the SES collection, and so a quantity of unstamped material was collected from the centre of the town, near the 19th century Forum excavations.

1.4.4 Modern Bricks

Brick manufacturing continues today in the Tiber Valley, and now-defunct 19th century brick yards still can be found across the landscape. It is difficult to conceive that the major clay sources exploited in antiquity have been exhausted, given the geology of the Tiber Valley (3.1.1). Samples of both fired brick and raw clay were collected from the major modern brickyards, and from two 19th century yards. The distribution of these yards is given in Figure 1.6. They range from the Valle Aurelia behind the Vatican in the South to Narni Scalo on the Nera River in the North, an area known to have had ancient brick production (the *figlinae Narnienses*). The analysis of the fabric of these bricks provides the 'pegs' on which to hang the geographical distribution of the cluster analysis of the ancient brick fabrics (3.5).

1.5 The Means

1.5.1 Archaeometric techniques

A major problem with studies of the brick industry is that the presence of *figlinae* names in the stamps has led to a view that the locations of production are more or less certain (based mostly on Huotari's unpublished toponomastic study of *figlinae* and early medieval place names, cited in Steinby 1978: 1508-9). In this study I undertook to locate the likely clay sources archaeometrically, and to question the interpretation of the brick industry based on a presumed knowledge of the production locations. The principal archaeometric method employed for this study is X-Ray spectrometry (combined with visual analysis of the fabrics). This technique is relatively cost-effective and offers a good degree of precision, accuracy, and speed of analysis. The Vatican Museums, which house the principal collection of stamped bricks in Rome, has lately embarked on a programme of X-Ray analysis of their material (Albertazzi *et al.* 1994; Baldi, Bertinetti and Camilli 1999), and so the choice of method was also influenced by a desire to produce results which would be comparable to the Vatican Museums' study. The data created through the archaeometry are analysed via discriminant and cluster analyses. The resultant groupings are compared with the analyses of the modern brick and raw clays in order to deduce the likely geographic sources for the ancient materials.

This drawing out of the interrelationships between bricks as indicated by their sharing of the same clay sources, or the usage of the same stamp, or their consumption at the same site provides an analytical framework for interpreting the functioning of the industry. Figure 1.7 contains the same information as in Figure 1.5, but in this diagram the archaeometric relationships have been discovered. The relationships between manufacturers, landlords, and workshops are considerably more complicated than was the case previously. Not only can sites be connected on the basis of stamps, but also because they have access to the same clay source. Different manufacturers can be connected to each other because they exploit the same clay body (...and so on). The picture I am developing depicts both the patterns of consumption and production.

1.5.2 Statistical Analysis

Much can be accomplished simply by considering the published collections of stamped brick from a statistical rather than an impressionistic point of view. Accordingly, I examined the main collection of stamped brick -the *Corpus Inscriptionem Latinarum* XV.1 - with regard to the numbers of examples per stamp type, shapes, *signa*, and the occurrences of consular dates. Even very basic statistical explorations of this data answers some of the problems I identify with the traditional interpretation (2.1, 5.3). I combine this information with the distribution patterns of brick. I adapt a gravity settlement model, developed by Rihll

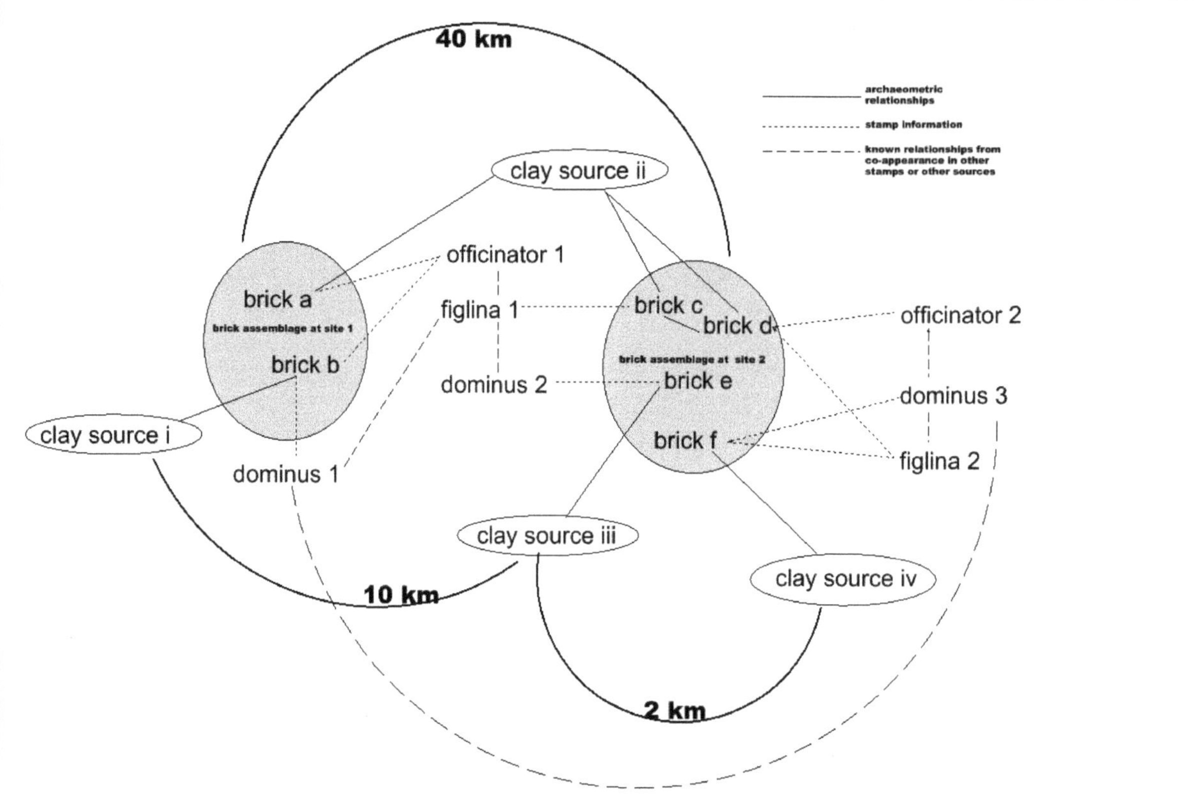

Figure 1.7 Intersite relationships as evidenced in brick assemblages and completed by archaeometric data. The diagram depicts relationships between two hypothetical sites, located 40 km distant from each other, and four clay sources. The relationships are based on co-appearance of names in different stamps, use of the same clay sources, and so on.

and Wilson (1991) for understanding the role of geography for city formation in Archaic Greece, so that I can study the geographical interrelationships between sites using stamped brick. This permits an explanation of puzzling aspects of the marketing and transportation of brick in the valley.

1.5.3 Social Networks, Evolving Networks, Dynamic Networks

Finally, I plot out the information regarding *domini*, *officinatores* and other players in the trade and analyse the information using social networks analysis. This is an approach familiar to sociology but hitherto never employed for Roman prosopography. The shapes of the social networks (the pattern of interconnections between individuals) are studied to determine whether they resemble 'small-worlds', a very particular configuration which allows spontaneous self-organisation in networks (which grow and evolve) to emerge from the dynamics of the network. The archaeometric data is also analysed from a networks analysis point of view, to understand the dynamics of the industry at the level of the manufacturers themselves, unclouded by expectations built on the information in the stamps.

It is in the interactions between the two levels - manufacturers and landlords- that I am able to examine the overall dynamics of this rural industry in the hinterland of Rome. Figure 1.8 and Figure 1.9 recast the information from the diagram in Figure 1.7 into pure network terms. In Figure 1.8 the brick production network (where the ties are between bricks which share the same source clays or are found at the same sites), is more complicated in terms of the number and strength of ties than the social network in Figure 1.9 (which records ties between individuals appearing in the same stamp or known to have been related from other sources). The task at this point would be to explain the differences, similarities, and interactions between the two. Because brick stamps can be dated, in understanding the interactions within and between the different networks we necessarily move to a historic mode of discussion: explaining the changing dynamics of the networks allows us to write history from archaeology.

1.6 Structure of the Study

The historic, economic, geographic, social, and academic issues considered in order to reach an understanding of the complex dynamics of the brick and tile industry are discussed in Chapter 2 'The Brick Industy, the Tiber, and the Hinterland'. Here the current knowledge of the industry is set out, and gaps and inconsistencies are identified. Because the brick industry in relation to the Tiber may be usefully compared to any extractive industry which uses a river

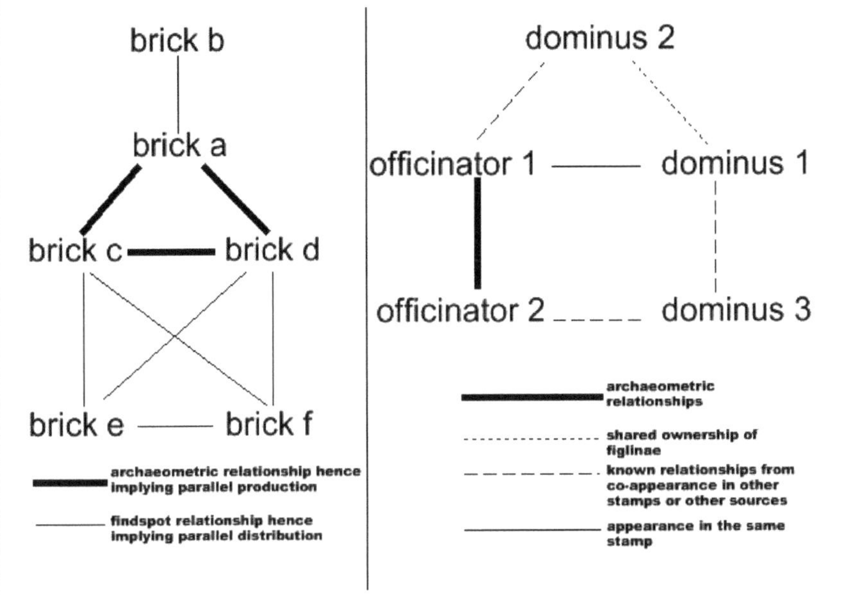

Figure 1.8 (left) and 1.9 (right). Archaeometric relationships between brick assemblages at different sites represented in pure network terms; social relationships in stamps from brick assemblages at different sites, represented in pure network terms.

as its principal means of transportation, the chapter illustrates by reference to an ethnographic parallel with the 19th century timber trade in the Ottawa Valley in Canada how the dynamics of a riverine industry might be interpreted (Graham 2005a). The other main issue discussed in this chapter concerns the relationship between a city and its hinterland. It introduces the concept of different levels of social complexity, and how individuals are crucial to formulating the interaction between different levels.

In Chapter 3 'Sourcing the Brick Industry' the archaeometric methodologies employed are described and the results presented. The archaeometric results are put into the context of earlier scientific studies of ceramics in the area, as well as the current Vatican Museums' project. I suggest a way of visualizing the results of the archaeometry which makes it easier to understand the geographical and geophysical relationships between individual manufacturers.

Chapter 4, 'An Industry in the Hinterland' and Chapter 5, 'Bricks to Rome, Bricks to the Valley' draw out the implications of the results of the archaeometry. The single biggest implication concerns the patterning of land exploitation and tenure. A new interpretation of the brick industry is offered in these two chapters, combining the statistical results of the study of the *Corpus* with the distribution patterns and archaeometric results.

Chapter 6, 'Dynamic Social Networks in the Brick Industry' looks at the dynamics within and between the ranks of the manufacturers and landlords, implied naturally in brick stamps. The shape of the networks of relationships between individuals (whether by blood, patron and client, sharer of the same clay resource and so on) allows me to study where the sources of social power lay. An understanding of these dynamics indicates how the manufacturers and landlords may have interacted, providing an alternative understanding of rural-urban relationships.

1.7 Summary

Studies of the brick industry in central Italy (the Tiber Valley) have focussed almost exclusively on the epigraphic evidence of brick stamps. Consequently, in this region, brick is useful only in the study of prosopography and chronology. Given the ubiquity of brick and tile in the architecture of the city of Rome, there would seem to be an opportunity missed for understanding Roman society and culture. Each stamped and unstamped brick represents the interactions of individuals. It is from these interactions that larger-scale phenomena arise (a simple manifestation of a larger-scale phenomenon for instance may be the homogeneity of stamp types arising without government control). The process could well be indefinite, with new, macroscopic phenomena emerging from the interaction of lower-level phenomena, feeding back into those lower levels. Using the brick industry therefore as an indicator for wider activities in the hinterland, the necessary first step to understanding rural and urban interaction is to tease apart how individuals work within, and create, the brick industry.

Chapter 2: The Brick Industry, the Tiber, and the Hinterland

2.1 Introduction

In order to understand the city's exploitation of the hinterland during the first three centuries AD, as represented by the brick industry, there are three issues which must be examined. First, how is the brick industry organised? Second, how might the Tiber work as infrastructure? Third, how can the relationship between the hinterland and the city be defined and explored?

This chapter examines the brick industry as it is currently known, and points out where there are gaps, problems, and inconsistencies. Then, given the frequent emphasis put on river transportation for heavy materials in the ancient world, it explores what the use of a river for infrastructure actually entails, and how that might affect our understanding of the brick industry. The river, and of course the roads, firmly tie the hinterland to the city; the hinterland is the focus of the final section of this chapter. Models of how the hinterland works do not in effect, exist; the hinterland is usually understood in opposition to the city. The main failing of models of the ancient city (cf. Finley 1985; Morley 1996; Horden and Purcell 2000) and the processes of urbanization is that these models confuse who or what is doing the acting: the people who live in the city, or the city itself? The city is taken as the basic unit of action, although what a city actually is, is not defined.

This chapter argues instead that we should concentrate on the interactions of individual people. Using a model from modern geographical thought, the chapter argues for urbanization as a process which takes place over physical and social networks, where the 'city' as such emerges from conscious decisions to intensify interactions at particular places, in what then becomes defined as the hinterland. What differentiates the 'city' from the 'town' or the 'village' or the 'villa' is in part the degree of intensification of these interactions. What this means for the brick industry is relatively straightforward. In brick, there is evidence for social and physical networks which overlie simultaneously the hinterland and the city. Sites which use brick, especially stamped brick, represent points of intensification in the hinterland. Understanding how the brick industry works, therefore, will allow us to examine the flows of material, capital, and social investment throughout the region from point to point (whether villa, village, town, or city).

2.2 The Brick Industry in the Tiber Valley

2.2.1 Nature of the Collections

One problem in understanding the brick industry that frequently goes undiscussed is the nature of the collections. Until fairly recently, brickstamps were only collected for their antiquarian rather than their archaeological value. It was not until the work of Bloch (1947) that their chronological usefulness became fully recognised. It was the names on them, for what they could say about the lives of the men and women mentioned on them, which were of particular interest. While it may be true, as Helen (1975: 13) argues, that all of the different types are now known, in even the recent past when stamps were found only the 'new' ones might be published. The ones already known often would not be recorded, or if they were, the numbers found might not be recorded either. In the search for a perfect representative of a stamp, or for new people named in stamps, information regarding frequencies and provenances was regarded as less important. Consequently it would seem that a study attempting any numeric analyses of brickstamps and stamping practice would be on shaky ground, open to charges that its dataset was unreliable.

If the collections are not representative of actual stamping practice, then there is no reason to suspect that different collections would have the same distribution patterns. Yet the four collections studied in this study, representing over 15,000 individual stamped bricks, *all have the same particular distribution pattern of number of examples per type.* They are power-law distributions. This means that there is something going on that is not an artefact of how the collections were created.

Power-laws in distributions of phenomena are significant. Essentially, a power-law distribution means that the stronger the event (however measured), the rarer it is seen (Buchanan 2002: 84-5). The occurrence of earthquakes for instance follows a power-law distribution. On any given day there are hundreds of almost undetectable tremors. Very rarely there will be a single earthquake capable of causing widespread damage. Human heights on the other hand follow a normal distribution. Most humans are about 5 ½ feet tall (ca. 1.7m), with a few taller and a few shorter individuals. If human height followed a power-law distribution instead, most people would be exceedingly short, but there would be one or two extremely tall people in the population as well (Barabási 2002: 67). *With stamped bricks, a power-law distribution means that we will only ever find a handful of examples for hundreds of types, but occasionally we should find hundreds of examples for only a handful of types.*

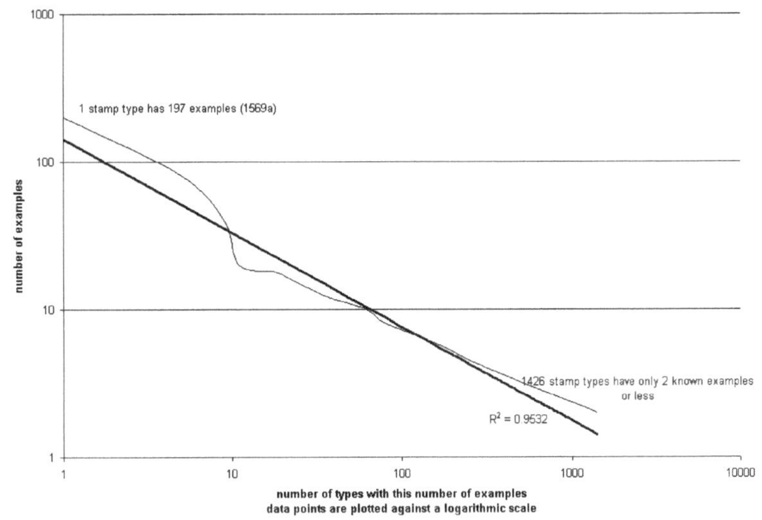

Figure 2.1 Power law distribution of stamp types in *CIL* XV.1 Solid line represents actual distribution; dotted line indicates ideal power law distribution

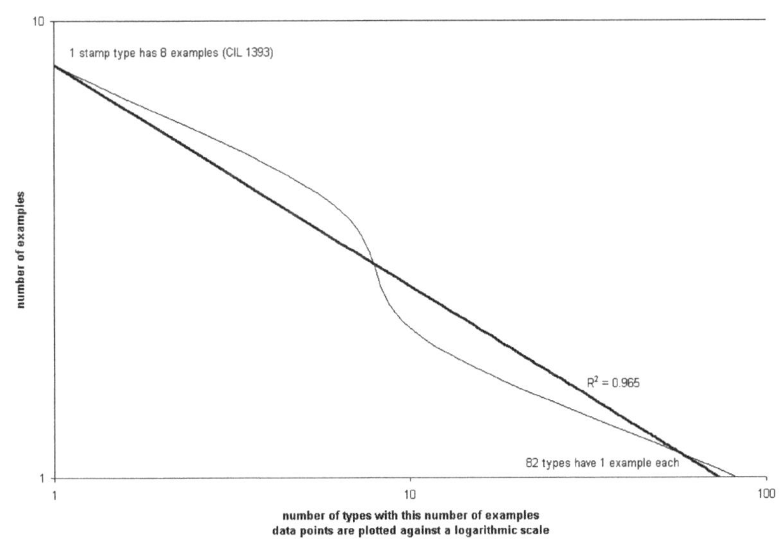

Figure 2.2 Power law distribution of stamp types the SES collection. Solid line

Yet the hit-and-miss nature of brick stamp collecting would lead one to suspect that there ought to be a reasonably normal distribution (i.e. bell-shaped curve) of stamp types to numbers of examples. It can be shown mathematically (Shennan 1997: 73) that if many independent factors come into play in the creation of a particular variable (as in individuals' heights) then the cumulative effect is to produce a normal bell-shaped distribution. Here the decision to record a stamped brick could have been influenced by any number of unrelated factors in the investigator's mind, e.g. Is this a new one? Is it interesting from an aesthetic point of view? Is it important to record these stamps carefully? *ad infinitum*. That is, over the hundreds of years that stamped bricks have been noted and collected, with all the biases and levels of professionalism involved, the number of examples of most stamp types recorded should vary little about the average. (It should also be noted that the number of examples of any particular type would have been influenced by whatever was standard in the productive outfit at the time of making the stamp. Again however we would expect a bell-shaped curve as a result of all the cumulative factors at play at the kiln site). The distribution should be normal.

In all of *CIL* XV.1 there are some 12,000 examples of stamps, notwithstanding the instances where it notes only that there were 'pluribus exemplis' or 'multis exemplis'. The distribution of these examples according to type does not, however, resemble a normal distribution (Figure 2.1) The BSR collection is much smaller (177 examples), and it does not resemble the normal distribution either (Figure 2.2). In DeLaine's database of stamps from Ostia (representing 3516 examples culled from the bibliographic sources on Ostia) the normal distribution is not found (Figure 2.3). In brickstamps recorded *in situ* in Ostia (DeLaine 2002), the distribution of individual *domini* named in stamp types by number of buildings their stamps are found in does not resemble a bell-shaped curve (Figure 2.4). This distribution it should be noted has more to do with how brick is used than how it is produced (to discuss production the stamping information should be noted at the level of *type* rather than *dominus*). Nevertheless, at these different orders of magnitude there are power-law distributions.

If investigator bias was significant, we would expect that these collections, created at different times over the last 150 years, should have bell-shaped curves (if a bit

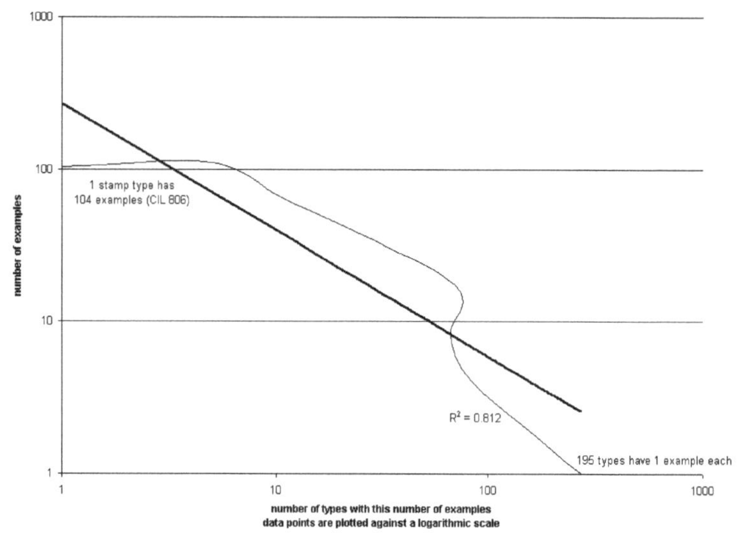

Figure 2.3 Power law distribution of stamp types found at Ostia. Solid line represents actual distribution; dotted line indicates ideal power law distribution

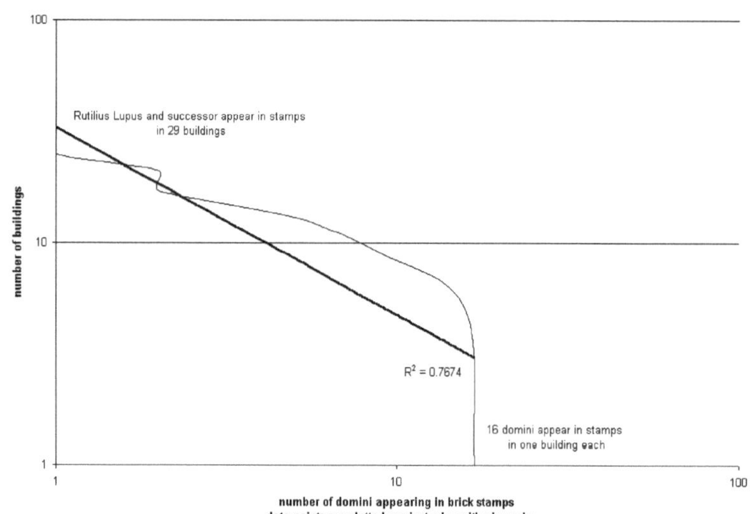

Figure 2.4 Power law distribution of different *domini* by numbers of buildings in which their stamped bricks are *in situ*. Solid line represents actual distribution; dotted line indicates ideal power law distribution

skewed one way or the other). That three are power-laws and one is very close to being a power-law is rather remarkable. That the *in situ* examples from Ostia also do not occur with a normal frequency is telling, for they at least should be free from any significant investigator bias and most closely reflect ancient realities (in fact, the distribution closely follows a power law as well although it is not a complete fit). Power laws are not an irrelevant oddity nor are they pervasive (Bentley and Maschner 2001: 42), and their presence in stamped brick therefore has important ramifications for understanding the industry (they are an indicator of self-organized criticality in complex systems or small worlds in social networks cf. Chapter 6). For the time being though it seems reasonably certain that the frequencies of different stamp types as recorded today have some basis in actual stamping practice, and therefore statistical analysis of the collections should provide real insight into the industry.

2.2.2 The Purpose of Stamps

In the mid second century, when the brick industry seems to be at its height, brick stamps carry several pieces of information. An example is *CIL* XV.1 861 (Figure 2.5, photograph of SE 42, *CIL* XV.1 861). The transcription of this stamp reads as follows:

EX FIG ASINIAE
QVADRATILLAE O D C NVN
NIDI FORTUNAT LVCIO
QVADRATO COS

Sometimes there is a small figurative device in the centre of the text, called by epigraphists the *signum*; in the case of *CIL* XV.1 861, there is a pine nut. A free translation of *CIL* XV.1 861 reads, '*opus doliare* [brick] of C. Nunnidius Fortunatus from the [clay district] owned by Asinia Quadratilla, in the year Lucius Quadratus was consul'. These elements (the name of the manufacturer (*officinator*), the name of the landowner (*dominus*), the brick yard or clay-district (*figlina*), the year, the *signum*, even the shape of the stamp itself) are present in various combinations throughout the corpus of stamp types. In the 1970s and 1980s members of the Finnish Institute in Rome studied the museum collection of the epigraphic stamps from Ostia, which comprised well over a thousand individual examples. Minute attention to grammatical forms allowed Helen (1975) to discern the broad outlines of the industry. Subsequent elaboration by Setälä (1977) and Steinby (1978, 1981, 1982) allows

Figure 2.5 Photograph of SE 47, carrying stamp *CIL* XV.1 861

the following generalization: *figlinae* or *praedia*, which literally mean 'potters' workshops' and 'landed estates' (*Cassell's Latin Dictionary*, Simpson 1991) are in this context both translated as 'brick yards'or 'clay-districts'. The expression used on stamps *ex figlinis illis* is taken to indicate the place of production, whereas *ex praedis huius, ex figlinis huius* is taken to mean the owner of the land (Helen 1975: 37, 82-83). The *figlinae* or *praedia* were owned by one or more landowners - *domini* - who either let or commissioned a second party - usually called *officinator* - to produce bricks on their land. Whether or not the *dominus* took an active role in the production of brick and tile is debated, based on finely-cut understandings of the grammar of the stamp and Roman contract law (Steinby 1982: 233-234; Aubert 1994: 232-233).

From what can be deduced epigraphically the stamps could serve a wide variety of purposes: distinguishing the output of different *officinatores* working side by side; for compensation or verification that the work has been carried out; to indicate the products of different *figlinae* belonging to one *dominus*; or different *domini* who used the same *tegularia* (brick warehouse; Steinby 1993: 144; Raybould 1999: 75). Aubert, however, doubts an accounting function for the stamps since stamps do not appear on all bricks (1994: 234). Five main possible reasons for stamping a product are surmised by Manacorda (1993: 38, 44 - 5): ownership; compliance with standards or expectations of quality and quantity; government control over producers and/or product; promotion; information for the organization of production or sale. Insofar as brick is concerned, the stamp as quality mark is rather unlikely, because as has been pointed out before, the stamp is applied before firing and, therefore, before the quality of the brick is known (Steinby 1993: 141). Aubert suggests that the stamp on the brick could serve as a sort of 'tracer', whereby flawed bricks could be tracked back to their source although it is unclear what sort of legal liability there would be (Aubert 1994: 234; see also the various papers in Bruun 2005, which came out too late for me to consult for this study).

Locatio-conductio : stamp as contract

Steinby (1993) argues that the stamps of the second century (which frequently contain all three pieces of information, *figlinae*, *dominus*, *officinator*, including a consular date) represent an abridged version of the *locatio-conductio operis* contract between the *officinator* and the *dominus* (Steinby 1993: 139-144). *Locatio-conductio* contracts were one of the usual means of letting out building contracts. In order to be valid, the parties to the contract had to agree on all terms and activities (including supplying the materials and/or plant to be used), as well as a fixed price (Anderson 1997: 69). In this context there are two varieties. *Locatio rei* refers to the plant and property used, while *locatio operis* refers to the finished product itself (Pucci 2001: 149). Steinby's idea is important because it means that the *dominus* was involved in production, and was something of an entrepreneur. If the stamp is an abridged *locatio operis* contract, then the *dominus* paid the *officinator* to make a certain amount of bricks. If on the other hand, the stamp refers to *locatio rei*, then the entrepreneur is the *officinator*, contracting with the landowner to use his land for the *officinator*'s own profit (Steinby 1993: 140-1; Pucci 2001: 149).

In the medieval period architects had a hand in controlling brick production through commissioning certain types of bricks and cutting their own moulds (Aubert 1994: 239). Buildings such as the *Horrea Epagathianae et Epaphroditiana* at Ostia with its moulded brick columns and capitals suggests that Roman contractors and architects might have had a similar interest in brick production. It is hard to imagine a landowner taking such an interest in such things as speciality brick, given the numerous social and political responsibilities, obligations, and other commercial schemes the *dominus* may have been involved in, which may suggest that *locatio rei* is the better interpretation. However, most commentators favour Steinby's *locatio operis* interpretation (cf. Manacorda 1993; Pucci 2001). Manacorda (1993: 46) believes Steinby to be right in identifying the stamps as abridged *locatio-conductio operis* contracts, but he is puzzled by the use of juridical language, if the stamps are only about the internal organization of production. He then asks why bother stamping at all when the stamp-form is not of these putative *locatio-conductio operis* contracts, and why are there stamps of just *dominus*, or just *officinator*?

This interpretation of stamps only describes what brick stamps are. It does not explain the 'why' of brick stamps. I find this interpretation unsatisfactory for a number of reasons. Firstly, it only actually applies to the 'fully developed' stamps of the second century. More importantly, the interpretation does not account for all of the elements in the stamps. The *signa*, the figurative devices often adorning the centre of the stamps, are thought to have a sort of heraldic function acting as a family badge, in keeping with the *locatio-conductio* interpretation. Steinby (1974: 22) relied on the appearance of *signa* to connect otherwise undifferentiated stamps to her chronological framework. If a *signum* and a named individual appeared in the same stamp, then the *signum* was taken to represent that person. Therefore stamps which carried only a *signum* could be connected to named individuals in more complete stamps on the grounds that two stamps with the same *signum* were made by the same person. This assumption has not been proven. A heraldic-device interpretation holds when only certain families can be shown to be using particular symbols. The prime example is that of Rutilius Lupus (*figlinae Brutianae*), whose stamp dating to AD 110 (*CIL* XV.1 21) carries the signum of the wolf. His is the first example of a stamp with a signum, and in this case there is clearly a connection.

Contrary evidence abounds, however. For example, in *CIL* VIII -North Africa- there are recorded 123 examples of Tiber Valley brick stamps. The pine nut device appearing in the *Fortunatus* stamp discussed above is found on 30% of the stamps with *signa*. If *signa* are heraldic devices, then we must conclude that all of the individuals in these pine-nut stamps found at Carthage are related to *Fortunatus*. However, we know that there was no familial relation between *Fortunatus*' patron, *Q. Asinius Marcellus* and the *gens Domitii*, yet the device appears on stamps connected to both families. As for Rutilius Lupus, of the 25 stamp types in *CIL* XV.1 which use a canine *signum*, nearly one third are from *figlinae* not connected with his *Brutianae*. It may be objected that the pine-nut is too common a symbol, and so that particular illustration is not significant. Nevertheless, regardless of the relative 'commonness' of a symbol, the same phenomenon obtains. *Locatio-conductio operis* requires that the participants be clearly named; the use of the same *signum* by a multitude of people perhaps demonstrates that the *signum* itself might not be a proxy for one of the named people in the text.

That is not to say, however, that the internal arrangements of brick production could not be arranged through *locatio-conductio operis* agreements; it was after all fairly normal practice in any number of undertakings (Anderson 1997: 69). However, the text of brick stamps does not include all the necessary elements found in *locatio-conductio* contracts. Recall the translation of *CIL* XV.1 861, with which we opened this section. The text of this stamp does not contain any of the elements of a legal contract, other than a pair of names (by which criterion a tomb-stone displaying the name of the deceased and the name of the person who erected it could be interpreted as a contract). A legal contract would include a mention of payment, but that is never the case in brick stamps. A legal contract would contain clauses to deal with the potential failure of one or the other parties to uphold their part of the bargain (Anderson 1997: 69). Besides which, there is no reason why a *locatio-conductio* contract, if indeed stamped on a brick, should conform to a limited range of very particular *shapes*. The importance of a contract surely must be in the *content*, not the *format*. Even if brick stamps were contracts, the question becomes, why record this information on a brick? Traditionally, it is regarded that such information is necessary for the successful running of the kiln, to sort out who has made how much (Steinby 1993: 140). If the information on the brick helps to differentiate individual production within the kiln, then the stamps might be *locatio-conductio operis* contracts; but if they are *locatio-conductio operis* contracts, then they must differentiate the individual production in the kiln. This is circular logic. At La Graufesenque, France, differentiating information was recorded on kiln dockets (lists scratched onto potsherds), not onto the fired vessels themselves (Parca 2001: 68). One wonders if a similar situation could not have been used in the brick kilns. It certainly would have been easier than sitting down and whittling a new stamp die from a block of wood each time kiln arrangements changed.

The idea that the stamps reflect a *locatio-conductio operis* contract perhaps has become conflated with the idea that the stamps are *locatio-conductio operis* contracts. Yet the stamps themselves are not *locatio-conductio operis* contracts because they are missing the necessary, and probably most important part- the *merces*, or payment (Aubert 1994: 232). Even if for a moment we accepted that the stamps do reflect *locatio-conductio operis* contracts, then it must be asked in addition to Manacorda's questions regarding this issue, why were anepigraphic stamps used? Anepigraphic stamps do not have any written text in them, and usually take the form of circles made from impressing the ends of staffs. More ornate stamps are made by impressing belt-buckles or other bits of metal-work (Broise 2000: 120-1). If these 'illiterate' stamps are used because they serve a different purpose from the epigraphic stamps, then the multi-purpose nature of stamps is admitted and there is no one interpretation. Conversely, if the anepigraphic stamps do serve the same purposes as the more verbose stamps, then the literate stamps are not *locatio-conductio operis* contracts, and their similarity (such as it is) to such contracts is coincidental.

Anepigraphic stamps

The purpose of anepigraphic stamps is a difficult subject, and the relevant bibliography is fairly sparse. Coates-Stephens and Parisi (1999:92) list the sites from which the primary collections of these stamps have been collected: from the City walls at Porta San Giovanni, the Largo Argentina, the Crypta Balbi, from the houses beneath Santa Prisca, the Palatine, the Lacus

Figure 2.6 Example of a *litteris cavis* stamp (*CIL* XV.1 563, from Filippi 1992: 236). Scale in cm.

Juturnus and the Lateran, from Ostia, from Santa Cornelia in South Etruria, and from the Arval sanctuary. Anepigraphic stamps are thought to date from the Severan period (AD 193-235) or later, when traditional epigraphic stamps seem to fall out of fashion and are largely replaced by these anepigraphic ones (92). However, work by Proietti (1990: 562-3) has noted that anepigraphic stamps can be found on the *same* bricks as epigraphic ones, from about the Trajanic period (AD 98-117) onwards. Coates-Stephens and Parisi therefore suggest that anepigraphic stamps have a very different function. The limited range of forms suggests that they were meaningful only in a limited context, and for a short period of time, with the only purpose being one of differentiation amongst bricks destined for the same kiln. They hypothesize that these stamps distinguish different batches with different drying times (1999:92). If Coates-Stephens and Parisi are correct, then we may imagine that epigraphic stamps are related to needs and uses beyond the production phase, that is, to distribution and consumption.

Stamping rates

Steinby (1974: 86) writes that the *figlinae Salarese* and *Quintanensia* specialised in the production of *bessales* bricks (the 8 inch variety; 9 of these arranged 3 x 3 equal the surface area of one *bipedalis*, a two-footer), *all of which* (so it is said) carry a stamp. Obviously, any theory of stamps and stamping practice has to be able to account for the various permutations we observe; if all stamps served the same purposes (whatever those purposes may be) then they should be stamped at roughly the same rates, and carry more or less the same information.

However, the stamps on *Salarese* and *Qintanensia* bricks are rather different from the 'normal' stamps, like the example of the Nunnidius stamp with which this section opened. Besides being on every brick, stamps on *bessales* from these *figlinae* are odd in that they usually carry a consular date and a slave name, impressed into the stamp using what is called '*litteris cavis*', or 'hollowed letters' (Figure 2.6 depicts an example). In normal stamps, the stamp die is usually carved so that the letters and any ornament are raised up against the background of the stamp. *Officinator* names are usually given in a more or less complete form, not just the slave *cognomen*, which is what happens in *bessales*. There is a further peculiarity: anepigraphic stamps appear at their earliest on *bessales* bricks *alongside* the *litteris cavis* stamps, but never (apparently) on *bipedales* (Broise 2000: 113). We might ask therefore if these *litteris cavis* stamps are not somehow related to anepigraphic stamps. It is still an open question, but it could be that *bessales* stamps are more akin to anepigraphic stamps than regular epigrahic ones. They would in that case serve the same purpose as anepigraphic stamps, of differentiating internal arrangements at a particular kiln. The specialization (and perhaps the scale of production?) of *Salarese* and *Quintanensia* in *bessales* bricks may have necessitated a slightly more developed form of stamp than the regular anepigraphic stamp, yet a form that was still only concerned with the internal workings of the productive unit. If this is correct (which is not beyond the realm of the possible although it has not been proven), then we might be able to ignore the apparent differences in stamping rates between *bessales* and other kinds of bricks, for these stamps would be in effect little different than anepigraphic stamps in scope and purpose.

Shapes and dates

There is also the problem of stamp shape. Commentators have always noted that there seems to be a rough 'chronological' development in the form of stamps, beginning with rectangular stamps (first century), moving on to semi-circular (second century-early third), and finally a completely circular stamp form (fourth-sixth centuries) (Dressel, 1891: 9; Bloch 1947: 23; Steinby 1974: 19-20). Yet there exist many instances where the same individuals appear in different shaped stamps, e.g. *CIL* XV.1 806a (orbicular) and *CIL* XV.1 806d (rectangular) of Zozimus; *CIL* XV.1 819a (rectangular) and *CIL* XV.1 819b (circular) of Antonius Marionis (5.3.1 for a fuller discussion). In fact, there are instances where on the same brick two different stamps have been used (Figure 2.7, a rubbing of a case of overstamping of *CIL* XV.1 800 and 801). It is argued that this is remarkable only for the fact that somebody made a mistake: stamps with incised letters were only to be used on *bessales*, so the person was wrong to use the semi-circular stamp (Filippi, pers. comm. 2002). Overstruck stamps do not seem to have ever been a subject of discussion; the illustration used here comes from the publication of a computerized typology scheme (Filippi 1992: 238). This example demonstrates that there can be a contemporaneity of stamp shapes, and that different stamp shapes were employed for different purposes.

Figure 2.7 Example of overstamping (*CIL* XV.1 800 and 801, from Filippi 1992: 238)

2.2.3 Summary: Problems of Interpretation

The epigraphic approach to stamped bricks places emphasis naturally enough on the text in the stamps. Consequently, there is an implicit belief that the Romans who used these stamps approached them in the same way - through literacy. If we assume that a certain proportion of illiterate people came into contact with the stamped bricks, and needed to understand what the stamp was about, then the figurative devices and standardised stamp shapes ought to convey the same meanings as the text. The question ought to be asked, for whom were bricks stamped? There is only *one* inscription in the Tiber Valley where a *portitor*, a carrier, names himself (*CIL* XI 4175- found in Terni, it names a man who is described as '*ocrisina*'- 'of Otricoli'). The implication is that the people who make bricks have a certain level of literacy, but those responsible for shipping are less literate. Because *signa* may not necessarily indicate the same information as the text of the stamp, there may be at least two levels of meaning in brick stamps. I will return to the question of literacy and the meaning of *signa* in 5.3.2.

Aside from the likelihood that stamps conveyed a multiplicity of meanings depending on who was reading or using the stamp, there are several other difficulties in interpreting brick stamps. These include: the nature of the collections; the purpose of *signa*; interrelationships between the different elements within a stamp's text; the relationship of stamps to contract law; the purpose of anepigraphic stamps; and impressionistic associations between stamp shapes and dates. If brick stamps are indeed merely some sort of abbreviated contract, of interest to only two parties, and concerned merely with internal arrangements of production, then they are meaningless when we find them on a site. 'Stamp-as-contract' does not refer to anything beyond itself. If this is our interpretation then the study of brick stamps on its own is not overly useful, other than being a source for prosopography and a criterion for dating buildings. This is not the case however. While it would be difficult to prove or disprove that *locatio-conductio operis* or *rei* may indeed have been used for the internal arrangements of brick production, the stamps themselves are not *locatio-conductio* contracts. The seeming correspondence between stamp shape and chronology is secondary (of course stamps will change with time; but why particular shapes were used has nothing to do with chronology). If anepigraphic stamps are related to the internal arrangements within the productive unit, I would argue that epigraphic bricks are connected instead to the problems of distribution, of getting the right product from point A to point B.

2.3 Some Logistics of, and Parallels to, the Brick Industry

2.3.1 Location, Demand, and Distribution

Location

Another ancient brick industry which has been the subject of considerable interest is that in Roman Britain. For the Romano-British industry the conventional view was that, owing to the bulky nature of brick and tile, transportation costs would have limited production to the immediate area of consumption (Darvill and McWhirr, 1984: 240, citing Hodder 1972; 1974). However, comparisons with later periods in British history (indeed, up until periods as recent as the 19th century) have shown that there were many different ways of meeting demand. There is no reason why these cannot be projected backwards to earlier periods: itinerant brick-makers moving from place to place; product being moved from place to place; and both bricks and brick makers moving (Darvill and McWhirr, 1984: 240).

Locations of production in central Italy, the *figlinae*, are on the other hand assumed to be static. On the basis of toponyms recorded in the medieval Farfa register, Huotari (cited in Steinby 1978: 1508-9) connected the names of Roman *figlinae* to various medieval *fundi*, which by and large are on the left bank of the Tiber in the Sabina. Steinby (1981: 237) distinguishes three categories of *figlinae*: those which can properly be called 'urban' since their stamps are found only within Rome and sometimes Ostia; *figlinae* the stamps of which are found in a limited suburban area (where the production centre is) but also to a degree in Rome; and those whose stamps are not found in Rome (but still within Central Italy). For all three categories it is presumed that the *figlinae* are not very far from Rome.

Demand

The key to understanding production and the location of production sites is to understand demand, according to Darvill and McWhirr (1984:240-3). They point out that demand in heavy industry is rather different than in light industry. For commodities such as pottery, the output of any particular manufactory is broken apart into single items and sold individually or in small batches to the consumer. Bricks on the other hand are needed in quantity and therefore

> bricks would behave in a similar way to pig-iron, lead ingots or timber with demand coming from a particular source that has fairly immediate need for them.
> (Darvill and McWhirre 1984: 243)

At different kinds of site the degree of demand would vary. Individual villas would need brick only from time to time, but small villages and towns would demand building materials in phase with the waxing and waning of their economic fortunes. Large settlements and cities (of which Rome of course was the largest) would have a very high level of sustained demand, with sporadic peaks. The greater the demand, the higher the price and the better a product is able to absorb the costs of production and transportation to get to market. Over a million *bessales* (the 8 inch variety) were required on the construction of the 10 km long stretch of the Aqua Anio Novus riding on top of the Aqua Claudia outside Rome in the mid-first century AD (by my own calculations); well over four million were required for the foundations, substructures and central block of the Baths of Caracalla in the third century (DeLaine 1997:124). These were only two Imperial building projects; private construction could easily have equalled that demand annually in Rome, if not also in all of the communities of the Tiber Valley as well, with obvious ramifications for cost, profits, and distribution. How was the distribution of these bricks effected?

Distribution

Epigraphers have viewed the information in the stamps as being related to the production aspect of the industry. Very little attention, therefore, has been given to the problems surrounding brick and tile after they have been produced, questions of transport, storage, and sales (Steinby 1982: 230; Steinby 1981: 237). There are, however, occasional references in the stamps of the second century to *horrea*, *porti*, *negotiatores*, and *actores* (warehouses, ports, businessmen, 'doers' or managers) which suggest that there was infrastructure in place for marketing and distribution (Steinby 1974: 74; 1981: 239). Cassiodorus (Var. 1.26) mentions a *tegularium* called the *Portus Licini* (still operating during Theodoric's reign, though known from second-century brick stamps *CIL* XV.1 139,226,408a-d,630) from which it has been argued that *portus* in the context of brick stamps means a place where produce is brought together and then redistributed (Steinby 1981: 239). What sort of change in meaning there would be between the second and fifth centuries is not discussed. DeLaine suggests that the *tegularium* meaning is secondary and originally such a place ought to be a harbour (1997: 90).

Steinby asserts that all the produce of the *figlinae* and *officinae* under the control of the *Domitii* family passed through the *Portus Licini*, and that *domini* in general (including the Emperor) were concerned, therefore, with transport and sale (Steinby 1993: 142). Helen (1975: 20) on the other hand deduces from the stamps that the manufacturer and the consumer had no contact. When stamps from the Tiber basin are found in far-flung places such as North Africa the traditional explanation is to posit the idea of transport as ballast (Aubert 1994: 240). Yet, Aubert points out that the high stowage factor (ie. high mass-to-volume ratio) of tile, occupying 2.13 - 2.27 m^3 per tonne, would make it a rather unappealing cargo to take on board merely as ballast (1994: 241, McGrail 1989: 356). Setälä (1977: 35-37) tied the appearance overseas of urban brick stamps to the particular governor (who also happens to be the *dominus* in question), which would suggest that *domini* are indeed involved in distribution. Stamps from the Tiber valley are also known along the Italian seashore as far north as Civitavecchia, as far south as Anzio, and on some islands (Giglio, Giannutri, Ponzi; these seem not to have had any local production) (Steinby 1981: 239). A recent shipwreck found off the coast of Sardinia (the Capo Carbonara C; Parker 1992 wreck 221) with a cargo of solely bricks, plus a certain amount of circumstantial evidence for the hoarding (for lack of a better word) of *bipedales* (the two-foot variety), has led Thébert (2000: 345-6) to conclude that certain types of bricks had a greater value than what we would traditionally ascribe to them. If Thébert is correct, then there would be good reason for *domini* to be interested in questions of distribution and marketing.

In Ostia, because stamped brick with stamps of different *figlinae* belonging to, or *officinatores* working for, the same *dominus* are regularly found together in the same building, it is supposed that they came from the same warehouse (Steinby 1981: 239; 1993: 142). Influenced no doubt by the deterministic cost of land transport, Steinby says that the distribution of material was limited to the river and sea for transport, while allowing overland distribution for the produce aimed at a strictly local market (Steinby 1981: 239). However, the supposed over-riding costliness of land transport, while enjoying long employ in ancient economics (first developed by Yeo 1946) ought to be reconsidered. Laurence (1999: 95-7, 99) demonstrated that the costliness of land transport as calculated by Yeo is faulty, and more importantly not an over-riding determinant of human behaviour.

How rivers work as infrastructure has not been well explored in Roman economics (cf. Laurence 1999: 109), and the traditional reliance on cost-ratios to explain trade patterns neglects the impact a river can have on society. Since the work of LeGall (1953), and latterly, that of Mocchegiani Carpano (1984) and Quilici (1986), the Tiber has curiously been absent from the scene of archaeological research. The Tiber Valley project acknowledges the role of the Tiber both as highway and as barrier (Patterson and Millett 1998) but it does not explore what that means in anything other than a symbolic fashion, for there is no actual exploration of the river itself as a piece of infrastructure created by humans. The Tiber has figured in discussions about the logistics of the brick trade (Helen 1975: 20, 44-5; Steinby 1981: 238-9), but again there has been little consideration of what it actually means to ship something by the river. In this section (which reprises some of Graham 2005a) how the river functions as infrastructure is considered first of all. The dynamics of using a river for the trade and transport of heavy goods are then discussed in the light of an ethnographic parallel with the 19th century timber trade on the Ottawa river (Canada).

2.3.2 The Tiber as Infrastructure

How we represent space affects our interpretation of how space works (Graham 2006; Montello *et al.* 2003: 316-331). When we look at a map, we conventionally place north at the top. There is no real imperative reason, from a planetary point of view, for why we do this. It is simply our habit. Maps of the Tiber Valley therefore have Rome at the bottom, which creates in the mind of the viewer an assumption that everything must eventually make its way directly to Rome. This has ramifications for how we understand the economic geography of the Tiber Valley, but the situation is more complicated than a straightforward settling of every good produced there in Rome. To see this complexity, to understand how the Tiber functioned as infrastructure, we need to shift our perspective away from our customary two-dimensional cartographic point and its attendant assumptions, down to the level of the water.

In what could be considered an early work of phenomenology, Louise and Leicester Holland (1950) took a rubber raft down the Tiber to explore the experience of the landscape from the point of view of those who would have worked on the river. They found that it was almost impossible for them to gain their bearings because the view from the river was hampered by vegetation along the shore. Elevated landmarks and the confluence of other streams became the markers by which they navigated. They were only able to keep track of where they were by tracking the confluence of other streams and by sighting the occasional hills in the distance which would appear and disappear as they meandered along the oxbow loops. We might easily imagine that these particular hills and confluences may have taken on a similar significance for the ancient boatmen. Such familiar 'companions' on any river journey may even have been named. These places, if they could be identified, might provide the evidence for understanding the usage of the river.

Some of these places might be where there are backwaters in the stream. These backwaters can form where there is an object in the river, disrupting the flow. Other places are where there is a confluence of rivers, or where curves in the river bed send the current from one side of the river to the other (Gabler *et al.* 1999: 474). The obvious place therefore to look for ports, landings, and other evidence for how the river was used as infrastructure is along those stretches where particular hilltops are visible, in the backwaters where other streams enter the Tiber, and also where meanders in the stream send the current from one bank to another. For instance, the town of Otricoli which was an olive-oil-exporting port also named in brick stamps (*CIL* XV.1 389a,b) sat on such an inside curve (although the change in the channel's course since antiquity has placed Otricoli on the outside curve, underlining the importance of understanding changes in the river's course over time). The port structures downstream from Rome in a bend of the river near Pietra di Papa (close to the suburb of Eur; Mocchegiani Carpano 1984: 34-5) also demonstrate the usage of the back-water phenomenon. There is also some evidence from geophysical survey at Forum Novum for port and warehouse structures on a curve of the river Aia (Gaffney *et al.* 2001).

Brick stamps *CIL* XV.1 917,1227-1230; S. 325-326, S.328 record *a(b) pr(aedia) a pila herculis*, which seems to be referring to an estate which sits in relation to a particular landmark, the Pila Herculis. The formula 'a pila...' is not confined to brick stamps, however. *CIL* IX 4121 a+b is an inscription recovered from a bridge on the Nera. It reads:

(a) A PILA SECUNDUM/ VIAM P L P [...]
(b) A PILA LO[...]/ NAR P LE[...]

The piers of this bridge, in being named, point to a role in the landscape beyond being merely a bridge. They are a 'place', or 'node' in a Favro-Lynch sense (Favro 1996: 13; Lynch 1960: 47, 72), where other things happen, a destination in themselves, perhaps a small port on the river or way station along the road. For Favro, the naming of places orders the landscape in the same way the ancient orator used the mnemonic device of the 'house of memory'. In each 'room' in the house, the orator 'stores' the parts of his speech, to be remembered as he 'walks' through each room in turn (Favro 1996: 7). In the same way, the associations a 'place' in the landscape has, the 'organizational clues' enabled an individual to successfully navigate the environment.

From that point of view, the Pila Herculis of brick stamps ought to be similarly a small port or settlement related to the piers of a particular bridge. (It may be in fact that the bridge referred to in the Narni inscription is the same place known in the 6th century as the 'Pile Augusto', the actual river port at Narni, below the city

(Quilici 1986: 209)). Other *praedia/figlinae* which are named with the preposition "ab" include the *figlinae: ab Apollini* (*CIL* XV.1 2156), *ab Iside* (*CIL* XV.1 248-255), and *ab Neptuno* (*CIL* XV.1 355). Perhaps these bricks are referring to landmarks in the same way the *a pila* formula does. Farmers ploughing their fields at Seripola (near Orte) in the 1970s uncovered the base of an alter dedicated to Isis; the rescue excavations conducted at Seripola during the construction of the Autostrade del Sole also uncovered some evidence for a cult centre connected to Isis (Nardi 1980: 235-6). Temples and sanctuaries would make excellent landmarks while navigating on the river; the temple of Venus for example above the so-called 'Sarno Baths' at Pompeii may be acting as a landmark for the river-port of Pompeii (E. Curti, pers. comm. 2002), while at Ostia there is a temple at the river harbour, facing the mouth of the Tiber (Heinzelmann 2002) which is probably serving the same purpose. The cult centre of Isis at Seripola therefore might conceivably be connected to the *figlinae ab Iside*.

With regard to the brick stamps which refer to '*portus*' (2.3.1), Albertazzi *et al.* (1994: 368) have shown that some bricks from the *Portus Licini* are composed of material from the banks of the Aia, some way upstream from the Tiber itself (cf. 3.1.3). Similarly, the *Portus Parrae* (*CIL* XV.1 409-412, S.103-104) ought to be at the confluence of a river with the Tiber, but perhaps the *figlinae* proper are situated up the lesser river some distance. Mocchegiani Carpano (1984: 39) suggests that the *porti* and *a, ab* names could refer to particular stretches of docks and wharves named after nearby warehouses and other facilities. Perhaps these landing places are indeed named after certain *figlinae*, but the *figlinae* themselves do not necessarily need to be physically placed on the Tiber itself. The text in stamps, by indicating quite specifically where they originate ("*ex figlinis...*") could be read backwards (as it were) to indicate the stopping places along the river, the points in the network where the river communicated with the outside world.

In *Portus Licini* stamps, *officinator* names are never mentioned. Sub-types are distinguished by four distinct *signa*: Mars (408a), Mercury (408b), Aries (408c), and Victoria (408d). Steinby uses these *signa* to connect stamps from other *figlinae* where the *officinator* is named to the *Portus Licini* (1974: 73-74). However, if the names of *figlinae* may be connected with the landing/unloading place, and *signa* do not necessarily correspond with named individual persons (2.2.2), could *signa* refer to particular docks (the Mercury docks; the Victoria docks) or wharves at the *Portus Licini* complex? During the 18th and 19th centuries at the Port of London, certain named docks were given over to the trade in particular commodities, a practice that resulted out of the logistical complexities of conducting world-wide trade from a narrow river port (Port of London Authority 2002). Perhaps on the Tiber we are viewing a similar solution to a similar problem (cf 5.3.1 and 5.3.2).

I have argued that some *figlinae* might take their names from notable local landmarks, and being landmarks in their own right some stopping places along the river might be named after the *figlinae*. This is not a circular argument but rather points to the fact that *figlinae* were part of the landscape, and their presence helped create a coherent order to it (cf. Favro 1996: 7-13). I have argued where we might find some of these landmarks, and stopping places, but I have not considered where the *figlinae* themselves might be in relation to the river. There is an assumption that *figlinae* were on the banks of the Tiber itself, and so are now buried under metres of alluvium (cf. Quilici 1986: 213); this assumptions seems to be based on the idea that bricks had to be shipped by water (cf. Steinby 1981: 239). Yet there are other sites accessible by water but high enough to have avoided the problems of Tiber flooding. Despite what the variability in the water regime in Central Italy would lead one to expect, there are indications that many lesser streams might have been suitable for trade and transportation (Laurence 1999: 109-114). Some of the tributaries of the Anio (especially the Fosso dell'Osa and the Fosso di Grotta Oscura, leading to the sources of the Gabine building stone) were canalized and fitted with locks (Quilici 1986: 210). The locks would hold back the water until there was enough to float the boats, barges, or rafts to the main river; the same system was used on the upper Tiber above Orte (211; Pliny N.H. III.V,53). The nine-day cycle of holding back and releasing the water corresponds with the market-calendar, the *nundinae* (LeGall 1953: 124) and indicates that these locks were tied to the necessities of trade (and not, say, irrigation). This suggests that as long as there was some sort of manageable stream, it was fully possible for places distant from the Tiber and the Anio to float their produce to the major river highway. Whether they did so or not is another matter, but we should not automatically assume that *figlinae* were fast against the banks of the Tiber (nor should we imagine that they are now forever lost to us, buried under metres of alluvium).

Effective Parallels for the Brick Industry

Brick stamping is often discussed in reference to other categories of stamped materials in the Roman world, usually *fistulae*, *amphorae*, and *terrasigillata* (cf. the variety of papers in Harris 1993). Comparing brick stamps to these other classes of material seems to treat the stamp itself as an entity entirely unrelated to the material on which it is found. The similarity between up-market pottery such as *terrasigillata* and common brick has not been demonstrated to the point that we are able to equate the meaning/purpose on one with the other. Besides which, Darvill and McWhirr (1984: 240-1) have a good point when they draw attention to the fact that the dynamics of a heavy industry will be quite different from those of a light industry, simply by virtue of how the product is sold in batches rather than in individual units; they draw a direct parallel with mining and timbering. Comparisons to those sorts of activities would be more appropriate.

Stamps have been found in fact on other building materials. In 1868 a piece of bitumen was recovered in the contrada Pignatara (Lettomanoppello) in Abruzzo with a rectangular stamp reading: [...]ALONI C(aii) F(ilii) ARN(iensis) SAGITTAE (Agostini and Pellegrini, 1996:57-58). A stamped squared-timber has been recovered from the Thames. Dendrochronology dates the felling of the timber to around AD 63 (Brigham *et al.* 1996: 36). The investigators record that the stamp mentions just a single name- TRAEGAUG or perhaps TRAECAUC (Brigham *et al.* 1996: 36). They have interpreted this as the stamp of a Thracian auxiliary unit, but stamps are notoriously difficult to read, especially TR's, C's, and G's. The stamp could be read as PRAECAUG or (ex) PRAE(dis) C(aesaris) AUG(ustus), a formula similar to that found on Tiber valley bricks. Assume that that is correct, and assume also that these timbers were floated downstream, then that would make them cut from an imperial estate belonging to Nero somewhere in the Thames valley watershed. There is another connection between brick and timber for according to Collingwood and Wright (1992: 125) the earliest stamped tiles found in Britain were the products of 'an imperial tilery of Nero's reign near Silchester'.

Timbers stamped with full Roman nomenclature have also been discovered *in situ* in the foundations of the circus of Arles at the mouth of the Rhone (Fuegere, M., pers. comm. 2000). A direct connection between brick and timber can be illustrated by the activities of the Roman Army in Britain. Stamped brick displaying the legend of the legion involved are well known (Collingwood and Wright, 1992; Peacock, 1979: 8). A wooden bung for a barrel was found in 1983 at Annetwell Street in London in a second century AD context. This bung had been cut from a piece of wood which had previously been branded: [...] LEG [...]/[...]EG XX [...] or *Leg(io) XX [V(aleria) V(ictrix)]* (Collingwood and Wright, 1992: 4, 2442.11). This is the same formula used in military-stamped bricks in Roman Britain. In all of these examples, the key points are the shapes, standardised abbreviations, the clear indication of origin, and the indication of the end user (eg, *Legio XX*). These timber stamps and the bitumen stamp are closer relatives to brick stamps than to the pottery which informs most models of stamping practice, and therefore the mechanics of these trades ought to be more directly relevant for our understanding of the brick industry.

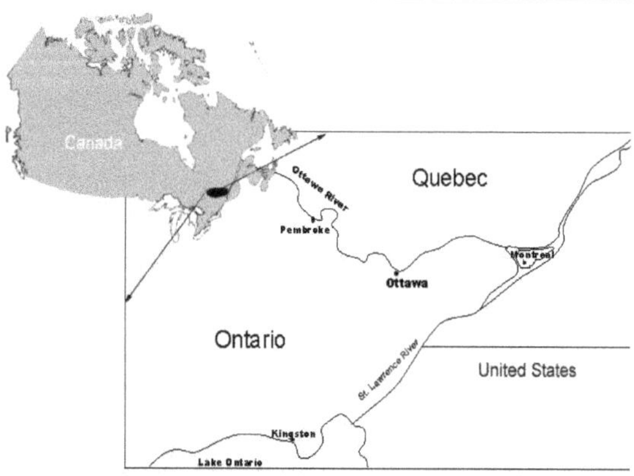

Figure 2.8 The Ottawa Valley, Canada

2.3.2 The Ottawa Valley Timber Industry in the 19th Century

Given these similarities between brick stamping and timber branding, an understanding of timbering practices could help us understand how the brick industry worked. The standard reference is Meiggs *Trees and Timbering in the Ancient World* (1982), but his time-scale is so broad that it does not deal with the period we are interested in overly much. Nor does it go very deep into the details. However, there are only about 75 generations which separate modern from ancient practice (Adam, 1999: n.151) and fewer than six of those have been industrialised. Until the advent of mechanisation, the logistics of exploiting primary resources are largely similar to ancient practice (cf. DeLaine 1997: 105-107). Ethnographic comparison with well-documented though importantly non-mechanised timbering practices in 19th century Canada provides an appropriate model for a riverine economy of the same scale and level of development as the ancient brick industry (cf. Dyson 1992: 16 and Dyson 1979 where he argues for the suitability of colonial North America as an effective parallel for the Roman world).

The Ottawa River in Eastern Canada (Figure 2.8) was used as a conduit for trade and as an integral piece of infrastructure by Europeans for less than two hundred years. The Ottawa valley was rich in stands of tall white pine, and this material was heavily in demand by the British Navy to build their ships, but also by the Americans to build their homes. The appearance, flourishing, and decline of the timber industry (in the form which relied on the Ottawa to get the timber to market) happened quite quickly. From one operator to hundreds in the space of twenty years, the industry organized itself without the imposition of formal government regulation. It experienced rapid consolidation and concentration in the hands of powerful well-connected men.

Little capital was needed for a man to get started in the trade, and since the actual trade was largely seasonal it complemented the agricultural calendar. Indeed initially it formed an important aspect of a farm's income. In its early days there were huge numbers of small producers in the forest. Profits could be considerable. At its height, one raft of timber could be worth C$14000, which in today's terms is a very considerable amount of money (Carlisle and Cheliak: 1984: 20-23). However, poor communications with the main markets meant that the lumberjacks could not accurately predict demand. Consequently there were repeated cycles of over-supply and price collapse (Reid 1990: xlvii).

These shocks drove out the many small operators, who were replaced by a restricted number of men who had the considerable capital, credit, and market information to be able to absorb the frequent crises. Consolidation took place in two forms. In the first, the larger operators subcontracted or provided financial backing to the smaller firms for a set amount of timber to be delivered to the water's edge (Reid 1990: lxiv). This shifted the risk of transportation, and the potential non-sale of goods from the little operator for whom it could be disastrous to the timber baron (or larger operator) whose social and political contacts and clout reduced that risk considerably for himself. In the second form, the larger firms simply continued to expand their operations at all levels, buying out their competitors (Reid 1990: lxiv).

Ownership of the logs was indicated in two ways. In the first, a heavy hammer with a particular pattern on the end was driven against the end of the log. The resulting imprint would be taken up by the grain of the wood and so even if the stamp itself were to be cut off, the ownership of the log would still be visible. In the second, a simpler mark would be scribed into the bark to be visible while the log was in the water (Hughson and Bond 1987: 88, 104). As the trade became more complex, so too did the uses of these stamps.

Production of timber by the large operators was centred on the camp or shanty, which acted as a base for the men. These were usually located no more than four miles from a serviceable body of water, which was the greatest distance felt to be economic. The shanty remained in use for as long as there were suitable stands of trees to exploit (Hughson and Bond 1987: 83).

Timber slides were used to get the squared timber around rapids or waterfalls where they would otherwise be damaged, or have rock splinters embedded in them, a grave danger to the men in the sawmill (Hughson and Bond 1987: 104). At the timber slides, the rafts were broken up into smaller units and run down the slide one unit at a time. The owners of the slides were able to charge a toll on the other operators for the logs that went through, based on close attention to the stamps (Theilheimer 1984: 33).

By the 1860's the largest operators on the river had banded together and formed a cooperative venture for the improvement of navigation and commerce on the Ottawa. The company undertook to build the necessary infrastructure on the river, the towing of logs over flat water, and the like. The stamps on the logs formed the basis of payment for the services offered by the company (Hughson and Bond 1987: 107-108). The stamps also served in the measurement of wood volume and later on, taxation by the government. As the trade became more complex and the timberers were farther and farther afield, some lumbermen used different stamps to indicate timber cut in different watersheds, and to indicate at which sawmill that particular log was to be cut (Stiell 1984: 33). This informal but complex system it will be noted came about without the intervention of any legal body.

In 1870, at the behest of the largest timbermen Parliament passed a law that required each log to be stamped according to a mark already registered with the Government. Registration entailed recording the mark, a description of it, serial number, and the name and address of the operator. Over 2000 different stamp types are known, formed from ligatures of the operator's initials, to depictions of beavers and turtles, to quite abstract shapes (Stiell 1984: 33).

Implications for the Brick Industry

The crucial points of comparison between the Ottawa Valley timber industry and the Tiber Valley brick industry are that -

- originally the trade was part of regular, seasonal farming activities
- low start-up costs, combined with the inability to predict production or demand, created repeated cycles of over-supply and price collapse
- uncertainty and crises drove the tendency towards consolidation into the hands of the larger players
- the cheaper the price of the product, and the further the distance from point of sale, the more improvements that were necessary to bring that product to market
- large operators banded together for the improvement of the river, giving them the concomitant right to charge others for the use of these improvements
- stamps were used for ownership, indication of destination, indication of origin, calculation of volume shipped, taxation, tolls
- stamps developed informally in response to the difficulties of shipping on the river; their codification in law happened later and was at the behest of the large operators, likely to their advantage

On a basic level, putting a mark on something is about control, about differentiating it from something else. The dynamics of the timber industry were tied not only to seasonal fluctuations in demand and poor communications, but also to the need to control access to the infrastructure which made it possible to use the river. Stamps developed in response to this need. With regard to rivers in general, the lesson is that not everybody has equal access to the river or its infrastructure, and in this fashion, river transport is much different to that by road. Whereas a road can transform the 'space-economy' of a region, making places closer together by shortening travel time (Laurence 2001a: 596, 598), a river requires an *intermediary*, a port. Consequently, a river's role in the 'space-economy' is much more complex.

Along the Tiber, the indication of access points implied by *figlinae* names suggests a limited number of places where the Tiber could be joined. Whoever controlled the ports controlled the river. The Tiber therefore may have not merely facilitated trade but rather enabled the *social control* of trade in a way that roads could not. In the distribution network which connected the hinterland to the city, the river offered a short-cut into the heart of the city which was not available to all (the crucial importance of short-cuts or long-distance links in network structures is a point to which I return in Chapter 5). The economic geography of the hinterland therefore owes at least as much to social constraints as geographic ones.

2.4. Hinterlands and Networks

2.4.1 What is the Hinterland?

What actually constitutes hinterland, and how does it function, economically? Models of the hinterland *per se* do not exist, although the hinterland is necessarily implied in models of the ancient city. The one cannot exist without the other, and countryside only becomes hinterland in relation to a city, but the hinterland as such is rarely discussed. Note that 'hinterland' is a slightly different concept from 'countryside'; one could discuss the countryside in isolation from the city, but hinterland always has to be conceived of in relationship to the city. The city economy, in antiquity, is now usually thought of in terms of Finley's 1985 re-working of Weber's 1958 'Consumer City' model. This is not the point for another long discurssus on the Consumer City for the model is well known (taxes and rents on the countryside draw off surplus, to be consumed in the city). I will however point out that the central point of the Consumer City model is that the city-hinterland relationship is in fact one where *social* relationships enable the City to feed itself by continually extracting surpluses from the hinterland (Wallace-Hadrill, 1991).

The historiography of the Consumer City model has been discussed at length, most recently in Erdkamp (2001: 332-340), Horden and Purcell (2000: 105-8), Laurence (1997) and Morley (1996: 13-32). Opponents have delighted in pointing out flaws in Finley's (1985) formulation of the model (cf most recently Mattingly and Salmon 2001; Parkins, 1997). The two main flaws in the model are that it is based on an artificial distinction between the undefined 'city' and 'hinterland', and secondly that it regards the hinterland as an 'undifferentiated economic entity' (Erdkamp 2001: 342). Whittaker (1995) asked if theories of the ancient city mattered, and concluded that despite the serious flaws in the Consumer City model, it was still the best model available (1995: 22). On a more pessimistic note, Horden and Purcell (2000: 108) asked whether the model, even if the best available, spurred scholars to 'ask important questions'.

Many still find a good deal of use in the concept. Morley (1996: 185) maintains that despite covering just about 'everything you might wish to know about the ancient city', the model does not necessarily imply parasitism, and indeed he goes on to show that this model allows for the stimulation of economic development in the hinterland: "the belly's hunger is what forces the rest of the body to rouse itself and exercise its strength and ingenuity". Morley's work is important in that he explicitly examines the hinterland on its own terms in relationship to the city. However, his discussion of the hinterland is ultimately a very traditional approach, novel only for considering the hinterland side of the equation rather than the city side.

Horden and Purcell (2000: 45-6) move beyond the traditional model, but their focus on 'micro-regional ecologies' goes too far in the opposite direction. Whereas the Finley model seemed to make the city a basic, self-evident unit capable of independent action, Horden and Purcell reject the notion of 'city' as a unit of analysis or indeed object of analysis entirely (helpfully defining a town as 'what each age takes it to be', 93). They argue,

> if we are intent on abandoning 'the town' as a distinct settlement type, we should not reasonably be expected to produce an urban theory of our own.
>
> (Horden and Purcell 2000: 109)

They prioritise 'micro-regional ecologies' as the basic unquestionable concept capable of independent action. However their definition of 'micro-regional ecology' is as equally vague as their town definition (45-9). Their discussion of 'ecologies' translated into an historical context is a confused mixture, with the literary deconstruction of Rappaport's 1968 anthropological study of the Tsembaga people of New Guinea, Sallares' *The Ecology of the Ancient Greek World* (1991), and 1960s 'New Archaeology' all major ingredients. That being said, one important aspect to their work is the idea that these micro-regions are connected to each other, and much of what matters in Mediterranean history happened because of these connections (Horden and Purcell 2000: 90). Yet strangely, they have forgotten something: people.

The Role of People

From Finley (1985) to Horden and Purcell (2000), *individual people* are the missing elements. Cities, towns, villages, and hinterland all existed, and the common denominator in all of them is people. The problem is that these concepts are really part of a continuum; there is no one point where a dividing line can be drawn and said to be the discrete point where to that side there are cities and to this side there are towns. Because it is a continuum, the same processes at play at one point along the continuum are at play at another.

This is what is implied by the rank-size rule, the observation that the rank and size of cities in a country observe a power-law distribution. Horden and Purcell (2000: 94-105) misunderstand what the rank-size rule indicates, though they come around to the correct conclusion in the end: they take the example of the city of Mantinea, which by administrative whim in the 4th century BC was reduced from a city to a collection of villages, and argue that if this could happen, then the rank-size rule has no use. However, power laws do not predict what any one individual case will be, rather they indicate that very specific complex system states are in operation in the background (cf 2.2.1). What Horden and Purcell (2000) fail to notice is that, despite any one city's ranking and its own historical travails, the rank-size rule obtains regardless over time. Mantinea may have dropped to the bottom rank, but somewhere else is now at the top and the power law remains. In fact, when Horden and Purcell arrive at what the rank-size rule implied all along, they have (as already noted) dispensed with towns:

> It is therefore the common processes by which micro-ecologies interact, rather than the presumed distinctions between one kind of settlement and another, or one period and another, that should hold the Mediterranean historian's attention
>
> (Horden and Purcell 2000: 122).

I agree, but when they argue that a 'proper conception' of towns and routes (meaning, their ill-defined micro-ecological conception) 'will resist the kind of systemic and mathematical approach that is [...] fundamental to rank-size analysis' (Horden and Purcell 2000: 104) they are belittling any approach other than their own as somehow less sophisticated. In fact, a 'mathematical' approach has much to recommend it, because it throws the essential patterns of interaction into sharp relief enabling them to be studied and compared across periods and cultures (cf 6.2, 6.3).

'Mathematical' arguments aside, it does not seem advisable to exclude towns from the analysis simply because settlements are of all different types and sizes. However, Horden and Purcell's focus on processes of interaction (2000: 122) is welcome. But what is doing the interacting, if 'cities', 'hinterlands', and 'micro-ecologies' are all found wanting? The answer is of course, people. Parkins (1997: 89) points to the obvious importance of élite households in the economy, and writes that the city 'can be regarded as comprising many mini-economies: the economies of individual households'. An analysis using her definition would have to pay close attention to the countryside holdings of these households because these are the source of their political legitimacy. This definition encompasses the town-country continuum. Indeed, using this definition, the City could be considered to be a manifestation of processes evident throughout the countryside, differentiated from them only in their intensity.

Levels of organisation

Parkins' definition is important because it shifts the discussion away from some vague ill-defined notion of 'the city' to lower levels of social organisation. This question of levels of complexity is important. Ocean waves are a classic example of different levels of complexity. Waves move across the surface of the water, but it is the energy which moves, not the water molecules themselves. The wave exists at a higher level of complexity. The same is true of traffic jams (with respect to individual cars), or rope, or anthing else in which waves manifest: 'In all of these waves, the motion of the wave is very different from the motion of the constituent parts'(Wilensky, 1998:3). What Finley and his disciples are describing is akin to the ocean in terms of waves rather than water. The discussion is pitched at too high a level, and the relationships between the different levels are misunderstood or misconceived, leading to the sorts of confusion decried by Davies (1998) in 'Ancient economies: models and muddles'. This may also be why the debate over the nature of the ancient economy goes round and round in circles. It is the central problem of archaeology in general: how to move from the individual bits and pieces (water molecules, i.e. archaeological evidence) to higher-level phenomena (waves, i.e. society)?

Wilensky (1998:4) describes three potential ways of understanding the concept of levels. The first may be called 'the hierarchy view'. This view understands levels in terms of control, with each level being subject to the one above it, as in the army (privates, corporals, sergeants, and so on). A second approach, the 'container view', sees levels as being parts of a whole, as for instance with units of time ie a year is 52 weeks, which are seven days long each, which are 24 hours each, which each have 60 minutes (contrast with: 12 corporals do not equal one sergeant). It is evident from her discussion that Parkins' understanding of the city (1997: 88-9) is similar to the 'container view' (e.g., 300 household economies = one city economy), but her definition could be understood in a third fashion: the 'emergent view'. Wilensky (1998:4) writes:

> ....we are focussing on yet another meaning of levels, which we call the 'emergent view' of levels. Our focus is on levels that arise from interactions of objects at lower levels- like the traffic jam that emerged from the interactions among the cars. These levels might seem similar to the part/whole levels: just as a year is made up of months, traffic jams are made up of cars. But the jam/car relationship is different in some very important ways. For one thing, the composition of the jam keeps changing; some cars leave the jam and other cars enter it. Moreover, the jam arises from *interactions* among the cars; it is not just a simple accumulation of cars. Months do not interact to form a year; they simply accumulate or "add up." A year can be viewed, essentially, as a long month. But a traffic jam is not just a big car. It is qualitatively different.

When Whittaker wrote that he found all theories of urban economics unsatisfactory, his concluding reason was because 'the study of cities is only an imperfect way of studying the operations of power in society' (Whittaker, 1995: 22). If, in taking an emergent view of the problem, we consider cities as a higher level of the 'operations of power in society', as can be understood from Whittaker's statement, then no wonder he found the theories unsatisfactory for it would be similar to trying to understand the nature of water molecules from ocean waves.

Power

Which operations of power? In Erdkamp's (2001) discussion of the consumer city model, he points to the fact that cities are not actually the 'heart of the concept'. Rather it is the relationship between cities and the countryside, the means (the *power*) by which agricultural surplus, primarily food, are redistributed, which matter (Erdkamp 2001: 340). Much debate has focussed on *where* production took place, whether in cities or in the countryside (cf. Whittaker 1993; Mattingly and Salmon 2001). But the geographical spot where production happens is, economically, irrelevant (Whittaker 1990: 116-7). Erdkamp's point is that, wherever production happened (be it rural non-food producing activities such as pottery-making or urban manufacturing like shoe-making), it depended on surplus production of foodstuffs. For Erdkamp, we must first distinguish between 'economically relevant entities' rather than 'city' or 'countryside', and then examine whether we are dealing with reciprocal or non-reciprocal exchanges. It is a question of access. If it is social or political entitlement which guarantees access to food surpluses, then we are dealing with a non-reciprocal relationship. If on the other hand those who produce the food surplus receive payment in kind, money, or services, then it is a reciprocal relationship. Erdkamp argues that because non-reciprocal exchanges dominated ancient life to a far greater extent than did reciprocal exchanges, virtually all ancient cities may be classified as 'consumer cities'. More importantly, the economy as a whole is a 'consumer economy' and in this regard, not much different than in any other period (Erdkamp 2001: 353).

While Finley wanted to distinguish ancient economies from modern economies, Erdkamp's reformulation of the model erases the distinction. His model does however point to certain features of the Roman economy which are of interest here. Erdkamp's model allows for subsistence agriculture while at the same time permitting a complex market economy (spin-offs from reinvestment of income from food surpluses and so on). It provides a framework for analysing élite wealth and investment in both rural and urban productive activities (Erdkamp 2001: 354). The key is to ask whether we are dealing with reciprocal or non-reciprocal exchanges between individuals. This question implies the existence of networks through which the relationships, the exchanges, were mediated.

2.4.2 Urban Geography and Networks

If the question with which this section opened is reversed, to ask what are cities and how do they function, Massey, Allen, and Pile. (1999:vii) reply that it is necessary to consider how cities are created in the context of social relations which extend beyond the conventional boundaries of the city, but also intersect within the city. Considering the social aspects first of all, we are compelled to remember that above all else, cities are places where large numbers of people are confined to relatively small areas. Hence, whatever it is that cities may be, they are definitely *intense* (Massey *et al.*1999: 42). The way people live, go about their daily work (or try to find work), buy and sell, seek out companionship, are entertained, are all social processes which in cities are far more intense (faster, differentiated) than they would otherwise be. In cities the individual experiences of people and the physical features which result from, constrain or empower those experiences, are combined or kept apart in particular ways (Massey *et al.* 1999: 49).

Massey *et al.* argue that examining how social processes extend beyond and also intersect within cities is to say that social processes work across various networks. Cities are the foci of multiple networks. These networks do not exist independently of the people who operate within them, but rather must be actively maintained and so are forever mutable. This perspective therefore

> ... requires an appreciation of the ways in which networks spread across, intersect with, or avoid, one another. It has been argued that networks both stretch beyond cities and also intertwine within them. (Massey *et al.* 1999: 49)

Networks extend *beyond* the city, linking different cities together in different ways (but also incorporating every point in between along the continuum of settlement types from humble rural farmsteads upwards). According to this model, it is cities themselves which are *nodes of social relations in time and space* (Massey *et al.* 1999: 100-136). At any given time a city will be a node in any number of different networks of power and influence:

> ...they do not simply map on to one another; these networks are of differing significance: they differ in terms of the kinds of social power which they carry as well as in the numbers influenced and in the degree and nature of that influence; individual cities have different balances of all of these, and different positions within them; cities are foci of such networks, but they are so in very different ways, giving them distinct kinds and ranges of influence. And moreover the position of any city within such networks may change over time [....] changes in a city's place within these networks can deeply affect its fortunes and its character (Massey *et al.* 1999: 117).

An Example from the Sabina

At all times, however, people must be taken into account: it is not enough that interconnections should exist. Rather, people must *make* something of these interconnections (Massey *et al.* 1999: 121). Intersecting networks on their own might only become a simple trading post, a seasonal market, or a transhipment point. Forum Novum, a settlement established by the Romans in the Sabina in the 1st century BC, never developed into a town as such; even today it is still little more than a church and a hotel. After the initial capitalization on the intersecting networks in the river valley where Forum Novum is situated brought the settlement into being, those networks were evidently not maintained. Laurence (2001a: 606-7) has discussed the establishment of *fora* (market-centres) in Italy as part of the process of establishing control in newly centuriated land, sometimes as part of a *viritim* land allotment (the process of finding land for de-mobbed veterans of the late-Republic civil wars). The establishment of *fora* represents direct élite intervention in the landscape, and so the direct decision to warp and intensify local network patterns.

An ordinary node such as Forum Novum does not become a city unless it acquires *intensity* (and many *fora* did not succeed as urban centres, Laurence 2001a: 604). Intensity *emerges* from the intersection of social processes and relationships in all their varieties in a localised space. How that space is organised is a reflection of how society is organised, and the two affect each other recursively. The same applies to interconnections over longer distances, and even disconnections have significant effects:

> we are arguing that cities may be understood spatially and the particular form of the spatial configurations which constitute them will affect 'what happens next'. (Massey *et al.* 1999:161-3).

Influence via Networks

Finally, there is the question of what flows through these networks. For Roman Britain, Laurence (2001b) has argued that the creation of the road network had a profound effect on the 'Romanisation' of that province. Laurence's model of cultural change (what could be labelled 'Romanisation') lies in the mobility of people, goods and capital along these new roads (Laurence 2001b: 67). Prior to the conquest, Britain was composed of various micro-kingdoms, ruled by *regules* or 'little kings'. Using the evidence of the Antonine itineraries for routes through Britain, Laurence shows that the newly reconfigured patterns of mobility in the province were tremendously different from previous indigenous patterns, but were not significantly different from the rest of the Roman world (for a critique and an alternative formulation using an agent-based model of information flow along the Antonine networks, see Graham 2005a). The disruption to previous modes of life by the imposition of the new patterns could produce hostility, but ultimately that hostility did nothing to change the new configurations (Laurence 2001b: 91-2).

'Urbanisation' might be best understood in the same manner as Laurence's 'Romanisation'; indeed they could be thought of as synonymous. The people, goods and capital which flowed over the intersecting networks I have been discussing could be considered realised in the fabric of a single brick, because a brick represents simultaneously the manufactured product itself and its travels, the capital required to make it, and the people who made it, who travelled to the clay sources, who consumed it.

2.4.3 Summary: From the Hinterland to the City

To discuss the hinterland necessarily implies an understanding of its relationship with the city; therefore one must come to grips with the ancient city to consider their inter-relationship. Earlier discussions, in pitching the terms of the discussion at such a high level of complexity as 'the City' without actually defining the terms ('what is a city?') has caused much confusion in the debate (exacerbated by the mixed definitions of Horden and Purcell (2000: 45-46, 93, 109)). Whittaker (1995: 22), however, identified the critical issue when he wrote that the study of urbanism is only an indirect way of understanding the 'operation of power in society'. This power I argue lies in the nature of economic exchanges. Erdkamp (2001) puts the focus on the nature of the economic relationships which take place within and across the city-hinterland boundaries, however those boundaries may be defined. For Erdkamp (2001: 342-3) the question is not about 'the city' and its relationship with 'the hinterland' but whether we are dealing with reciprocal or non-reciprocal exchanges of food surpluses. The question of reciprocal or non-reciprocal exchange of food can be seen as the basic operation of power in Roman society. The Roman economy is characterised by the dominance of non-reciprocal exchanges rather than reciprocal exchanges, hence it is a 'consumer' society. These exchanges do not necessarily map onto a division of town and country, but rather point to networks of relationships which overlap the traditional barriers.

Parkins (1997: 88-9) argues for an understanding of cities based on the agglomerated effects of the operations of élite household economies (which owe their political/social legitimacy to their landed holdings). Parkins' work implies the relationship with the countryside without having to cast the dialogue in stark city/hinterland dichotomy. If it were possible to identify the economic activities of households archaeologically, then we would be in an excellent position. Yet, most often all we have are the traces of individuals. To extrapolate from this basic level of the individual to more complex levels of social organisation (Graham 2005b), all the way up to that of the city so we can examine the relationship with the hinterland, is therefore quite difficult. It is however a project for which archaeology is well-suited provided we have an appropriate interpretative framework.

Such a framework may be found in the geographical writings of Massey, Allen, and Pile (1999). Here the city is an 'open intensity' (161-3), a node of social relationships in time and space open to other places and differing from other forms of settlement in the intensity (speed, number, and heterogeneity) of the relationships taking place. Individuals matter in their conception. The interconnected and intertwining networks of the city, created by and creating the possibilities for social relationships (and all that that entails) to take place, and the movement of people, goods, and capital along these networks, as in Laurence's discussion of Roman Britain (2001b), provide the key. This understanding of urbanisation can be seen to have been presaged by Millett (1982: 422) when he argued for an understanding of Roman Britain as a 'network of individual settlements and groups of settlements which were originally connected by a network of social relationships'. City and countryside are blurred together. Urbanisation takes place along a continuum, measured in the intensity of the patterns of mobility of people, goods, and capital (cf Urry 2000 : 49 - 76 on the sociology of mobility) through social networks over space both within and without the conventional borders of the city, regardless of formal administrative definitions. In this study, brick will be studied as representing these exchanges, so the distribution of brick in the countryside can be used to explore hinterland - city socio-economic dynamics.

2.5 Chapter Summary

In this chapter, we have considered certain issues regarding the nature of this industry, of the Tiber as infrastructure, and the relationship of a city with its hinterland, including:

- the problem of what a stamp represents: *locatio-conductio operis* or *rei*, or something else?
- the interrelationship of stamp elements (names, consular dates, *signa*)
- the logistics of the industry
- the problem of how people 'read' a stamp, whether literally or symbolically
- the payment of the *merces*
- how the practice of brick stamping could develop spontaneously in response to the usage of the Tiber for trade
- how the Tiber may have been an instrument for controlling trade, rather than simply facilitating it
- the special way in which Rome could thus be connected with the hinterland by the Tiber
- how hinterlands may be conceived
- how hinterlands and cities are interrelated
- how the movement of information, capital, and people along networks, and the intensification of relationships when networks intersect, can transform a region

In this chapter I have discussed the way the Tiber may have tied the hinterland to the city, and possibilities for a new understanding of the process of urbanisation. These understandings will allow the build-up of a picture of the dynamics of the economy from the traces of individuals. Brick stamps record one aspect of the household economies (which take place over the whole urban to rural continuum) of members of the élite of Rome. In brick there is evidence of an infrastructure for moving material from the hinterland to Rome, but also evidence for people moving to the hinterland for the purposes of the industry. One aspect of the lives of people named in stamps is of course that they are involved in brick, but some were also involved in politics, in other regions, in other industries. In this way, the people named in brick stamps can be considered as nodes in overlapping networks. In the same way that the city can be defined as an open intensity (i.e., where different networks, open to outside influences, overlap and intersect), we might also expect to find places in the hinterland where the networks in the brick industry intersect with other networks (of whatever kind) to create new nodes and urban-type places. The mobility of resources, whether human or capital, along these networks is part of the process of becoming urban (urbanisation). The lines of inquiry will therefore be followed within this general framework. I will need to explain how individuals worked or were prevented from working in the industry, who stood to benefit, and how the different networks enabled brick to be used in the valley, and in Rome. But this cannot be done until the locations of the production sites are discovered. The localisation of production sites through archaeometry is the subject of chapter 3.

Chapter 3: Sourcing the Brick Industry

3.1 Introduction

In this chapter, potential sources of clay used in the brick industry are hypothesized through an archaeometric study of stamped and unstamped bricks from the SES collection, as well as a number of samples from modern brickyards in the Tiber Valley. A programme of X-ray diffraction (XRD) and X-ray fluorescence (XRF) analysis has been undertaken. X-ray diffraction is a technique which describes the mineralogical composition of the sample, while XRF characterises the chemical composition. X-ray analysis (i.e. diffraction and fluorescence) was chosen for this study because it is both quick and cost-effective; it also does not destroy the sample during analysis.

This chapter first discusses in 3.2 the archaeometry undertaken to date in the Tiber Valley and what Roman-era sources of production have been localized. It considers the results of the Vatican Museums' project, and what their results along with the findings of the other studies would lead us to expect to find archaeometrically in the SES stamped bricks (3.2.5). The next section (3.3) discusses the methodology of X-ray analysis and the logic behind the sample selection. The results of the XRD and the XRF are then presented (3.4). In 3.5 our results are compared with what we would have expected, given the earlier work discussed in 3.2 The chapter then turns to a visualization of the geophysical interrelationships, with potential production areas included (3.6). Finally, A statistical technique, multi-discriminant analysis, is used to tease apart some of the interrelationships in the bricks, as well as the evolution of the exploitation of the brick fields (3.7).

3.2 Archaeometry in the Tiber Valley and the Locations of Production

While there has been a certain amount of work identifying the various raw building materials used in construction in Rome (cf. DeLaine 1995, 1997 for the most recent work), clays for brick production have been comparatively neglected until fairly recently. The first major archaeometric work on ceramics in the Tiber Valley was Peña's 1987 PhD dissertation. It was aimed at understanding the marketing of Roman pottery, which necessitated the identification of production centres (cf. 3.2.2). Peña examined the material texturally before submitting a further sample to petrographic and chemical analysis. The *Opus Doliare* Working Group[2] headquartered at the Vatican Museums, on the other hand, began in 1994 a program of investigation into the 'urban' (i.e. bricks found primarily in the city of Rome itself) brick and tile industry itself through X-ray analysis (reported in Baldi *et al.* 1999, cf. 3.2.3). Careful study of place names in the modern landscape and in historic documents by various scholars also suggests particular areas that were probably exploited in antiquity (Steinby 1978: 1508-9). Taken holistically, these various approaches allow us to build up a picture of where the production locations may be, to which the findings of the current archaeometric study can be compared for assessment.

3.2.1 Geology of the Tiber Valley

The highly variable and complex geology of the Tiber Valley is the result of different processes at different geological periods. In the east in the Sabina, the left wall of the Tiber valley (the Apennine mountain chain) was formed through the collision of the European and African tectonic plates at the end of the Cretaceous period. In South Etruria the right wall of the valley was formed towards the end of the Pliocene through a series of volcanic eruptions. There are four principal volcanic zones, from north to south: the Vulsino (the main crater of which is today's lago di Bolsena), the Cimino-Vicano (lago di Vico), the Sabatino (the Tolfa-Ceriti-Manziana district), and the Colli Albani to the south of Rome. The Tiber cuts to the sea between the last two of these volcanic zones (Constantino *et al.* 1993: 12-13, 50). Sedimentary deposits in much of the area lie below the tuffs expelled by these eruptions, although the drainage pattern cuts deeply in places, exposing outcrops of Pleistocene and Pliocene marine clays (Peña 1995: 69). Major clay deposits are exposed in the areas around Orte, while on the left bank there are large sources surrounding Narni and in the southern part of the Sabina. Around Rome itself, the major sources of clay are in the minor gullies and valleys to the south and west of the Vatican (DeLaine 1995: 556-7, 559) (Figure 3.1).

Peña has spent a number of years working on ceramics production in South Etruria on the right bank of the Tiber. His 1987 PhD thesis was the first systematic study which identified pottery workshops and put them into their wider economic context. Eleven clay areas in

[2]The Working Group brings together researchers from the Musei Vaticani, the Musei Comunali di Roma, the Soprintendenze Archeologiche di Roma e Ostia, ENEA (the national energy company), the Centro Ceramico di Bologna, and the Colorossia Italia di Sovigliana, and the universities of : Berlin; Napoli-Orientale; Oxford; Roma La Sapienza; Siena; Viterbo (Baldi *et al* 1999: 617).

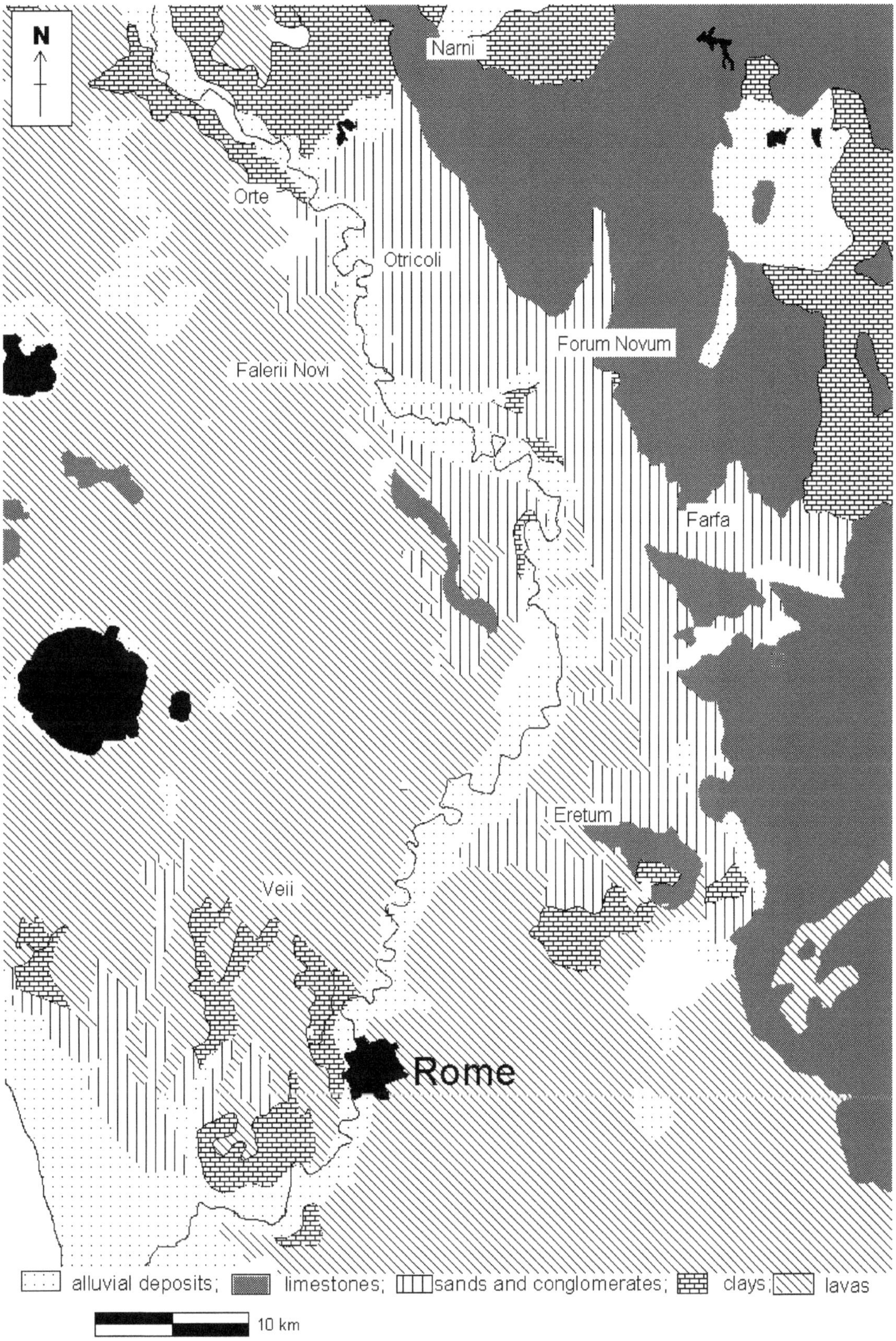

Figure 3.1 Geology of the Middle Tiber Valley

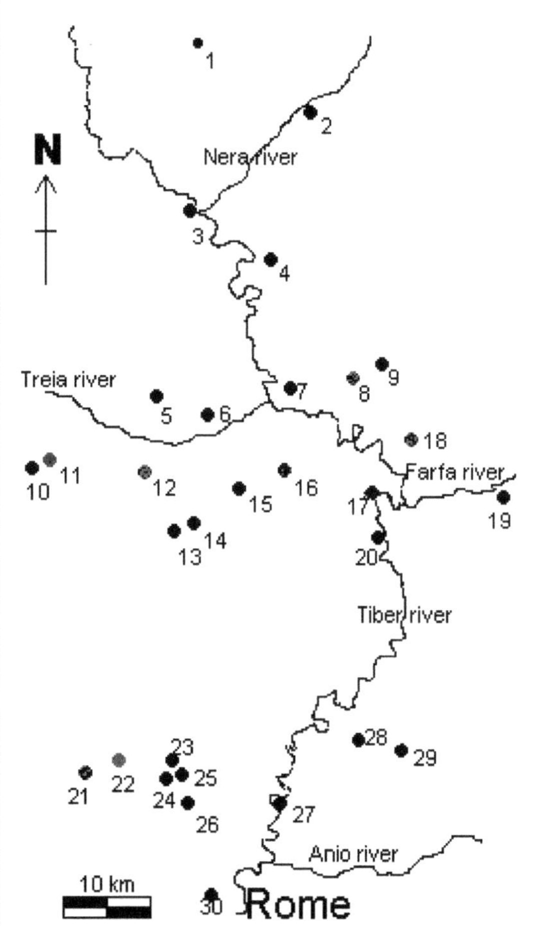

1 – Poggio Gramignano	7 - Grappignano	13 – Valle l'Abbate	19 – Castel Nuovo di Farfa	25 – Isola Farnese
2 – Narni Scalo	8 – Aia Roman kiln site	14 – Mazzano	20 – Fiano Romano	26 – La Giustiniana
3 – Orte	9 – Forum Novum	15 – San Biagio	21 – Fosso Arrone	27 – Castel Giubileo
4 – Otricoli	10 – Sutri	16 – Mt. Soracte	22 - Fosso Galeria	28 – Monterotondo Scalo
5 – Falerii Novi	11 – Fonte Vivola	17 – Nazzano	23 – Castel del Pino	29 – Mentana
6 – Cività Castellana	12 – Nepi	18 – Galantina	24 – La Storta	30 - Valle Aurelia

Figure 3.2 Places named in the text.

South Etruria have been explored by Peña (1987: 55-71) with reference to ceramic production. These are plotted in Figure 3.2, a map of places and areas mentioned in this chapter.

- Clay area 1 consists of grey clays and sand clays with interbedded conglomerate, from about Orte to 5.5 km northwards.
- Peña's clay area 2 is made up by blue clays with intermixed gravels and silts, ranging from Orte Scalo for about 6 km to San Michele Archangelo.
- Clay area 3 is along the lower Treia river near Cività Castellana, with grey and sandy clays and conglomerates.
- Clay area 4 is again in the Treia system, near the town of Nazzano. It is composed of sandy blue grey clays.
- The fifth clay area is to the immediate north of the town of Sutri. Here the grey clays are formed from marls and marly limestones.
- To the east and south of Mt. Soracte, extending along the western margin of the Tiber floodplain is clay area 6, where there are clayey sands and sandy grey clays.
- Clay area 7 runs for 7 km north northwest by south southeast to the south of Mt. Soracte, with sands having lenses of clay.
- The eighth clay area has the same sort of clay as clay area 1. It is in the valley bottoms of the Valle di San Sebastiano and the Valle la Fata, between the modern town of La Storta and the Isola Farnese (in the neighbourhood of Veii).
- Clay area 9 is a small zone near the settlement of La Giustiniana. The clay is similar to that of clay area 7.
- Clay area 10, with clays similar to that of area 3, is along the bottom of the valleys of the Fosso Arrone and the Fosso Galeria.
- Clay area 11 is in the Tiber floodplain to the east of Castel Giubileo. The clay is greyish and/or reddish brown in colour, and can contain gravels, sands and silts.

3.2.2 Ceramics in General

Peña conducted a field survey searching for evidence of ceramic production at a site (as evidenced by the presence of wasters), and then analyzed the ceramic fabrics from these sites along with specimens of raw clay from outcrops within the study area. Although primarily concerned with pottery, he did consider other ceramic products when he encountered archaeological evidence for their production. He located 12 pottery workshops, which is certainly a fraction of the total number active. The following five also produced brick and tile, although only the one at La Storta seems to have used stamps (see Figure 3.2 for location):

- At Fonte Vivola, 1.8 km NE of Sutri, a workshop active in the 1st - 2nd century AD produced tile characterized by having a highly ferruginous, low-calcium body, and medium to coarse volcanic sand.

- At San Biagio, 2 km ESE of Nepi, a workshop attached to a large villa produced brick/tile characterized by having a ferruginous, carbonate body and containing fine to coarse volcanic sand. This workshop was active from the late 1st century to the 2nd or 3rd century AD.

- 1.5 km west of Mazzano, at Valle l'Abbate a workshop at a large villa, dating from the second half of the first century to the second or third century AD made brick/tile in a non-ferruginous, carbonate body containing medium to coarse volcanic sand.

- One kilometre E of La Storta at the Casale del Pino a workshop attached to a small villa was active in the 1st century AD. Brick and tile produced here had a ferruginous, carbonate body, and contained coarse reddish tuff fragments.

- Again near La Storta (1.8 km SW) a workshop attached to a large villa (3rd - 5th centuries AD) made brick/tile in a highly ferruginous, non-carbonate body and contained fine mica, medium quartz, and medium to large volcanic sand grains.

(Peña 1995: 71-72)

Peña used neutron activation analysis (NAA) to differentiate fabric groups composed from clays drawn from different sources within a region, and to match the fabrics with the sources. He noted the wide variety in fabrics, discernible by eye, within the coarse-bodied fabric groups at these workshop sites. The principal inclusions in these fabrics were augite, flint, grog, mica, plagioclase feldspar, sanidine, shell, tufa, calcareous bodies and 'soft red bodies' which are probably granules of iron-oxide (Peña 1987: 99-100). In a later work he cautioned researchers against making too fine textural distinctions in fabrics from assemblages from non-workshop sites, lest they lose archaeological significance (Peña 1995: 74). Finally, Peña makes the point that a workshop could and often did employ a wide range of raw materials in order to make ceramics for different purposes; that is, the more utilitarian the ware, the more likely the use of local, non-imported materials (Peña 1995:75-76).

3.2.3 Brick and Tile in Particular

A Finnish scholar named H. Appelquist, who was part of the team from the University of Helsinki which studied the Ostian collections (2.2.2) performed the first archaeometric investigation of Roman brick and tile in the early seventies as part of his doctoral work (Steinby 1977: 20). That work has never been published. It was the only scientific study of the problems of brick and tile (aside from Peña's brief digression) until 1993 when Gloria Olcese of the Università di Siena published her initial results on the material from the Villa of the Quintili at mile V of the Via Appia (Olcese 1993; recent archaeometric work reported in Bruun 2005 appeared too late to be incorporated here).

Her objective in that study was to create an 'identity card' for the various *figlinae* present in her sample, using x-ray fluorescence. She tested forty-five samples carrying stamps of the *figlinae Sulpicianae* and twenty-one of the *figlinae Domitianae*. On a graph of the potassium versus magnesium values of the samples, there were two distinct clusters, divided between the two *figlinae*. She then averaged all the values for each element, and calculated the standard deviation for each. To know whether a particular unstamped brick came from these particular *figlinae* one would only have to test the brick and see if the values fell within one standard deviation of the mean (Olcese 1993: 123,8). However, this 'identity card' approach is problematic because it rests on the assumption that all stamped bricks carrying the same stamp are made from the same clay. This has not yet been proved. Until such time, it is impossible to demonstrate the significance of any correspondence between a particular brick and a 'chemical signature' for an entire *figlina*.

The *Opus Doliare* Working Group project at the Vatican Museums is a direct descendent from this earlier work. Its primary goal is to locate the clay pits and kilns, although its net is cast wide to include: obtaining knowledge of the production technologies; defining the structure of social relations both within and without the productive unit; inquiring into the economics of the industry; using *opus doliare* in dating archaeological contexts (Baldi, G, E. Gliozzo, D. Manacorda: forthcoming). There have been two papers published by members of the Working Group to date, one (Albertazzi *et al.* 1994) localizing the production of a particular run of stamped bricks, the other (Baldi *et al.* 1999) testing whether the epigraphy of the stamps matches the chemical signatures, à la Olcese, of the various *figlinae*.

Albertazzi and her colleagues (1994: 347-350) analysed 14 fragments of brick along with six samples of raw clay and one sample of a kiln waster from 13 sites within a 15 km^2 area along the Aia river in the Sabina (see Figure 3.2 for location). They used a barrage of techniques: thin sectioning, XRD, XRF, and firing tests. The study area was within the territory of the municipality of Forum Novum. This *municipium* was established sometime between the late Republic and the mid 1st century AD and its territory was roughly 330 km^2. It had a largely agricultural economic base, related to the large numbers of *villae rusticae*, in which the production of brick it is argued played an important part. The identified brick stamps came from nine

Fragment ID	Stamp #	*Figlina*	*Dominus*	*Officinator*	site	date
SAB 5	*CIL* XV.1 862			C. Nunnidius Fortunatus	1	ca. 142
SAB 6a-b	*CIL* XV.1 532	Sabin(iana)	Servilius Capito	Paccius Verna	2	100-110
SAB 7	*CIL* XV.1 2196		Servilius (who could be either a *dominus* or an *officinator*)		3	?
SAB 10	*CIL* XV.1 408 (variant not identified)	Portus Licini	the Domitii family		5	211-217
SAB 11	*CIL* XV.1 408 (variant not identified)	Portus Licini	the Domitii family		5	211-217
SAB 12	*CIL* XV.1 404	Ponticulanae	C. Fulvius Plautianus	Licinius Felicissimus	5	212-217
SAB 13	*CIL* XV.1 762a	Domitianae Maiores	C. Fulvius Plautianus	Felicissimus	5	212-217
SAB 14	*CIL* XV.1 438	Quintianae	Rubriae	M. Antiochus	6	mid 1st century
SAB 15	*CIL* XV.1 659c	Tonneianae de Viccians			6	Flavian

Table 3.1: Identified stamps found along the Aia
After Albertazzi (1994: 351), augmented with Steinby (1974-1975:38,66,73-77,80-82,94-96)

fragments of brick (Table 3.1). The remaining samples are listed in Table 3.2.

Thin-sectioning determined two major groups, with the second group being further divided into three. The main difference between the two groups was the presence of volcanic material in the second group, with the subdivisions being a function of the percentage abundance of volcanic material (Albertazzi *et al.* 1994: 354).

- Group 1: SAB 6,8,9,14, characterised by inclusions of quartz, feldspar, metamorphosed carbonates, and micas

- Group 2a: SAB 1,10,11 characterised by circa 40% volcanic material

- Group 2b: SAB 12,13,15 characterised by circa 30% volcanic material

SAB 1	brick fragment, from kiln site
SAB 2	brick fragment, from kiln site
SAB 3	waster, from kiln site
SAB 4	fragment of kiln furniture
SAB 8	brick fragment, from site 4
SAB 9	brick fragment, from site 4
SAB 16	clay from ancient clay pit, at kiln site
SAB 17	sand from ancient clay pit, at kiln site
SAB 18	pozzolana and tufa from quarry at Grappignano
SAB 19	clay from clay pit at Cerreta
SAB 20	clay from clay pit at Galantina
SAB 21	sand from clay pit at Galantina

Table 3.2: Unstamped Brick and Unidentified Stamped Brick Found Along the Aia River. After Albertazzi (1994: 351)

- Group 2c: SAB 5,2,7 characterised by circa 20% volcanic material

(Albertazzi 1994: 354-356)

Albertazzi and her colleagues (1994: 364) found that the chemistry of the Group 1 samples was similar to that of the clay from Galantina and the tufo from Grappignano, but the other group did not readily agree with any of the non-brick samples. They concluded that Group 1 represented a local production, but the inclusion of *CIL* XV.1 438 (from the *officinae Quintianae*) in the group puzzled them. They suggest that there is perhaps an 'alternate explanation' for this stamp's text (Albertazzi *et al.* 1994: 368); though not explicitly stated, it would seem that they expected this brick to fall in with the other 'urban' stamps of Group 2.

'Urban' in the context of brick stamps is a term which only refers to the fact that many of these types were originally found in the city of Rome, and later collected under that term in Chapter 2 of *CIL* XV.1. They state that the petrographic characterisation of the Group 2 bricks was not very good, leading them to conclude that neither the petrology nor the epigraphy allow them to localise the exact production of these bricks (Albertazzi *et al.* 1994: 368). However, by calling all of the stamps in Group 2 'urban' and putting them in opposition to the demonstrably 'local' production of Group 1, they give an impression that the bricks in these subgroups were made in the city. Group 2a is 'Portus Licini', 2b is 'Ponticulanae-Tonneianae', and Group 2c is headed by the two artisans 'Nunnidius and Servilius' (Albertazzi *et al.* 1994: 368).

The other paper by the *Opus Doliare* Working Group characterised 97 stamped bricks from the storerooms of the Vatican Museums, using XRF analysis. (Baldi *et al.*, forthcoming). They found that the samples divided into two broad groups. The first group corresponded to the *figlinae*: *de via Nomentana; de via Triumphalis; a*

Stamp #	*Figlinae*	Villa Production	Probable Villa Production	Imports
		impasto 4	*impasti* 1, 2, 6	*impasti* 3, 5
CIL XV.1 283	Macedonianae		1	
CIL XV.1 416	Propetianae		3	
CIL XV.1 427a	Publilianae		2	
CIL XV.1 430	Publilianae		6	2
CIL XV.1 433	Publilianae	1	2	3
CIL XV.1 526	Salarese	1	1	1
CIL XV.1 661a	Viccianae			1
CIL XV.1 795a	Sex. Annius Aphrodisius		3	
CIL XV.1 862	Asinii family		1	
Unidentified				1
Anepigraphic			1	

Table 3.3. Concordance between Martin (1999) and Monacchi (1999) fabric classifications.

Creta; L. Sestius. The output of each individual officinator in these *figlinae* was clearly distinguished within the cluster analysis. They suggest that the *figlina de via Nomentana* is probably at the point where that road crosses the Anio River, in today's Monte Sacro suburb of Rome (Baldi *et al*, forthcoming). In the past, the other *figlinae* tested were thought to be in the Vatican area; their association in the cluster analysis with the stamped bricks of the *de via Nomentana* suggests to the investigators that they should be relocated to this area instead (Baldi *et al*., forthcoming). It is worth pointing out that the identification of the location of the *figlina de via Nomentana* rests on the information in the stamp only; it has no independent corroboration. Why the *de via Nomentana* should trump the *via Triumphalis* (which passes through the Vatican area) is not stated either.

The second group discerned through cluster analysis contained *figlinae* whose names suggest geographic locations somewhere other than the city. The *figlina a pila Herculis* and the *figlina Isiacae* were clearly distinguished from the others, but the remaining *figlinae* were intertwined, suggesting to the investigators close geographical proximity. Interestingly, the chemical signatures of stamped bricks explicitly mentioning the name of the *figlina Camillianae* were substantially different from stamped bricks attributed to this *figlina* on the basis of common *signa*, known patron-client relationships, and other family ties (Baldi *et al*., forthcoming).

Finally, a recent excavation at a villa near Poggio Gramignano (see Figure 3.2 for location) found strong evidence of production of both pottery and other ceramics. Martin (1999: 373-4) found that there was a correlation between the type of architectural ceramic and whether it had been produced at the villa or not (local production was assumed based on the presence of wasters). All the column bricks, voussoirs, flooring bricks (*opus spicatum*) and pipes used the same clays as the numerous coarse ware wasters. The *tegulae mammatae* (a type of wall tile with a few small projections on the back, creating an air space in the wall, Adam 1999: 269) and the *suspensurae* (the tiles which make up the floor, Adam 1999: 269) in the baths

Stamp #	*Figlinae*	*Dominus*	*Officinator*	*Others*	*Date*
CIL XV.1 283	Macedonianae	L. and P. Cassius			end of 1st C AD.
CIL XV.1 416	Propetianae	Egnatius Clementis		from the *praedia* of Hortensius Paulinus. *Negotiator* Valerius Catullus	end of 2nd C AD, beginning of 3rd
CIL XV.1 427a	Publilianae	Aemilia Severa			end of 2nd C AD, beginning of 3rd
CIL XV.1 430	Publilianae	Aemilia Severa		*Negotiatrix* Iuniae Antoniae	end of 2nd C AD, beginning of 3rd
CIL XV.1 433	Publilianae	Aemilia Severa			end of 2nd C AD, beginning of 3rd
CIL XV.1 526	Salarese	Flavius Titianus	Iulius Theodotus		end of 2nd C AD, beginning of 3rd
CIL XV.1 661	Viccianae	Apollinaris			mid 1st C AD
CIL XV.1 795	?	Sex. Annius Aphrodisius			Domitian
CIL XV.1 862	Asinii family	Asinia Quadratilla	C. Nunnidius Fortunatus		ca. 142 AD

Table 3.4. Epigraphic Information Associated with Stamps listed in Table 3.3

were made in a much different fabric, suggesting that they had been imported (Martin 1999: 380).

There were thirty stamped bricks recovered in this excavation (Monacchi 1999: 382; Tables 3.3 and 3.4). Monacchi studied and classified the fabrics of the stamped bricks using a different scheme than Martin did for the architectural ceramics, so it is somewhat difficult to make a concordance between Monacchi and Martin's work. 'Villa production', 'probable villa production', and 'imports' are Martin's concordance with Monacchi's various *impasti* (fabrics). Only *impasto* 4 agreed completely with what Martin deduced to be Villa production, while *impasti* 1, 2, and 6 were highly probable. Nevertheless, it appears, on a quick consideration of the problem by Martin that at least two stamps and probably another twenty are of the same fabric as the ceramics produced at the villa (Martin 1999: 374 n.6).

Martin describes the villa-produced fabric as being reddish-yellow, smooth and hard. The breaks were moderately clean, and the main inclusions were tiny abundant 'micaceous grits'. There were also red, white, and black inclusions which ranged in size from small to large (although these sizes are not defined). The imported material was pink in colour, broke irregularly and was rough to touch. The inclusions were frequent in abundance, ranging in size from medium to large (again, sizes are not defined). They were colourless, or red, or dark grey (Martin 1999: 374).

3.2.4 Toponyms and Other Inferences

During the major push towards understanding this material by the Finnish Institute in Rome in the 1970's and 1980's, Huotari (unpublished; cited in Steinby 1978: 1508-9) studied the location of *figlinae* on the basis of toponomastic similarities between the names of *figlinae* and early medieval estate names (*fundi*) recorded in the Farfa register. Steinby summarised Huotari's work along with some other topographical indications in an article in the *Reale Encyclopaedie* (Steinby 1978: 1508-9; Figure 3.3) :

- *fig. Brutianae* - Rome, Vatican
- *fig. Mucianae* - Trastevere
- *fig. Quintianae* - Rome, Vatican
- *pr. Quintanensia* - Colonna
- *fig. Subortanae, pr. Suborta* - Orte
- *fig. Sulpicianae* - by the Baths of Caracalla
- *fig. Varianae* - Vicovaro
- *fig. Naevianae* - Vicovaro
- *fig. Bucconianae, Vocconianae* - Ficulae-Fidenae (*contra*, Huotari who connects this figlinae with a *fundus Buccunianus* close to modern Bocchignano on the north side of the Farfa river just below Poggio Mirteto)
- *fig. Albianae, Buccinianae, Caninianae, Domitianae, Furianae, Germanicae, Licinianae, Macedoniae, Marcianae, Occiasae, Platanianae, Ponticulanae, Terentianae* - Ager Sabinus
- *de Narn, fig. Narniensis* - Narnia
- *de Ocri(culo?)* - Otricoli
- *fig. Pagi. Stel(latini)* - Soriano

Helen (1975: 80-2) identified Aemelia on the left bank of the Tiber across from Orte at the confluence of the Tiber and the Nar as the site of the *figlinae Caepionianae*, *Marcianae*, and *Oceanae*. This identification is based on the appearance of the same people mentioned in brick stamps as are named in funerary monuments in the region, and who are also known to have worked in the *figlinae Subortanae*, which was presumably just downstream from Orte.

The *Portus Licini* first appears in stamps in the Severan period, but it was later mentioned by Cassiodorous (*Var.* 1.26) and was, therefore still operational during the reign of Theodoric. *Portus Licini* was in the *Praedia*

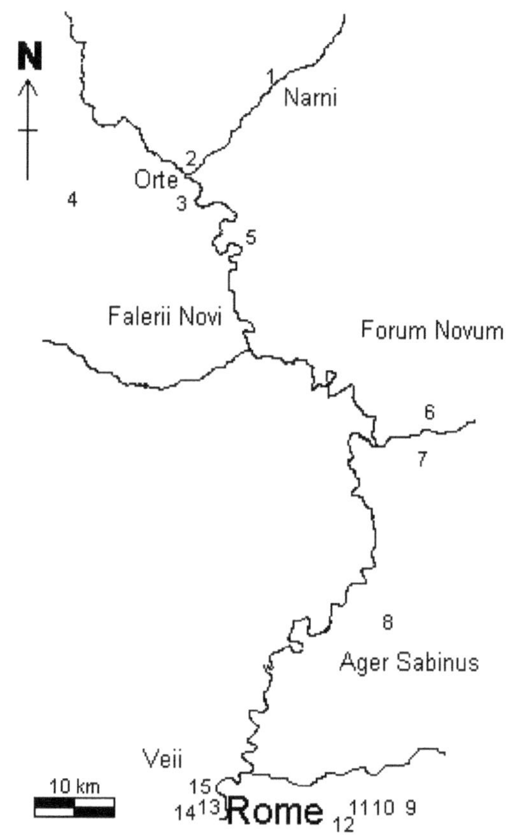

1 Narnienses
2 – Caepionianae, Marcianae, Oceanae
3 – Subortanae
4 – Pagi Stellatini
5 – de Ocri(culo)
6 – Bucconianae
7 – Portus Licini, Ponticulanae
8 – Ager Sabinus group
9 – Naevianae
10 – Varianae
11 – Quintanensia
12 – Sulpicianae
13 – Quintianae
14 – Mucianae
15 – Brutianae

Figure 3.3 Locations of *figlinae* based on toponyms and other inferences

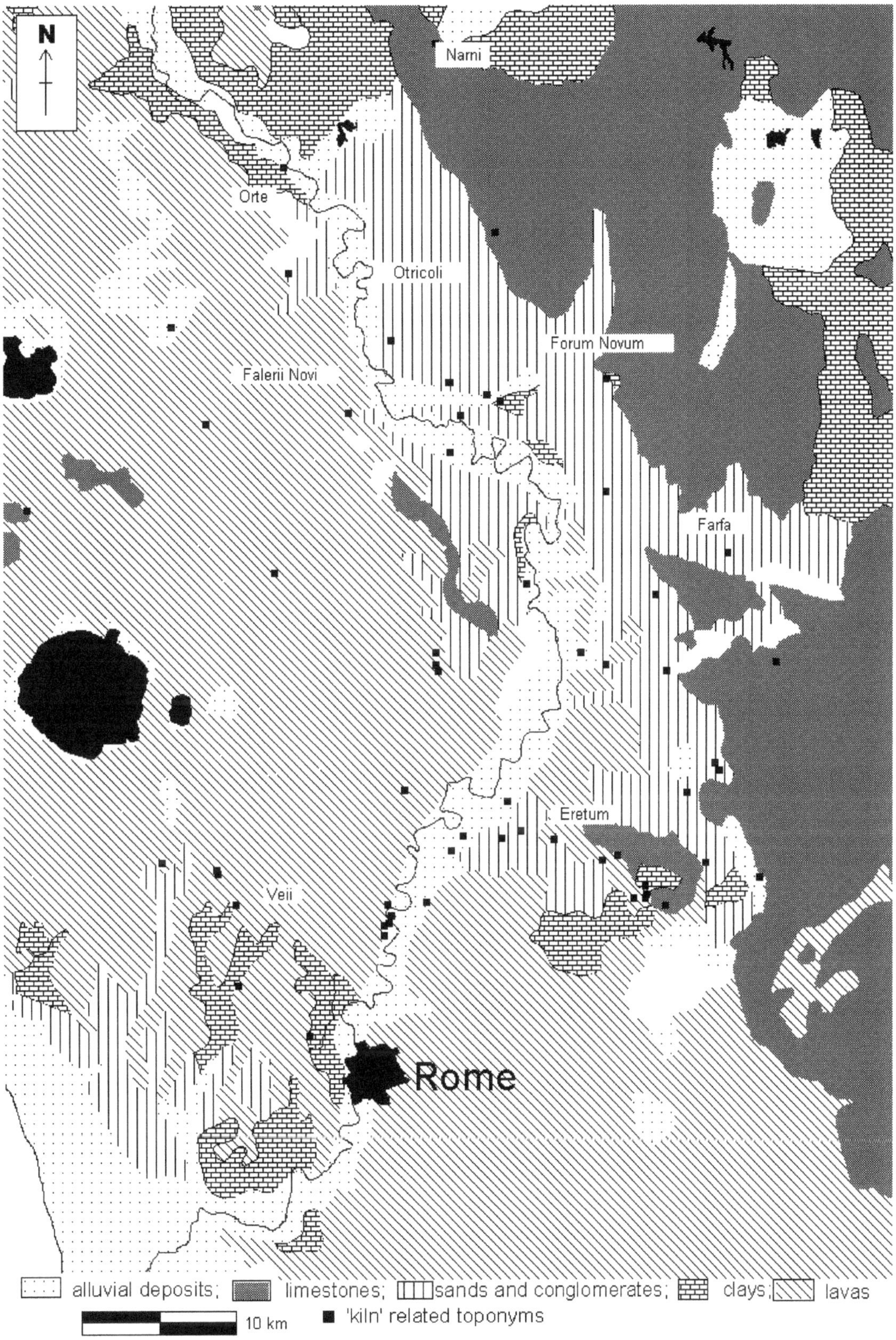

Figure 3.4. Locations of *furnace* place names in relation to the geology of the middle Tiber Valley

Licini as was the *figlinae Domitianae Veteres* (Steinby 1974: 73-4). By linking the *officinatores* appearing in these and related stamps and those of the *figlinae Ponticulanae*, and tying them through toponyms to an identified medieval site (Castello Tribucum, in the *fundus caesarianus* near Ponticlum, which ought to be the site of *figlinae Ponticulanae*) DeLaine (1997: 90-1) argues that this brick-making locus ought to be found somewhere along the Farfa river, about four kilometres upstream from the Tiber. The *figlinae* making up this locus are the *Ponticulanae, Domitianae Maiores, Bucconianae, Oceanae*, and *Genianae*.

Toponomastic evidence for the production process itself, rather than the locus of production (degenerate names from ancient *figlinae*) can also be found on any detailed map of the region. In 3.2.1, the geology of the Valley was discussed in fairly broad terms. Close study of the available geologic maps indicates that in volcanic South Etruria, clays are present at the base of Monte Soratte, a limestone massif. On the sedimentary Sabina side, basaltic lavas are present in large pockets around the Eretum area, and above the confluence of the Nera and the Tiber in Amelia. In the landscape today there are many place names which recall the Italian for 'kiln', *fornace* (Patterson, Witcher, and Di Giuseppe, Tiber Valley Project database CD-ROM, Version 1). When these are plotted on the map, (Figure 3.4) they seem to occur around the edges of these pockets on either side of the Tiber, especially where rivers and streams have cut through the various layers, exposing them. There seems to be a correlation between *fornace* sites and the interfaces between geological zones, suggesting that there is something particular about the materials available in these areas which made them attractive to those who worked in clay. The question of interfaces will be taken up again in 4.5.2.

3.2.5 Expectations: The Physical Nature of Roman Brick and Tile and the Locations of Production

There is rather a lot of information already available concerning the locations of *figlinae* in the Tiber Valley, as has been recounted in the previous section. We may summarise these findings, and what they imply for the archaeometric study of the British School collection as follows:

- Brick and tile in South Etruria seems to have been produced at a relatively small number of sites. Ceramics from these sites are characterised by the presence of flint and sanidine. The production site of stamped bricks of Aristanius held at the BSR has been confirmed by Peña in his clay area 8 (cf 3.2.1, Peña 1987: 55-71).

- In the Sabina, production sites have been located at Poggio Gramignano and along the Aia River near Forum Novum. The brick and tile at Poggio Gramignano were made on site. The inclusions which characterise this site are frequently occurring small micaceous grits, and also frequent small to large white, red, and black flecks. Peña's descriptions of minerals may permit a tentative identification of the inclusions in the Poggio Gramignano fabrics as some sort of calcite or plagioclase feldspar (white), grog or iron oxide (red), and augite (black). Imported materials at Poggio Gramignano are characterised as having medium- to large-sized colourless, red and dark grey grits (Martin 1999: 374, fabric 2). These perhaps may be identified using Peña's descriptions as calcite or plagioclase feldspar (white), grog (i.e. crushed pottery) or iron oxide (red), and sanidine (colourless)(Peña 1987: 99-100).

- The brick and tile at Poggio Gramignano are also differentiated according to function. By having brick with the same stamps in both the local and imported groups demonstrates that the term *figlina* can encompass more than one clay source. At the same time the situation at Poggio Gramignano demonstrates that one clay source can accommodate more than one *figlina*.

- Along the Aia, a site where wasters (i.e., tiles that were warped, cracked, or were otherwise deformed during firing) were found is believed to be a production site for the *Quintinianae* and the *Ponticulanae*. This kiln demonstrates one of the same situations as at Poggio Gramignano. One kiln produces for two different *figlinae*. The minerals present in brick from this kiln are quartz, feldspars, carbonates, and micas. The 'urban' stamped brick found in the area were shown not to be local. These bricks tended to be differentiated on the proportions of volcanic clastics (Albertazzi *et al.* 1994: 354).

- Olcese's work (1993: 123) does not attempt to locate production centres. She attempts a different tactic, the identification of 'chemical signatures' for each individual *figlina*. By averaging all values together for samples presumed to come from the same *figlina*, she inadvertently masks the possible situations discussed above: a *figlina* encompassing more than one source, and a production site accommodating more than one *figlina*. Later work by the *Opus Doliare* Working Group finds that many *figlinae* can share a common source, and that individuals connected to this or that particular *figlina* by scholars on epigraphic grounds often do not produce brick from the clay used in that *figlina* (Baldi *et al.* forthcoming).

- The various studies based on toponomastic evidence (Helen 1975: 80-1; Steinby 1978: 1508-9; DeLaine 1997: 90-1), when considered altogether, do not support the mono-locale interpretation of *figlina*. Rather than seeing the various locations posited by various scholars for various *figlina* as evidence of erroneous reasoning, we could see this multiplicity of answers arrived at through a multiplicity of deductive chains as *confirming* the geographically scattered viewpoint.

Given these findings we might expect to find evidence for multiple clay sources for bricks from the same *figlina*, when that *figlina* is explicitly named. The stamped brick of individuals known at some point in their careers to have worked in a particular *figlina* may not have the same chemistry as stamped bricks where the *figlina* name is given. Bricks which originated in the Sabina will have chemistry and mineralogy similar to those bricks sourced to Poggio Gramignano and the Aia River. Produce from South Etruria will have characteristic inclusions such as sanidine, biotite, and volcanic material. The evidence of *fornace* place names at the interface between volcanic and sedimentary zones suggests however that there might be an overall homogeneity in the coarse mineralogy of Tiber Valley brick.

3.3 Methodology for the Archaeometric Study of the SES Collection

The archaeometric study considers a subset of brick and tile from the SES collection, plus samples from currently operating or recently defunct brick yards (Appendex A for the SES collection). The 75 stamped examples come from sites where at least two or more stamped bricks were found; by selecting these stamped bricks for study I hope to increase the chances that they do represent a real building event at the site, and that any relationships discerned between this site and others are meaningful. The remaining 20 examples are unstamped brick from Forum Novum, Falerii Novi, and modern brick. The 75 stamped bricks account for 14% of all the stamped bricks recorded as having been found in the Tiber Valley to date (total: 523 examples, Filippi and Stanco catalogue; at the time of writing I examined the draft version of Filippi and Stanco 2005). The probability of a sample of this size of being representative is just under 95%, nine times out of ten. In comparison to the total collection recorded in the *CIL*, some 12 000 examples, a sample of nearly 400 cases would be necessary to have a 95% chance of being representative, 19 times out of 20. There have been enough archaeometric studies conducted on stamped brick, if every researcher pooled his or her data, to create a sample of this size, but until such time, if we are willing to relax our margin of error, a sample of 75 cases does have a slightly better than 90% chance of being representative of the overall corpus as recorded in *CIL*, 9 times out of 10. The virtues of studying stamped bricks found in the Tiber valley are that smaller sample sizes are statistically valid and can be studied by one individual in a reasonable amount of time. Also, the known geographical relationships between findspots will help to clarify relationships in the clays (4.3.1).

It has been found in Britain that it is often possible to use only a hand lens or a binocular microscope to tell apart the different fabrics of Roman brick and tile (Betts, 1994: 51). Given the geology of the Tiber Valley (3.2.1), it may be the case that it is possible to explore the problems discussed in 3.2.5 by eye alone. If this could be confirmed, field methods could be developed where only a hand lens would be necessary to determine the origin and so on of the brick and tile. As a first step, the SES collection was examined visually to identify different fabrics. Inclusions that are volcanic in origin (e.g. mica, augite) might indicate an Etruscan origin, while those that are sedimentary (e.g. limestone) could indicate a Sabine provenance.

X-ray analysis was used to investigate the chemical and mineralogical composition of the samples. Statistical methods used to determine the significance of the results were discriminant analysis and cluster analysis.

3.3.1 The Nature of the Sample

The tested examples from the South Etruria Survey collection are noted in Appendix A. Seventy-five of these carried stamps ranging in date from the Julio-Claudians to Diocletian. Nine of the examples tested were gathered from modern brickyards in the region (over an area roughly 2500 km^2)- from the Val Aurelia, Fiano Romano, and Orte in South Etruria; and from Monterotondo Scalo, Aia valley (Forum Novum), and Narni Scalo in the Sabina and Umbria. Eleven examples were from unstamped bricks.

The SES collection itself does not contain any unstamped bricks (indeed, it is a very rare thing for unstamped bricks to be collected or studied at all in the Tiber valley, a notable exception being Potter and King 1997: 230-6). At Forum Novum, along the Aia River in the Sabina, the on-going excavations of a villa on the edge of the town (Gaffney *et al.* 2001) presented an opportunity to study unstamped bricks. Unfortunately, it is unknown whether the field survey conducted in 1997 collected or quantified the brick and tile at the site (Patterson, pers. comm. 2000), so the possibility was lost of correlating that which was found on the field with that which was found underneath. During the 1999 excavation season, due to pressures of time and resources, it was not possible to quantify the amounts of unstamped material found. Nevertheless, it was possible to determine the presence of at least five fabrics (3.3.2). Samples of these fabrics were taken in order to place Forum Novum in the wider networks of production and consumption of stamped brick through archaeometric comparison with the SES collection (cf. 3.6 and 5.4). Samples of unstamped brick were obtained from the town of Falerii Novi, to be used in the same way as those from Forum Novum. This town was born in a forced re-settlement of the Faliscan people of Falerii Veteres (Cività Castellana) by Rome during the Republic (Patterson and Millett 1998: 13). The samples were recovered from the Forum area.

3.3.2 Textural Analysis

Thirteen different fabrics are distinguishable by eye in the SES collection, using an x10 hand lens. Significant aid in determining these differences was provided by Dott.sa Helga di Giuseppe and Dr. Helen Patterson, both of the British School at Rome. In the visual determination of fabric, personal preferences play as great a role in how many fabrics seem present as any 'objective' criteria. This manifests itself in the tendency for ceramics specialists to fall into two broad

Fabric	Description
BSR 1	large ferrous bodies, augite, with lots of very small limestone
BSR 2	ferrous bodies
BSR 3	large ferrous bodies and augite
BSR 4	clean fabric, occasional large and small ferrous bodies frequent very small limestone
BSR 5	coarse, packed fabric containing many ferrous bodies, augite, limestone and other inclusions
BSR 6	quartz, augite, shell
BSR 7	notable for clean fracture- lots of medium sized inclusions, ferrous bodies less obvious, quartz/feldspar, limestone, possibly augite
BSR 8	small occasional ferrous bodies, frequent augite
BSR 9	small occasional ferrous bodies, very crumbly
BSR 10	little ferrous bodies, flecks of limestone, frequent quartz/feldspar clean fabric, moderately clean fracture
BSR 11	few small ferrous bodies, some with opaque inclusions (quartz/flint?) possibly augite
BSR 12	clean fabric with some rounded tufa inclusions, occasionally very large
BSR 13	coarse, very densely packed fabric with Aeverything@, medium sized ferrous bodies (smaller and less frequent than 5) quartz/feldspar, augite

Table 3.5. A "splitter's" fabric description. In this scheme, fabrics were classified by eye (using a x10 hand lens). The terms used are qualitative rather than quantitative. 'Tufa' is used in its general archaeological sense rather than its precise geological sense. 'Small' ~ 250 – 500 μm; 'medium' ~ 500 – 1000 μm; 'large' ~ 1000 – 2000 μm.

Supergroups	constituent fabrics
"peppery"	5, 13
"quartz/feldspar"	7,10,11,6
"limestoney"	1,4
"ferrous"	2,9
"ferrous + augite"	3,8
"tufa"	12

Table 3.6 Fabric "supergroups". In this scheme, the most similar fabrics from Table 3.5 were grouped together and given a descriptive name to capture the most apparent aspect of the combined fabrics. Hence 'peppery' is a fabric densely packed with many different kinds of inclusions, giving it a 'peppery' look.

FNV A	coarse texture, ferrous bodies, limestone; corresponds with BSR 1
FNV B	fine texture, quartz, voids; corresponds with BSR 4
FNV C	coarse texture, augite, quartz, limestone, ferrous bodies; corresponds with BSR 13
FNV D	fine texture, clean; does not correspond well with any particular BSR type
FNV E	powdery texture, occasional augite; corresponds with BSR 11

Table 3.7 Forum Novum fabric types and their correspondence with BSR fabric types in Table 3.5

methodological camps: those who may be called the 'clumpers' (those who only see a few broad differences) versus the 'splitters' (those who see many fine differences). Tables 3.5 and 3.6 define the different visual fabrics first at a fine level of resolution and then by clumping them together into broad 'supergroups'. The names used for the different groups, and their descriptions, are impressionistic guides to visual groupings. The term 'tufa' is used in this study in its archaeological sense of any volcanic ejecta, rather than its more precise geological meaning.

Over 170 kg of brick and tile was examined at Forum Novum to determine the different fabrics present. Because fabrics FNV A and FNV B more or less correspond together with the same 'limestoney' group in the SES collection, it was decided to test FNV B, FNV C, and FNV E to capture the variety in the sample (three different SES fabric 'supergroups'). Clearly, the situation with regard to the FNV material was not ideal but the solution adopted here (to try and work out the source clays used and study Forum Novum's position in the wider context of brick production), was better than ignoring the opportunity altogether. The samples from Falerii Novi are closest in type to BSR 1 and 4. Only three samples could be collected. What little brick and tile there was, visible in the section walls of the excavation of the Forum area, seemed little differentiated.

3.3.3 X-Ray Analysis

Given that the Vatican Museums' collection of stamped brick is probably the largest in the world I opted to use similar analytical methods to the *Opus Doliare* Working Group to allow easy comparison of the results. The differences lie in the statistical treatment of these results.

X-ray fluorescence (XRF) and X-ray diffraction (XRD) have had a long use in geological studies. Powdered-pellet XRF has the advantages of a good degree of precision and sensitivity; a wide variety of elements can be tested; and a large number of samples can be tested fairly rapidly. 'Powdered pellet' refers to the way the sample is prepared. XRF bombards a sample with X-rays, then measures the resultant energy yield. As the ionising radiation hits the electrons surrounding the nucleus of an atom, the electrons will jump to a higher energy level, or 'shell', of the atom. The electron then returns to its normal shell emitting x-rays characteristic of that element.

XRD on the other hand measures how a sample's crystalline structure diffracts a beam of X-rays. While XRF can give us the chemical makeup, XRD gives us the mineralogical (Herz and Garrison 1998: 222). The atoms which make up particular minerals are arranged in characteristic lattices or planes of atoms at characteristic spacings, which diffract the x-rays at particular angles, either cancelling out or reinforcing the beam. By measuring the angles at which the maximum reinforcement occurs, the interplanar or d-spacings of the mineral are recorded and hence the particular mineral can be identified.

There are difficulties however. For every d-spacing there are thousands of minerals; identifying which one requires comparison of the different intensities at each d-spacing known for a particular mineral. The problem becomes compounded in archaeological ceramics because of the act of firing the clay and the subsequent physical and chemical weathering alter the minerals; hence what laboratory measurements for a pure specimen indicate as normal may not be easily identifiable in an archaeological sample (Velde and Druc 1999: 273). The coarseness of the brick fabrics tested here mitigated this problem, enabling secure identifications of the minerals present.

Although the stamped bricks tested here were originally recovered from field survey, they have spent ca. 20 years in storage at the BSR. Samples were taken from the cores of the bricks to counter as far as possible the effects of weathering and post-depositional physical and chemical change. In some cases, duplicate samples were taken from elsewhere within a single brick to ascertain the variability within a single brick, but the samples appeared visually homogeneous, and the differences between individual bricks appeared greater than those within a single brick. It is also apparent from striations in the fabric of the bricks that during the manufacturing process nothing more complex than simple folding and mixing of clays with tempers took place. Since we are looking to group together bricks of similar composition (and therefore manufacture), the relationship between clay and temper is not as important as it would be if we were trying to identify the precise actual clay source used.

I do have to assume that manufacture of brick within the Tiber Valley was a fairly standardised process, both within individual manufacturing units, and on a larger, regional scale (cf. Shennan 2002: 49 on cultural transmission and uniformity of knowledge). Since brick is a highly utilitarian product, the clays and tempers used were likely quite local (even today, modern brick makers in the area use only local sands and gravels, available within a 500 m radius of the kiln sites as additives). The relationship therefore between a brick's fabric and the original source material is sufficiently good therefore for my purpose. Firing conditions do affect the mineralogy of the bricks to a certain extent, but the basic differences between clay and temper sources are still preserved. Except for a loss of volatile elements such as water the chemistry of the bricks is also not all that affected.

	Mean	Std. Deviation	Range	Minimum	Maximum
Augite	1.30	1.34	7	0	7
Haematite	0.24	0.2	1.13	0	1.13
Gehlentie	0.32	0.46	3.13	0	3.13
Calcite	0.98	1.38	7.94	0	7.94
Analcime	0.16	0.26	1.39	0	1.39
Muscovite	0.17	0.21	1.17	0	1.17
Dolomite	0.42	0.44	1.64	0	1.64
Anorthoclase	0.64	0.86	4.25	0	4.25
Sanidine	0.30	0.72	4.57	0	4.57
Albite	0.76	1.16	5.5	0	5.5

Table 3.8 Relative amounts of minerals in the tested SES collection examples, expressed as a ratio to quartz

	Augite	Haematite	Gehlenite	Calcite	Analcime	Muscovite	Dolomite	Anorthoclase	Sanidine	Albite
MOD1 Orte Scalo	0.5	0.0	0.5	0.0	0.1	0.1	0.0	0.9	0.0	0.0
MOD2 Monterotondo Scalo	3.2	0.2	1.2	0.0	0.3	0.0	0.7	0.6	0.0	0.0
MOD3 Fiano Romano	0.2	0.2	0.6	0.9	0.1	0.2	0.0	0.6	0.0	0.0
MOD4 Val Aurelia	0.0	0.0	0.3	7.9	0.0	0.0	0.0	0.0	0.3	0.0
MOD5 Aia Valley	1.8	0.2	0.9	0.0	0.2	0.0	0.5	1.0	0.0	0.0
MOD6 Aia Valley	1.8	0.1	1.1	0.0	0.1	0.0	0.4	0.6	0.0	0.0
MOD7 Aia valley raw clay	0.0	0.0	1.0	3.4	0.0	0.4	0.0	0.0	0.0	0.7
MOD8 Narni Scalo	0.0	0.0	0.2	1.0	0.2	0.2	0.0	0.3	0.0	0.0
MOD9 Orte Scalo raw clay	0.0	0.3	0.0	4.9	0.0	0.3	0.4	0.0	0.0	0.3

Table 3.9. Relative amounts of minerals expressed as a ratio to quartz in samples from modern brickyards

	Mean	Std. Dev	Range	Minimum	Maximum
Na_2O	0.99	0.31	1.89	0.09	1.98
Mg_2O	3.72	0.79	4.71	1.85	6.56
Al_2O_3	14.59	1.65	8.56	9.55	18.11
SiO_2	53.59	5.15	24.81	36.60	61.41
P_2O_5	0.23	0.10	0.56	0.10	0.66
K_2O	2.77	0.60	2.61	1.33	3.94
CaO	14.96	5.02	26.48	6.27	32.75
TiO_2	0.70	0.02	0.32	0.51	0.83
MnO	0.12	0.02	0.10	0.07	0.17
Fe_2O_3	6.38	0.65	4.44	3.36	7.8

Table 3.10 Major elements in tested SES collection examples

	Na_2O	MgO	Al_2O_3	SiO_2	P_2O_5	K_2O	CaO	TiO_2	MnO	Fe_2O_3
MOD1 Orte Scalo	0.99	3.63	14.13	54.67	0.13	2.67	16.49	0.65	0.1	6.21
MOD2 Monterotondo Scalo	0.99	4.67	12.62	48.66	0.14	2.32	24.02	0.59	0.09	5.57
MOD3 Fiano Romano	0.91	2.62	13.75	53.49	0.14	2.34	15.7	0.67	0.14	6.08
MOD4 Val Aurelia	0.2	3.72	9.55	36.6	0.14	1.75	30.91	0.53	0.11	4.89
MOD5 Aia Valley	1.22	5.67	14.05	50.35	0.14	2.55	18.83	0.67	0.11	6.04
MOD6 Aia Valley	1.07	6.56	13.47	47.69	0.13	2.46	21.31	0.63	0.1	5.87
MOD7 Aia valley raw clay	0.66	3.66	11.33	44.48	0.13	2.5	20.88	0.59	0.09	4.98
MOD8 Narni Scalo	0.61	3.45	13.42	51.44	0.13	2.56	17.96	0.64	0.12	6.08
MOD9 Orte Scalo raw clay	0.09	1.85	9.85	37.69	0.1	2.22	32.75	0.51	0.09	3.36

Table 3.11 Major elements in samples from modern brickyards

Element	Mean	Std. Dev	Range	Minimum	Maximum
V	114	16	81	80	161
Cr	123	24	97	77	174
Co	17	3	16	8	24
Ni	71	13	61	46	107
Cu	35	6	28	22	50
Zn	103	13	89	66	155
Pb	43	13	63	25	88
Rb	227	84	422	81	503
Sr	505	110	561	268	829
Y	38	6	39	26	65
Zr	191	49	255	82	337

Table 3.12 Trace elements (ppm) in tested SES collection examples

	V	Cr	Co	Ni	Cu	Zn	Pb	Rb	Sr	Y	Zr
MOD1 Orte Scalo	128	142	18	82	35	115	32	142	449	36	149
MOD2 Monterotondo Scalo	129	125	13	58	24	98	35	135	526	30	124
MOD3 Fiano Romano	127	105	23	70	35	99	39	155	387	34	211
MOD4 Val Aurelia	111	88	8	46	45	77	36	81	829	27	108
MOD5 Aia Valley	132	130	17	62	32	94	32	158	341	33	163
MOD6 Aia Valley	127	120	14	64	28	101	29	147	336	37	141
MOD7 Aia valley raw clay	104	92	11	46	28	70	30	119	268	26	151
MOD8 Narni Scalo	142	125	19	74	35	117	64	135	419	32	167
MOD9 Orte Scalo raw clay	102	103	16	56	34	90	26	112	528	28	82

Table 3.13 Trace elements (ppm) in samples from modern brickyards

Each sample was ground down to the consistency of flour in an agate ball-mill grinder. For XRF, *ca.* 5 g of each sample in turn was put into a mould and surrounded with boric acid. The mould was placed in a press, and subjected to 12 tonnes of pressure for thirty seconds. The pressure changed the boric acid into a neutral holder for the sample.

The remaining powder was used for XRD. A small amount (*ca.* 5g) was placed in plastic trays, where the powder was cut up using a glass slide so that the mineral crystals would be as homogeneous as possible in their alignments. This increased the chances of the crystals of diffracting the X-rays, making the result more reliable. The actual analysis of the samples was performed by the technicians in the Postgraduate Research Institute for Sedimentology (PRIS) at the University of Reading, Mike Andrews and Franz Street.

3.4 Results

3.4.1 X-Ray Diffraction

It was possible to identify a suite of 11 minerals present in the samples: quartz, augite, hematite, gehlenite, calcite, analcime, muscovite, dolomite, anorthoclase, sanidine, and albite. While XRD does not necessarily indicate absolute amounts of the minerals present, it is possible to compare the relative amounts of the identified minerals to a high degree of precision. The quantification methods of Alexander and Klug (1948) and the later refinement by Hooton and Giorgetta (1977) were used in this study. Their methods assume that the peak intensity is proportional to the mass fraction of the mineral within a sample. Using the ratios of peak intensity allows a relative scaling to the mass fraction ratios. Mineral peaks were selected to minimise overlap of different minerals and maximise intensity.

Here, because I want to identify groups of samples with similar composition I expressed the amount of each mineral as a ratio to that of quartz for each sample. With quartz being one of the most common minerals in the Earth's crust, it can be argued that if two samples were made from the same clay, they will have minerals in similar proportions to each other than to samples made from a different clay. Tables 3.8 and 3.9 indicate the results by mineral; for individual sample results, please see Appendix B.

3.4.2 X-Ray Fluorescence

The XRF analysis measured the following elements in the samples: majors (% wt), Na_2O, MgO, Al_2O_3, SiO_2, P_2O_5, K_2O, CaO, TiO_2, MnO, Fe_2O_3; trace elements (ppm), V, Cr, Co, Ni, Cu, Zn, Pb, Rb, Sr, Y, Zr (results indicated in Tables 3.10 to 3.13; for individual sample results, please see Appendix C).

3.5 Critical Evaluation: Expected versus Observed

3.5.1 Textural Analysis Evaluated

In order to determine the validity of the visual fabrics, the XRD data were compared against their visual fabric type via a discriminant analysis. This technique (cf. Shennan 1997:350-352) searches for patterns in the mineralogical data which can also be distinguished in the visual fabric (e.g. if the samples said to have visual fabric 1 could in some fashion be differentiated from the others on the basis of their mineralogy). Since the mineralogy is a 'macro' characteristic of the brick fabric, something which can to a degree be studied by eye, it stood to reason that there would be good agreement between the visual fabric groupings and groupings determined by the discriminant analysis. The 13 'splittist' visual fabric categories were used. Surprisingly, the visual fabric could be discriminated from the mineralogical data for only 36% of the samples.

However, when the same analysis using the 13 'splittist' visual fabrics (where many distinctions were made between bricks) was performed against the entire available data (minerals, major elements, trace elements), there was an overall success rate of 89%. That is, nearly nine times out of ten, the analysis found that a certain visual fabric category could be distinguished from the others within the mineralogy and chemical composition of the tested bricks. When the discriminant analysis was performed using the 'clumpist' fabric categories (where far fewer distinctions were made), the success rate fell somewhat, to 84%. While still good, this suggests that the 'supergroups' do not correctly group together similar fabrics; there is a loss in resolution in exchange for ease of use. The success rate of the 'splittist' categories suggests that the classification by eye of the fabrics was somehow tuned to a greater subtlety in the clays and tempers in the brick than was expected. What may account for this subtlety? Chi-squared tests of association between the fabrics and the date of the stamps found very high probabilities of weak associations between certain fabrics and chronological periods. The strength of the associations ranged between 0.04 and 0.14, as measured using Cramer's V statistic. Fabric BSR 1 had a 99.9% chance of an association with the Julio-Claudian period; fabric BSR 4 had a 90% chance of an association with the Flavian period; Fabric BSR 11 had a 99.9% chance of an association with the second half of the second century (Antoninus Pius - Commodus); and fabric BSR 13 had a 95% chance of an association with the Severan period It appears as if it were the subtle differences in clays (perhaps represented by the trace elements in the discriminant analysis) and the tempers (as represented by the mineralogy) to which the textural classification of fabrics was responding. These differences seem to have a slight chronological aspect, and might be related to the gradual shifting from one clay body to another by brick producers over time. We will take up in some detail the question of chronological aspects of brick production in 3.7 and 5.4.2. In the meantime it seems that the discriminant analysis can be the basis for learning to recognise some real differences in the fabrics of brick by eye, an ability which should be a useful tool in the field.

3.5.2 Mineralogy

Cluster analysis was used to determine like groups of bricks. There is a variety of techniques but the appropriateness of which one to actually use is judged by the archaeologist based on a sort of impressionistic 'feel' given the nature of the data. In this case, our expectations are that the use of modern samples of known sources should provide a control. Monterotondo Scalo and Narni Scalo, being nearly 100 km distant from each other, would not be expected to appear in the same cluster. If, for instance, they do in the results of a particular clustering technique, then a different technique should be explored to find a better fit. Broadly speaking, there are two main groups of techniques: the first is 'agglomerative', the other 'partitionist'. 'Agglomerative' methods (Shennan 1997: 235-49) start with an individual observation, group together the most similar observations to that one, and then continue to add observations to these initial groups. The result is a number of different hierarchical levels of clustering at various degrees of similarity. The 'partitionist' methods on the other hand require the user to decide before starting how many clusters are appropriate. It then assigns individuals to the closest cluster, the definition of which changes slightly each time a new observation is added (Shennan 1997: 249-53).

A variety of agglomerative and partitionist methods were attempted. A number of bricks always clustered together using whichever methodology (in particular but not limited to, the C. Nunnidius Fortunatus bricks SE 47 and SE 42; the Ostorius Scapula bricks SE 170, SE 174, SE 151, also SE 166 and SE 153; and the Q. Sulpicius Sabinus bricks SE 104 and SE 116). However, when using these methods the modern samples did not cluster in any sensible way. Only 'Ward's Method' (which is agglomerative) seemed to produce results which were intuitively correct, for it grouped geographically close modern bricks into neighbouring clusters, whereas the other methods did not do this. 'Ward's Method' groups individuals into clusters with the idea that the clusters should be as homogeneous as possible- *homogeneity* being defined as the smallest difference between the summed deviations for all the members of a cluster and the mean of that cluster. The clusters are kept homogeneous by keeping the differences as low as possible in each step in the process when a new case is joined to a group (Shennan 1997: 220-241).

Broadly speaking, the cluster analysis identified two main groups, with the first group being further subdivided in two again (Figure 3.5). The inclusion of modern, sourced samples in the analysis allows the clusters to be assigned a likely geographic provenance. There is a rough east - west partition of the clusters in the two main groups. The eastern group (clusters 1-8), seems to come from the Sabina, while the western

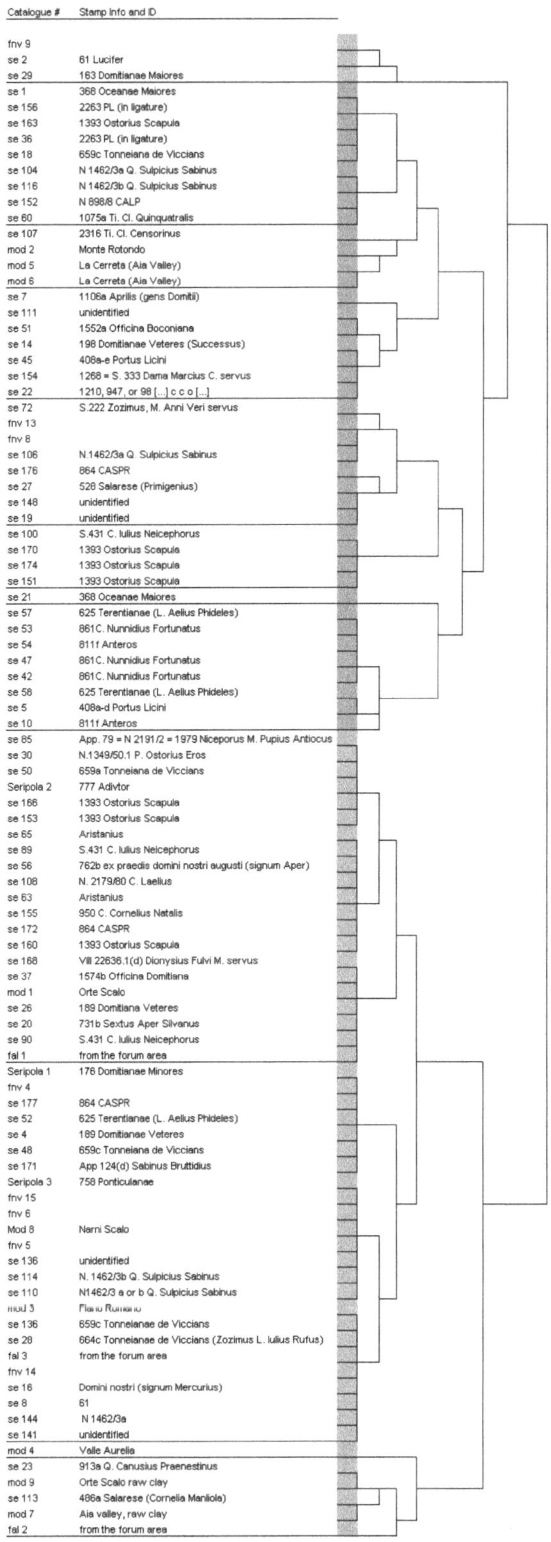

Figure 3.5 Dendrogram of the cluster analysis using Ward's Method on the XRD results

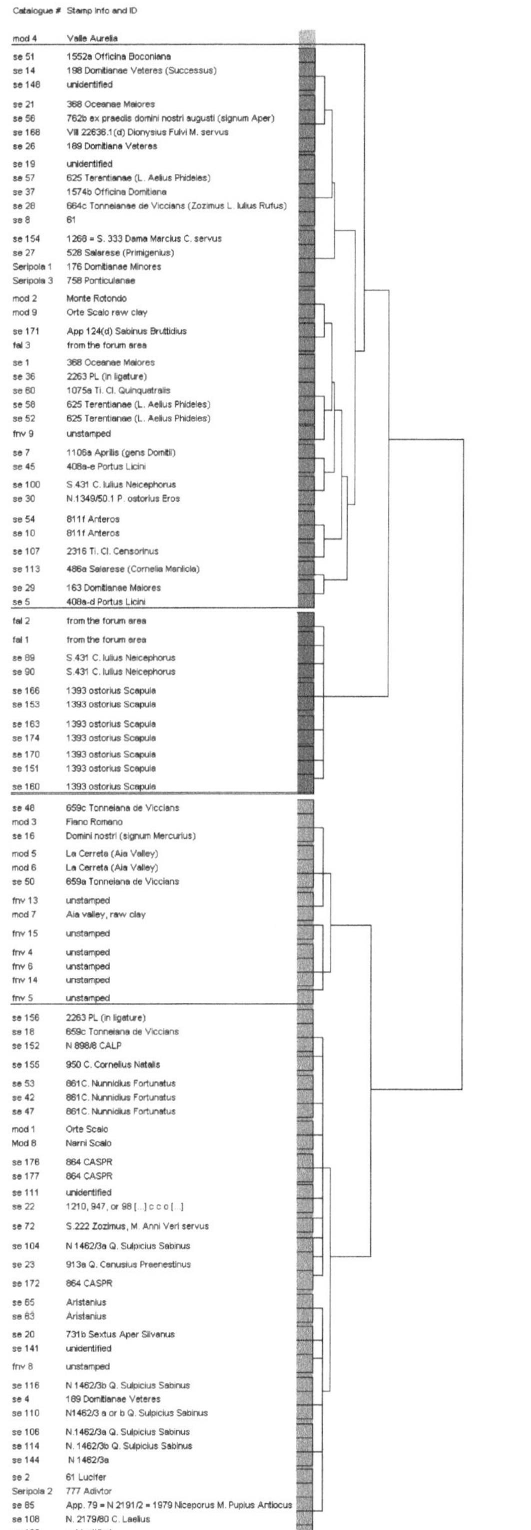

Figure 3.6 Dendrogram of the cluster analysis using Ward's Method on the XRF results

group (clusters 9 - 12) by and large comes from South Etruria, although the pocket near Narni Scalo is also a part of this group. This rough division agrees with what had already been deduced concerning some production sites. The Aristanius stamps fall in the same group as Orte, Fiano Romano, and Valle Aurelia. One site where wasters of Aristanius' work was found is just to the SW of Veii, suggesting nearby production (Kahane 1977: 182-3). On the other side of the river, Q. Sulpicius Sabinus stamps are known on the evidence of wasters, to have been produced near Eretum (Ogilvie 1965: 108-9). These fall in the same group as the modern samples from the Aia valley and Monterotondo Scalo.

3.5.3 Chemistry

'Ward's Method' was again used to group the samples, but only using the trace elements data. Meaningful results (using the same criteria as in 3.5.2) were difficult to produce using the data on the major elements. 'Ward's Method' has been used extensively for trace elements analysis (Shennan 1997: 241). It may be that the trace elements are 'locked' into the mineral structure of certain minerals which do not suffer great changes in the firing process, nor in the post-depositional history of the brick.

The dendrogram plot (Figure 3.6) of the cluster analysis reveals a number of patterns. First, there is the assignment into individual clusters of individual samples (e.g., se 47,42,53) carrying the same stamp (*CIL* XV.1 861). There is an overall dual division, as was seen on the cluster analysis of the XRD data. Again, this broad division may be interpreted as representing real geographic difference between the clay sources, while perhaps the individual clusters can be thought of as representing individual brick-making traditions. The second group can be further divided in two. By the same logic that permitted the rough division of the XRD cluster analysis into eastern and western groupings, the different clay groups of the trace elements cluster analysis can be connected with a particular region of the Tiber Valley. The division however is one of north and south, with the positioning of the modern samples in the cluster analysis running from Narni Scalo at one end of the dendrogram to Valle Aurelia at the other.

While the cluster analysis and dendrograms of the results for both the XRD and the trace elements data seem to indicate a broad bipartite division, they seem flatly to contradict each other in their assignment of different samples together into the same clusters. The question becomes which cluster analysis do we accept as the *right* one if both seem to make a bit of sense, yet suggest contradictory groupings? How can we reconcile the fact that the modern samples from the Aia group with that from Monterotondo in one instance, but with Orte and Fiano Romano in another?

3.6 Possible Sources in the Brick Industry

In fact, this is only a problem if we reason that since the mineralogy is necessarily connected to the chemistry (molecules to atoms) the cluster results ought to be similar. This would be the case if both techniques were measuring the same things. As it happens, the mineralogical data considered in the cluster analysis are connected to the *major* elements, not the traces. The trace elements are generally measured in parts per million, and do not therefore determine the nature of the major minerals or mineral groups (e.g. augite or pyroxenes). It seems to me that the cluster analyses are *complementary*, not contradictory.

The complementary relationship between the mineralogical and chemical data, as explored through cluster analysis, can be further examined through a visualization technique I call 'cluster mapping'. 'Cluster mapping' means here simply a representation of the position of a sample in terms of its positions in the original two cluster analyses. Cluster analysis really only gives one measurement, that of distance, between individual samples. It is one-dimensional. The dendrograms (Figures 3.5 and 3.6) connect the samples in a hierarchical 'staircase', a representational sleight-of-hand so that the results can be understood visually, rather than as a string of numbers. But because the two cluster analyses performed here measure complementary aspects of the same thing, the two analyses can be combined into a two-dimensional plot, cluster against cluster. My method for combining the two begins with the first observation in the first cluster in one dendrogram, and assigns it point 1 along the x axis. Each subsequent observation is positioned one discrete point higher. Every time a significant difference between clusters is encountered, the algorithm scales the co-ordinate up a user-defined amount proportional to the difference. The same process is repeated for the other dendrogram along the y axis. Each observation is thus assigned an x and y co-ordinate which can then be plotted. This can be performed by hand as easily as by computer, depending on the size of the data set. The process is illustrated in Figure 3.7.

The resulting 'cluster map' *by itself* is not to be taken as a schematised representation of actual geographical space. It merely represents in a two-dimensional way the distance between groups of bricks based on their mineralogy and trace-element chemistry, and therefore it can be used to understand the relationships between stamped bricks, and to assign provenances when sourced samples are included in the analysis. (Other statistical methods might be able to depict the same information, but as a way of illustrating the relationships between the bricks, the clustermap is more easily understood).

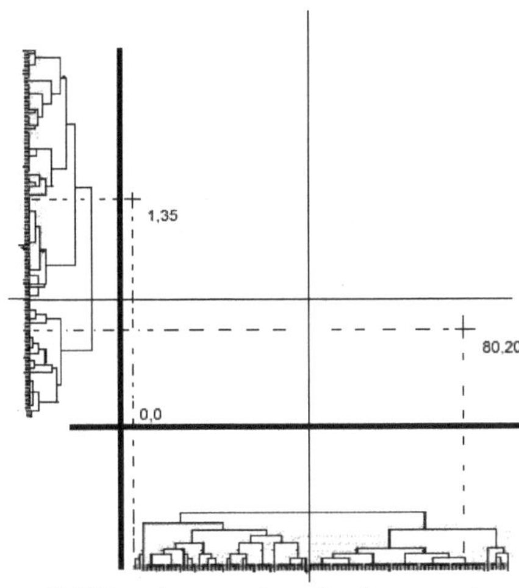

Figure 3.7 How to cross two dendrograms to produce a cluster map. The two broadest divisions in the original dendrograms, when crossed, produce a cluster map with four quadrants. Each sample then is plotted according to its position in each dendrogram. One dendrogram provides the x-coordinates, the other the y-coordinates.

The cluster map (Figure 3.8) is broadly divided into four quadrants, corresponding to the bipartite divisions in the original dendrograms, while the various blocks correspond to the original clusters at the same level of similarity in those dendrograms. In the different blocks several individual groups of bricks can be discerned (each assigned here an alphanumeric reference number). The inclusion of the modern samples does allow some rough geographic correlations to be made for the entire sample, and some quite specific provenances to be made for particular blocks. What is particularly striking is that the banks of the Tiber Valley are clearly differentiated: quadrants A and D correspond to the Sabina, while B and C correspond with South Etruria. This is probably due to the fact that South Etruria has primarily a volcanic geology while the Sabina is composed largely of sedimentary rock.

- Quadrant A : the hypothesized sources for these samples are in the area around the Aia valley (the town of Magliano Sabina), in the Sabina
- Quadrant B: the sources for these samples are hypothesized to be in South Etruria, particularly B2: Orte Scalo; B6: Fiano Romano. B3 would seem to be along the Nera river, at Narni Scalo.

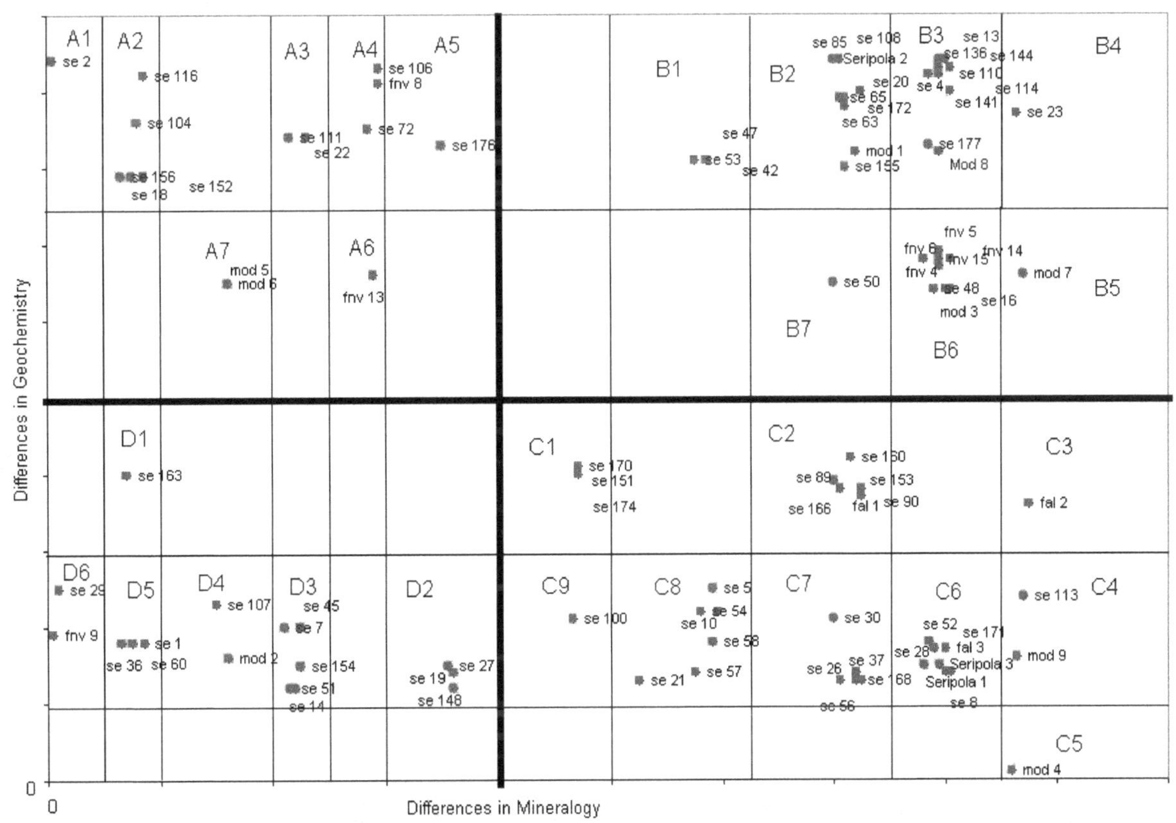

Figure 3.8 Cluster map of tested SES bricks. The inclusion of modern samples allows a generalised localisation for each quadrant: A – Aia Valley in the Sabina; B – Upper South Etruria (also a cluster, B3, around Narni); C – Lower South Etruria; D – Lower Sabina in the area of Monterotondo Scalo.

- Quadrant C: the sources for these samples are hypothesized to South Etruria also, but much closer to Rome, in the area near the Vatican

- Quadrant D: the sources for these samples (and in particular, D4) are hypothesized to the area around Monterotondo Scalo, in the Sabina

If we are studying a finished brick, which can be thought of as raw clay + temper, then the inclusion of raw clay alone for analysis in this methodology is not enough to determine an accurate provenance. The two samples of raw clay included in the study (MOD 7 and MOD 9) do not fall in the cluster map close to their associated manufactured bricks (MOD 5, MOD 6; MOD 1, respectively). This difference may be attributed to the *human* element, that is, the tempers added by the brick makers (cf. 4.5.3). In the modern samples studied the tempers added were quite local, and this tradition of using both local clays and local tempers is likely no different from ancient practice. In this methodology, the mineralogical data can be thought of as differentiating the samples on the basis of their tempers, while the chemical data tease apart the differences in the clays. Combining the two I would argue gives a complete picture whereas studying the raw clay (without the addition of temper) is necessarily only partial. Future studies which use this method of visualizing the interrelationships in an assemblage of brick therefore ought to take care that samples chosen for comparison to anchor the provenancing should be as similar as possible to the material under consideration. The success of this methodology can be estimated by performing a discriminant analysis of source quadrants against the total geochemical and mineralogical composition of the bricks. If the methodology is not valid, then there should be a very poor success rate in the discriminant analysis. As it happens there was a 94% success rate in discriminating source quadrants as determined by the cluster map from the geochemical and mineralogical data. For the 'errors', the suggested better groupings were between Quadrants A and D, and Quadrants B and C. The probabilities of the 'wrong' samples for membership in the other Quadrants were sufficiently minimal that it makes better sense to leave them as the cluster map suggested.

The different visual fabrics do not seem to be associated with any of the source quadrants. The only exception is BSR Fabric 4. This fabric has a 99.9% chance of being associated with Quadrant B, the upper South Etruria. The strength of this association is weak (Cramer's V = 0.22), but still stronger than any associations between fabrics and dates (3.5.1). However it may be possible to determine the source of a brick based on the presence of only a few key minerals. The relative amounts of the different minerals present do seem to differentiate according to provenance. Figure 3.9 plots the relative amounts of minerals by source quadrant. The amounts of augite, dolomite, and the feldspars in relation to quartz are the useful indicators. For instance if there seems to be three times as much augite present as quartz, the brick may come from near Monterotondo Scalo. Because there does not seem to be an association between any of the visual fabrics identified in 3.3.2 and these source quadrants, comparing the relative amounts of minerals in a fabric might therefore be a better qualification for fabric type (if we wish type to be related to source) than the subjective impressions which led to the creation of Tables 3.5 and 3.6, my 'splittist' and 'clumpist' fabric descriptions.

The cluster map and the relationships it illustrates form the bedrock for the reevaluation of the brick industry developed in the following chapter. However, the immediate implications of knowing where bricks were

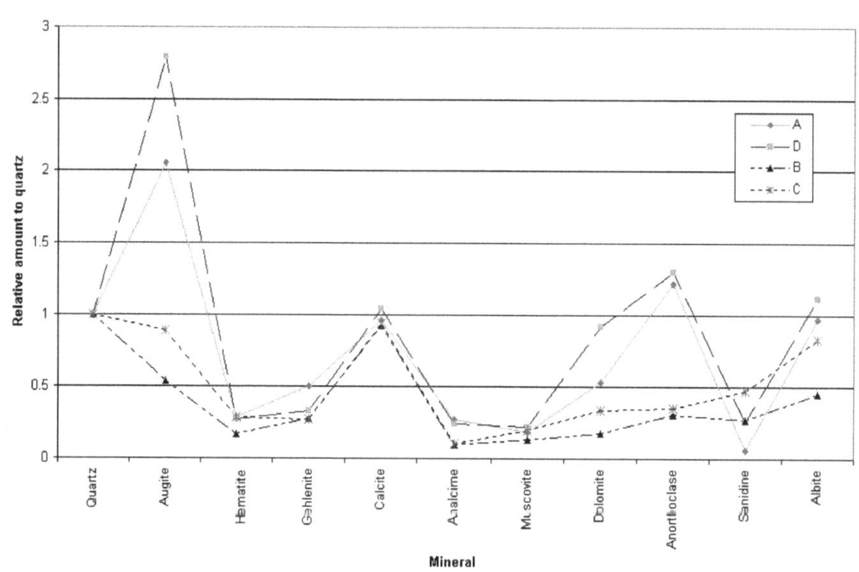

Figure 3.9 Mineral assemblages by source quadrants. Quadrants A and D correspond with the Sabina; quadrants B and C correspond with South Etruria.

produced in the Tiber Valley are that:

- certain *figlinae* had multiple establishments across the landscape: *Salarese*; *Viccianae*; *Portus Licini*, *Terentianae*, *Domitianae Veteres*, *Oceanae Maiores*
- certain sources were exploited by multiple *figlinae*
- certain individuals exploited different sources at the same time: Ostorius Scapula, St. Marcius Lucifer
- certain individuals used different stamps when concurrently exploiting different sources: C. Nunnidius Fortunatus (*CIL* XV.1 861 and 862)
- others used the same stamp when exploiting different sources: St. Marcius Lucifer, (*CIL* XV.1 61), L. Aelius Phidelis (*CIL* XV.1 625), Ostorius Scapula (*CIL* XV.1 1393).
- much of the unstamped brick recovered at the Forum Novum villa site (Gaffney *et al.* 2001) was imported across the river from the Fiano Romano area, an area also exploited by the *figlinae Tonneianae de Viccians*. A small amount was produced in the local area of the settlement
    - bricks were transported both up and down the Tiber valley. For example, SER 1, found at Seripola (across the river from Orte) is provenanced to the area near the Val Aurelia. The cluster at B3 is provenanced to Narni, but were all found in South Etruria

These results should caution one from letting what is written in brick stamps guide interpretation of archaeometric results (i.e., one should not argue that because these two bricks carry the same stamp they are from the same yard therefore any variation in the mineralogy/chemistry must be that present in the original source clay; cf. 3.2.3). Some *figlinae* do indeed exploit multiple clay sources, while other clay sources are exploited by multiple *figlinae*. It does not seem possible therefore to create 'identity cards' for the various *figlinae* mentioned in brick stamps (*pace* Olcese 1993: 123). In some cases it may be better to think of *figlinae* as not so much brick-yards, or clay districts, but rather as estates which owned numerous parcels of not necessarily contiguous land.

3.7 Refining chronologies of production and development

Multi-discriminant analysis can be used to refine this picture presented above (especially 3.5.1), where instead of discriminating the visually identified fabric types from the geochemistry of the bricks, I try to discriminate the information on the stamps themselves from the bricks' geochemistry. In most cases the differences between the groups are sufficiently large that there is little danger of misclassification. In the highlighted cells in the tables in this section, there is some danger of overlap. In these cases, where for example the predefined group was Julio-Claudian but the analysis suggests that Nerva-Hadrian or Antoninus Pius-Commodus might be a better fit one has to examine the probabilities carefully, and see which chronological reassignment makes sense given what else is known about that particular brick. It is important to note that the determination of stamp *types*, as recorded in *CIL*, has more to do with what the actual *text* of a stamp says rather than the use of a new *die*. To determine whether a different die is being used one needs to measure the size and depth of the letters and so on

If the MDA can discriminate patterns that correspond to groupings provided by the dates recorded on the stamps (indicated by the appearance of consular date formulae, or the names of otherwise known individuals), we might conclude that the clay sources used in the industry changed over time (which was suggested in 3.5.1).

Possible Outcomes

Should the MDA not be able to differentiate the groups, then a possible explanation might be that some clay sources remained in usage throughout the four centuries.

Another possible outcome might be that some periods can be discriminated and others cannot, which would indicate that multiple sources remained in use both within and across date periods.

We may be able to say something about the patterns of land-holding in the middle Tiber Valley if patterns in the geochemistry agree with family groupings suggested by the stamps.

Misclassifications are also significant, because the datings for many stamps are fairly secure. These are the proverbial 'exceptions which prove [ie., test] the rule'.

The MDA was run five times, grouping the stamped bricks together in different groups on chronological grounds. The first run, the 'fine-dating group', had a resolution of about 50 years. (In a much larger sample, it would be possible to group at the level of individual years, thanks to the practice of using consular dates on stamps). The second run collapsed groups together, so that the 'medium-dating group' had a resolution of about 100 years. For the third run the bricks were grouped and tested at the level of 'undated/dated'. Brick stamps also contain much information concerning the landlord and the brick producer themselves,

Group	1	2	3	4	5	6	7	8
1	0	2.9446	**9.6149**	**25.6729**	**8.9367**	**7.2836**	**10.2892**	35.0643
2	2.9446	0	**9.6892**	**25.6763**	**8.1452**	**7.8595**	**10.805**	30.9824
3	9.6149	9.6892	0	19.1295	13.0877	5.8403	16.411	40.4039
4	25.6729	25.6763	19.1295	0	23.4212	18.6745	**24.837**	25.0008
5	8.9367	8.1452	13.0877	23.4212	0	7.8472	4.6949	27.3759
6	7.2836	7.8595	5.8403	18.6745	7.8472	0	8.2881	34.8675
7	10.2892	10.805	16.411	24.837	4.6949	8.2881	0	24.1689
8	35.0643	30.9824	40.4039	25.0008	27.3759	34.8675	**24.1689**	0

Table 3.14 Squared distances between 'fine dating' groups. Shaded cells indicate 'overlap', where it is difficult to clearly discriminate groupings.

Group	1	2	3	4	5
1	0	9.0949	5.7714	9.8339	29.7064
2	**9.0949**	0	6.4376	14.5773	31.402
3	5.7714	6.4376	0	4.5988	26.5395
4	**9.8339**	14.5773	4.5988	0	22.2761
5	29.7064	31.402	26.5395	22.2761	0

Table 3.15 Squared distances between 'medium dating' groups. Shaded cells indicate 'overlap', where it is difficult to clearly discriminate groupings.

Group	1	2
1	0	5.54452
2	5.54452	0

Table 3.16 Squared distances between 'coarse dating' groups. There are no overlaps.

allowing the tested bricks to be grouped by various families (these groupings cut across the groupings by date), which is what was done in the fourth run of the MDA. In the final run, these family groupings were collapsed into 'Imperial House/non-Imperial House'. In all runs the predictors used were: Na_2O, MgO, Al_2O_3, SiO_2, P_2O_5, K_2O, CaO, TiO_2, MnO, Fe_2O_3, V, Cr, Co, Ni, Cu, Zn, Pb, Rb, Sr, Y, Zr.

In the discussion which follows and in particular the conclusions drawn from the discussion, one should understand that the number of conditional phrases which by rights ought to be used would make the text too unwieldy to read. Therefore I have tended to write this section in a positive manner to ease the reading.

3.7.1 MDA: Fine Dating

53 of 75 examples discriminated into the 'correct' groups, a success rate of 71%. See Table 3.14 for overlaps.

Group 1: Undated. 23 examples; 12 grouped correctly, a success rate of 52%.
Group 2: Unidentified. 5 examples; all 5 grouped correctly.
Group 3: 1st century. 13 examples; 8 grouped correctly, a success rate of 62%.
Group 4: Flavian. 2 examples; both grouped correctly.
Group 5: Nerva to Hadrian (1st half of 2nd century). 9 examples; 7 grouped correctly, a success rate of 78%.
Group 6: Antoninus Pius to Commodus (2nd half of 2nd century). 10 examples; 9 grouped correctly, a success rate of 90%.
Group 7: Septimius Severus et al. (end of 2nd, 1st quarter of 3rd). 11 examples; 8 grouped correctly, a success rate of 73%.
Group 8: Diocletianic. 2 examples; both grouped correctly.

Fine Dating Discussion

The overlap of all later periods (aside from the Group 8 Diocletianic bricks) with groups 1 and 2 suggests only that there are bricks within these groups which could be separated into the other categories, if only we had better information from their stamps. The overlap of Group 7 Severan bricks with Group 4 end of 1st century bricks and Group 8 Diocletianic bricks is more interesting. Since Group 7 is clearly discriminated from the immediately preceding periods, this overlap could be interpreted as a return to earlier-exploited clay sources, which continue to be exploited during the Diocletianic period.

Misclassifieds with greater than or equal to 90% probability

SE7 originally Group 5; MDA suggests reclassification into group 7 (group 5, 8%; group 7, 90%)

This was the only one with 90% probability. The information on this brick's stamp is as follows:

SE 7 **CIL XV.**11106a LSO 873
A[...]MITI/[...]BVLI/DOL
aprilis cn domiti / agathobuli / dol(iare)
Trajan/Hadrian (Steinby 1974: 55)

The dating is fairly secure, based on a known individual's name showing up in other stamps with the year 123 marked on them. The MDA suggests reassigning this brick to group 7, which dates to the end of the 2nd century, beginning of the 3rd (the reign of Septimius Severus). Aprilis might be an early user of a later-exploited source.

Overall, the MDA finds that 71% of the bricks fit with their assigned groupings, in this fine dating category.

3.7.2 MDA: Medium Dating

60 of 75 examples discriminated into the 'correct' groups, a success rate of 80%. See Table 3.15 for overlaps.

Group 1: undated/unidentified. 28 examples; 21 grouped correctly, a success rate of 75%.
Group 2: 1st century. 15 examples; 11 grouped correctly, a success rate of 73%.
Group 3: 2nd century. 19 examples; 15 grouped correctly, a success rate of 79%.
Group 4: end of 2nd century, 1st quarter of 3rd. 11 examples; all 11 grouped correctly.
Group 5: Diocletianic. 2 examples; both grouped correctly.

Medium Dating Discussion

The overlap here between Group 1 Undated and Group 2 1st century and Group 4 end of 2nd/beginning of 3rd suggests that the bricks in the undated group could probably be sorted into these other two categories.

Misclassifieds with greater than or equal to 90% probability

SE7 originally group 3; suggested reclassification group 4 (group 3, 1%; group 4, 98%)

This is the same brick that was misclassified above (in section 3.7.1). The analysis has slotted it into the same date group again.

3.7.3 MDA: Coarse Dating

65 of 75 examples discriminated into the 'correct groups', a success rate of 87%. See Table 3.16 for overlaps.

Group1: undated/unidentified. 28 examples; 23 grouped correctly, a success rate of 82%.
Group2: dated. 47 examples; 42 grouped correctly, a success rate of 89%.

Coarse Dating Discussion

Misclassified with greater than or equal to 90%

SE50 originally group 2; suggested reclassification group 1 (group 1, 91%; group 2, 9%)

SE 50. CIL XV.1 659a LSO 563
TONNEI DE FIGLIN VICCIANS
tonnei de figlini(is) / viccians
mid 1st century (Steinby 1974: 96)

Other bricks carrying this stamp type were correctly discriminated which might suggest that this brick was made from a different clay source. This reinforces the argument earlier that a single *figlinae* name could incorporate multiple sources.

3.7.4 Other Misclassified Tested Bricks

In this section, the various probability levels suggest the level of confidence one should put in the interpretations.

MISCLASSIFIED BRICKS IN ALL THREE DATING GROUPS

Key:
Brick ID# Coarse Dating original group - suggested reclassification probability; Medium Dating original group - suggested reclassification probability; Fine Dating original group - suggested reclassification probability
Archaeological information associated with this brick.

SE 18 2-1 60%; 2-1 65%; 3-6 52%
CIL XV.1659c LSO 565
...]ONNEI DE[...
[t]onnei de [figl(i)n(is) / viccians
mid 1st century (Steinby 1974: 96)

SE20 2-1 61%; 3-2 54%; 6-4 69%
CIL XV.1731b LSO 627
EX PR VM [... / ...] XAP[...]
op dol] ex pr vum[i qvad et an / faus ex fi se]x ap[silv
AD 140-160

SE50 2-1 61%; 2-1 79%; 3-1 44%
CIL XV.1 659a LSO 563
TONNEI DE FIGLIN VICCIANS
tonnei de figlini(is) / viccians
mid 1st century (Steinby 1974: 96)

SE156 1-2 83%; 1-3 71%; 1-6 84%
CIL XV.1 2263
PL

SE168 1-2 83%; 1-3 45%; 1-6 40%
CIL VIII 22636.1(d) = CIL XV.1 976
DION[...]/FVL[...]
dionysiu[s]/fulvi m s[er]

SE171 1-2 57%; 1-4 65%; 1-7 60%
App 124(d)
[...]VT/[...]AE
[sabinus br]ut[/tid volusian]ae

Two of these bricks have fairly secure datings based on the internal evidence of their stamps, SE18 and SE20. In the case of SE18, the suggested re-datings are not consistent across the board, which could be interpreted as a case where this brick may be made from a relatively minor clay source not otherwise present in the tested bricks. In the case of SE20, the re-datings are consistent, which might make this brick more similar to the late 1st century bricks than bricks made contemporaneously roughly 100 years later and so represent a manufacturer exploiting a previously expired source (perhaps because of better technology? Alternatively, it could represent a case where a source closer to the city of Rome was being used, if we assume that the earliest sources used would be closest to the city.

A third brick, SE156, and its brother, SE36 (see below), both carrying stamp CIL XV.1 2263, are thought on the basis of the stamp shape (PL in ligature, in a rectangle), to date to the 1st century (Kahane 1968). 'PL' is tentatively thought to refer to the 'Portus Licini', which is known from stamps dating to the Severan age, and also to still be active during the reign of Theodoric. The MDA reassigns both of these bricks to the 2nd half of the 2nd century, just before 'Portus Licini' starts being used regularly on stamps. The MDA suggests we move forward by roughly a century the presumed date of birth for this *figlinae*, which would make it likely that this *figlinae* was owned by the *gens Domitii*, and inherited by the Severans.

For the last two bricks, SE168 and SE171, the re-assigned datings are consistent (from undated to dated; from undated to end of 2nd century/beginning of 3rd; from undated to the reign of the Severans). These were thought to date to the 1st century because of the rectangular shape of their stamps; the MDA suggests that stamp shape and period might not be as closely correlated as traditionally thought.

MISCLASSIFIED BRICKS IN THE COARSE & MEDIUM DATING GROUPS

Key:
Brick ID# Coarse Dating original group - suggested reclassification probability; Medium Dating original group - suggested reclassification probability
Archaeological information associated with this brick.

SE22 1-2 68%; 1-3 64%
CIL XV.1947 (?)
[...] C C O [...

SE113 2-1 68%; 3-1 51%
CIL XV.1 486 a LSO 445
APR ET PAET CO[...]/CORM[...]
AD 123 (Steinby 1974: 83)

SE148 1-2 53%; 1-3 51%
Unidentified
C[...]/E[...]

The MDA for these bricks is equivocal; it is suggesting that they should perhaps be redated. Are these probabilities sufficiently strong for us to do so? The consular date in SE113 settles the matter for that brick, in which case the MDA should be interpreted as indicating another minor source in use during that period. As for the other two, both redatings for SE22 are to the second century, as are the redatings for SE148, which suggests for SE22 that perhaps the initial classification was incorrect, and for SE148, that the correlation between rectangular stamps and the first century is not completely clearcut.

MISCLASSIFIED BRICKS IN THE MEDIUM & FINE DATING GROUPS

Key:
Brick ID# Medium-Dating original group - suggested reclassification probability; Fine-Dating original group - suggested reclassification probability
Archaeological information associated with this brick.

SE7 3-4 98%, 5-7 90%
CIL XV.11106a LSO 873
A[...]MITI/[...]BVLI/DOL
a[prilis cn do]miti / [agathob]uli / dol(iare)
Trajan/Hadrian (Steinby 1974: 55)

The date for this stamp type is not in question (Aprilis is known in other dated stamps), so we should interpret the MDA result as suggesting that Aprilis was using a piece of land later exploited heavily under the Severans.

SE36 2-3 71%; 3-6 80%
CIL XV.12263
PL

See the discussion above in *MISCLASSIFIED BRICKS IN ALL THREE DATING GROUPS.*

SE48 2-3 75%; 3-5 59%
CIL XV.1 659c LSO 565
TONNE[...]
tonne[i de figl(i)n(i)s / viccians
mid 1st century (Steinby 1974: 96)

The tortured chronology of this *figlinae* is discussed by Steinby (1974), and this particular stamp type is fairly securely dated, so the MDA may be indicating another piece of land exploited heavily at a later date. Not all the tested bricks carrying this stamp type were misclassified, which suggests that the same stamp type could be used on products from different sources.

SE54 3-4 54%; 5-6 47%
CIL XV.1 811f LSO 684
DOL AN[...]
dol(iare) an[terot(is) Sever(iari) caes(aris)

The low level of probability in this case means that we can safely ignore this reassignment. Be that as it may, there is debate about the chronology of Anteros' stamps; on the basis of his name, he might be in the Severan household, but for this stamp type at least, the dating is fairly secure. Another brick carrying the same stamp, SE10, was not misclassified, so we might have Anteros as an early exploiter of a later source, and using the same stamp to mark bricks from different sources.

SE152 1-4 55%; 1-2 56%
N 898/8
[...]ALP [c]alp

This may have been reclassified by the MDA because its clay source is not otherwise attested in the tested bricks (given the conflicting groupings, and low levels of probability).

SE177 1-3 47%; 1-6 50%
CIL XV.1 864 LSO 714
[...]ASPR
caspr

Here we might have more evidence supporting the idea that rectangular stamps do not necessarily indicate a creation date in the 1st century.

3.7.5 MDA: Family Groupings

Brick stamps frequently carry the name of the land-lord or the brick manufacturer, which allows a re-group of the tested bricks into family groupings. The groups used here are the largest groups within the tested bricks, although other family names are present. These latter individuals and the tested bricks for which family information is not known are grouped under 'others'. It is also helpful to divide out Aristanius and Q. Sulpicius Sabinus, because the location of their workshops is reasonably certain (Aristanius- La Storta Peña 1987: 55-71; Q. Sulpicius Sabinus – Eretum Ogilvie 1965: 108-9)

Family
62 of 75 examples discriminated into the 'correct groups', a success rate of 83%. See Table 3.17 for overlaps.

Group 1: gens Domitii. 5 examples; 4 grouped correctly, a success rate of 80%.
Group 2: Severans. 11 examples; 10 grouped correctly, a success rate of 91%.
Group 3: Asinii. 3 examples; All 3 grouped correctly.
Group 4: Q. Sulpicius Sabinus. 6 examples; All 6 grouped correctly.
Group 5: Ostorius Scapula. 7 examples; All 7 grouped correctly.
Group 6: Aristanius. 2 examples; Both grouped correctly.
Group 7: Others. 41 examples; 30 grouped correctly, a success rate of 73%.

Group	1	2	3	4	5	6	7
1	0	7.445	**24.234**	47.687	111.659	55.69	6.364
2	7.445	0	**24.431**	36.143	137.65	38.656	5.922
3	24.234	24.431	0	53.379	147.217	48.073	18.946
4	47.687	36.143	53.379	0	174.536	52.308	33.044
5	111.659	137.65	147.217	174.536	0	167.013	117.285
6	55.69	38.656	48.073	52.308	167.013	0	32.271
7	6.364	5.922	18.946	33.044	117.285	32.271	0

Table 3.17 Squared distances between 'family' groups. Shaded cells indicate 'overlap', where it is difficult to clearly discriminate groupings.

Family Group Discussion

There is close overlap between the Asinii (Group 3) and the gens Domitii (Group 1) and the Severans (Group 2). These are aristocratic families; Groups 4, 5, and 6 are not (although Ostorius Scapula may have been the general who led the mopping up operations in Britain after the invasion in 43). This might suggest a pattern of land-holding where aristocratic properties are intertwined in a nucleus of productive lands, while lesser folk are isolated from this nucleus. The groups where the MDA analysis agrees completely suggests a single source each for the products of the Asinii, Q. Sulpicius Sabinus, Aristanius, and Ostorius Scapula. The production of Aristanius bricks was assigned by Pena to a site near Veii (La Storta, Peña 1987: 55-71); that of Q. Sulpicius Sabinus is assumed, based on the presence of wasters, to have been located near Eretum (Ogilvie 1965: 108-9).

MISCLASSIFIED
Greater than or equal to 90%
SE7 originally group 1; suggested reclassification group 2 (group 1 3%, group 2 95%)
CIL XV.11106a LSO 873
A[...]MITI/[...]BVLI/DOL
aprilis cn domiti / agathobuli / dol(iare)
Trajan/Hadrian (Steinby 1974: 55)

Given that the Severans inherited the imperial properties upon reaching the purple, and that a significant proportion of the gens Domitii's lands would have been included, it is not surprising that this brick should be reclassified. It was also reclassified by date to go with the Severan bricks, which suggests that Aprilis was exploiting a virgin piece of ground which later became part of the Severan industry.

SE141 originally Group 7, suggested reclassification Group 4 (group 7 1%; group 4 99%)
Unidentified
N[...]S.L[...]

On this evidence, it would seem that somebody else is exploiting the same clay source as Q. Sulpicius Sabinus. Another sample was reclassified, though with a lower level of probability, to the Q.

Sulpicius Sabinus group:

SE4 originally Group 2; suggested reclassification Group 4 (group 2 19%; group 4 73%)
CIL XV.1189
...]R[...]/[...]D D N
[... opus doliare ex prae]d d n[/ ex fig vete]r[es]
AD193-198 (Steinby 1974: 39)

The standard explanation for the name of this *figlinae* (Veteres) is that it may have been an older part of the *figlinae Domitianae*, brought back into service, and that its name is a contraction of the also known '*Domitianae Veteres*'. But perhaps, on this evidence, it is Q. Sulpicius Sabinus' estate, secured and renamed by the new Imperial owners.

OTHERS RECLASSIFIED:
Into the gens Domitii (Group 1)
SE13 originally group 7; probability 59%
CIL XV.1659c LSO 565
TO[...]/[...]CCIA[..]
to[nnei de figl(i)n(is)] / [vi]ccian[ns]
mid 1st century (Steinby 1974: 96)

SE21 originally Group 7 ; probability 84%
CIL XV.1368 LSO 347
[...]C OCEA MAI CAES N OP DO / [...]Q[...]DE[...]
[ex fi]c(linis) ocea(nis) mai(oribus) caes(aris) n(ostri) op(us) do(liare) / Q. Perusi Pude(ntis)
Antoninus Pius (Steinby 1974: 70)

SE36 originally Group 7 ; probability 66%
CIL XV.12263
PL

SE89 originally Group 7; probability 75%
CIL XV.1 S.431 LSO 1175
[...]LINEIS/[...]EICEPH
[fig]lineis / [c iuli n]eiceph

SE90 originally Group 7 ; probability 55%
CIL XV.1 S.431 LSO 1175
FIGILINEIS/CIVLINEICEPH
figlineis / c iuli neiceph

What is interesting about these reclassifications is that they suggest that lands owned by the gens Domitii are being shared by other brick producers. Also, SE89, SE90, and SE100 all carry the same stamp, but only SE89 and SE90 were reclassified here, which suggests that C. Iulius Neicephorus was using the same stamp to stamp products coming from different areas. The SE21 stamped brick carries the name of the *figlinae Oceanae Maiores*; it has been argued by T. Helen that the (a?) location of this *figlinae* is near the town of Orte. Not far from this town, in the neighbouring territory of Bomarzo, is a known estate of the Domitii. The *Portus Licini*, when named as such on brick stamps, belongs to the Severans. But if the PL stamps can be redated to the period immediately preceding the Severans, then these results suggest that originally it belonged to the *gens Domitii.*

Steinby (1974: 39-39) has worked out the development of the *figlinae* connected with the *Domitii* family from the known history of the family and the appearance/disappearance of *officinatores*, slaves, and freedmen named in the stamps. According to her reconstruction, in the time of Domitian, the only *figlinae* owned by the family were the *Figlinae Domitianae*. After this early appearance, there seems to be a lacuna in stamping bricks identifying these *figlinae*, although Cn. Domitius Tullus and Cn. Domitius Lucanus (and indeed all the members of the family) often appear in the stamps unconnected with a named *figlina*, so it is probable that production continued. The next stamp which names the *figlinae* appears in AD 138, but by the time of Marcus Aurelius and Faustina, the *figlinae* have split into two subdivisions, the *Domitianae Maiores,* and the *Domitianae Minores*. At this time there seems to be a connection based on an *officinator* (Sex. Publicius Consors) who transferred between the *Maiores* and the *Figlinae Ponticulanae*, and later under Commodus, to the *Bocconianae*. If, and it is by no means proven, transfer of *officinatores* between *figlinae* occurs only where the two *figlinae* are in close physical proximity, then *Maiores, Ponticulanae,* and *Bocconianae* could be neighbours. (The *Bocconianae*, it has been argued, give their name to the modern village of Boccignano, and so probably were located in that area. Steinby 1978)

In a similar way, the *Domitianae Minores* is connected in the latter half of the second century with the *Genianae, Caepionianae,* and the *Portus Licini* (via the *officinator* P. Aelius Alexander). Finally, under Commodus a new division of the *Domitianae* appears, the *Domitianae Veteres*. It seems to be connected to the *Minores* (on the grounds that a common *signum* appears in stamps of both; 37-39). If one can connect the *Domitianae Veteres* with the estate of Q. Sulpicius Sabinus as argued above, then it might be argued that the locus for these *figlinae* is somewhere between Eretum and the Farfa River.

Into the Severans (Group 2)
SE10 originally Group 7; probability 62%
CIL XV.1 811f LSO 684
DOL ANTEROT SEVER CAES
dol(iare) anterot(is) SEveri(ari) caes(aris)
ca AD 123

There is some debate about Anteros (SE10), concerning his precise relationship with the Severans and the dating of his stamps. SE10, carrying stamp CIL XV.1 811f is dated by Steinby (1974) and others to around 123; in the MDA on the date groupings (above section 5.2.1 – 5.2.4), SE10 is not reclassified, so that date probably holds (although the other example, SE54, has a low probability of dating to the end of the 2nd century; see discussion above) . In which case, Anteros might have been an early exploiter of a piece of land later brought into production by the Severans (much like Aprilis).

SE51 originally Group 7 ; probability 73%
CIL XV.1 1552a
OFSPOFBO
of(ficina) s(ummae) p(rivatae) of(ficina) bo(coniana)
Diocletianic

SE51 is a late stamp; perhaps suggesting that the exploitation of Severan lands continued through the intervening century?

SE152 originally Group 7; probability 73%
N 898/8
[...]ALP
[c]alp

SE152, when considered in the medium dating group, was reclassified to the end of the 2nd century and beginning of the 3rd; but in the fine dating group, it had a slightly higher chance of being in the undated group; the reclassification here might suggest a point in time at the opening of the 3rd century, and a point in space on the lands owned by the Severans.

SE171 originally Group 7; probability 56%
App 124(d)
[...]VT/[...]AE
[sabinus br]ut[/tid volusian]ae

SE171 was reclassified in each run of the MDA dating groups, from the undated group to the end of the 2nd century/beginning of the 3rd. Its appearance here perhaps confirms this dating, where this producer is sharing lands, or producing alongside those properties, of the Severans.

Into the Asinii (Group 3)
SE2 originally Group 7; probability 61%
CIL XV.161; LSO 92
]CIFER
[sta(tius) marcius lu]cifer
late Trajanic

Another brick carrying a copy of the stamp in SE2 is SE8; SE8 did not get reclassified, and so we have another instance of a brick producer using the same stamp on products from different sources.

3.7.6 MDA: Family Groupings Redux

Much is made of the concentration of brick production into the hands of the Emperor over the course of the second century. This grouping tests whether bricks stamped with Imperial nomeclature or formulae can be differentiated. If so, then that may suggest that the consolidation of ownership within the hands of the Emperor is an actual rationalisation of land-ownership patterns on the ground, a concentration of production on particular clay-sources (which perhaps led to economies of scale?)

Group	1	2
1	0	4.70271
2	4.70271	0

Table 3.18 Squared distances between 'Imperial and non-Imperial House' groups. There is no overlap.

63 of 75 examples discriminated into the 'correct groups', a success rate of 84%. See Table 3.18 for overlaps.

Group 1: Imperial Household (all periods). 26 examples; 23 grouped correctly, a success rate of 89%.
Group 2: non-Imperial House. 49 examples; 40 grouped correctly, a success rate of 82%.

Discussion of Family(2) Group

Misclassifieds greater than 90%

SER2 originally Group 1; suggested reclassification Group 2 (Group 1, 6%; Group 2 94%)
CIL XV.1 777 LSO 655
[...]TORIS/[...]GL [adiu]toris/[au]g l
1st century

Nothing much can be argued from this, other than that this individual worked lands not exploited by the later 2nd and 3rd century emperors (whose signal in the data probably swamps whatever signals are present from the 1st century imperial activities).

Dividing into 'Imperial House' and 'non-Imperial' House is in effect, due to the nature of the sample, not much different from dividing into '2nd century' and '1st-century', because of the evolution of stamp types: 2nd century and later stamps are much more likely to include longer individual and estate name formulas (which can indicate imperial ownership). But the clear ability of MDA to discriminate Imperial House bricks from non-Imperial House does indeed suggest some sort of rationalisation of brick production during the second century.

3.7.8. Refinement Conclusions

- Fine-Dating Groups: 71% chance of grouping correctly (roughly, 50 years/group)
- Medium-Dating Groups: 80% chance of grouping correctly (roughly, 100 years/group)
- Coarse-Dating Groups: 87 % chance of grouping correctly (dated/non-dated)

- Family-Groups: 83% chance of grouping correctly (6 families, other)
- Family-Groups Redux: 84% chance of grouping correctly (Imperial House/non-Imperial House)

This would suggest:

1. That *figlinae* can be fragmented bodies, with scattered holdings throughout the hinterland of Rome.
2. That different sources are brought into production at different periods;
3. That the differences in the clay and temper are sufficient that these can be distinguished, but occasional sources seem to be in production over a number of periods. There would also seem to be differences within periods:

   - aristocratic families seem to be clustered together, exploiting similar (yet distinguishable) clays,
   - lower social classes are using very different clays (which are perhaps on the marginal clay lands in the Tiber Valley).

These suggestions support the conclusions in section 3.6, which is heartening, since both approaches focus on different aspects and different statistical treatments of the data.

The pattern of misclassifications, especially where another brick carrying a copy of the same stamp does not misclassify, is especially interesting, because it suggests that the information on the stamp is not necessarily being used for indication of origin. Other misclassifications point to the eventual congregation of lands in the hands of the Emperor, a process surmised from brick stamp evidence, but here visible in the statistics.

Overall, both dates and family groups can be distinguished statistically using MDA on XRF data. This can be explained by arguing that since clay is a non-renewable resource (in the form that is useful for brick production, where its relative geographical position vis-à-vis the city, river, and road networks, and fuel sources, are all material considerations), different families exploit different sources at different times, but over time, the overall locus for the Tiber Valley brick industry is shifting (perhaps as easier to exploit sources are exhausted). The tendency for the ownership of brick production to concentrate in the hands of the Emperor also probably accompanies a rationalisation of brick production on the ground (an acceleration of the traditional patterns of aristocratic land exploitation?), perhaps replacing an earlier pattern of dispersed production. No doubt other grouping patterns can be deduced using MDA, since 'family' and 'dating' only enjoy a roughly 80% success rate.

3.8 Chapter Summary

There have been various archaeometric studies of Tiber valley ceramics over the last thirty years. When these were considered altogether, there appeared to be certain patterns which I therefore expected to find in this study. The most important of these patterns was that different products of the same *figlinae* might be composed of differing clays, and that different *figlinae* might produce products using the same clays as other *figlinae*. In due course this study did find that to be the case. This is a crucial finding, for it complicates the interpretation of the brick industry considerably. Of equal if not greater importance and of interest to Roman economic historians in general, is the generalised locating of the various clay sources used in the brick industry and the identification of the stamped bricks made from those clays. The MDA study drew out more of the nuances, showing how the exploitation of the brickfields developed and evolved. The implications of this are discussed in Chapter 4. That chapter revisits the brick industry, and reinterprets it in the light of this new knowledge.

Chapter 4: An Industry in the Hinterland

4.1 Introduction

This chapter re-interprets the Tiber Valley brick industry in the hinterland of Rome by correlating the results of the archaeometric study of the SES collection with the information on the stamps. The principal findings of Chapter 3 (that *figlinae* could mean one clay source in particular, but also one clay source could accommodate multiple *figlinae*; and the locating of the geographic areas where the clay sources for particular bricks may be found) complicate the picture enormously. It would seem that the meanings of stamps and their usage or purpose are two slightly different issues. The differences are mostly connected with perception. What meaning? The meaning to us as archaeologists? The meaning to an ancient brick layer? The meaning to the fellow who impressed it in the first place? The meaning to the person who decreed its use (whoever that may have been)?

This is not to discuss the various elements within a stamp, the *signum*, the *figlina*, the *officinator*, and so on. However, the discussion of usage is of course connected to the meaning of the particular elements. If the name of a particular *figlina* is stamped on bricks from a variety of dispersed sources, as is indeed the case for many tested here, then *figlina* cannot mean 'clay district' or 'brick yard' in the sense that Helen (1975: 82-83) understood. That is, the word cannot be pinned down to one geographic location, but rather refers to a collection of disparate parcels of land known by a generic name. Its use is reminiscent of the way a modern farm can be known by a single name even though its land may not be one contiguous whole. To return to the scene of my ethnographic parallel in the Ottawa Valley (2.3.2), 'Rusendale Farm' and 'The Russel Farm', owned by two different families, have become intertwined as each family has sold off different parcels of land to the other family. In Antiquity, similar examples are known epigraphically from North Africa, Egypt, and Italy, and can be detected in the letters of Pliny (4.2.2).

When we have an assemblage of bricks, the relationships which exist between them of same or different find-spots, stamp, and fabric are the key to establishing their wider meanings for *us as archaeologists*. However, by establishing what these relationships are we can begin to define some of the constraints in the industry, and work out the likely meaning for the ancient tradesman as well. The first issue then is to establish the geography of the brick industry, which *figlinae* were established where, and so on (4.2.1). The patterning of this geography leads to a consideration of how land was exploited in Antiquity in general in order for the landlord to achieve a suitable return (4.2.2). I then return to the archaeometric results and propose a framework for interpreting these results in 4.3 which enables a study of the modes of production employed. The different modes are placed against the backdrop of the discussion of land exploitation. Having established the *where* and the *how* of brick production, the *why* of brick stamping is reconsidered in this light (4.4). One level of meaning in the stamps is related to their value as a *commodity* and their ability to absorb the costs of being distributed far afield: bricks have value (4.5).

4.2 The Geography of the Brick Industry

4.2.1 The Locations of *Figlinae*

The generalised source locations of the various *figlinae* named in the stamped bricks tested are listed in Table 4.1 and mapped in Figure 4.1. Locations of *figlinae* have traditionally been deduced from the circumstantial evidence of toponyms, and also from tracking the transference of *officinatores* from one *figlina* to another assuming that for transfers to happen the two *figlinae* ought to have been fairly close together cf. 3.2.4. This sort of work has tended to produce contradictory results, with one researcher arguing that the location for *figlina* X is here, while someone else argues for it to be over there. The fact that certain *figlinae* exploited multiple sources throughout the landscape is a neat solution to those sorts of problems, but if my 'cluster mapping' puts a *figlina* in an area for which there is no other circumstantial evidence, some might take that as proof that the cluster-mapped location is incorrect. For the *figlinae* considered in this study, however, what circumstantial evidence there is for a given location is not inconsistent with the cluster-mapped location.

Figlinae / Dominus / Officinator	Production areas
Domitianae Veteres	Narni, Lower Sabina, Lower South Etruria
Domitianae Maiores	Lower South Etruria
Domitianae	Lower South Etruria
Oceanae Maiores	Lower South Etruria, Lower Sabina
Officina Bocconianae	Lower Sabina
Portus Licini	Lower South Etruria, Lower Sabina
Salarese	Lower South Etruria, Lower Sabina
Terentianae	Lower South Etruria
Tonneiana de Viccians	Aia valley, Narni, Fiano Romano, Lower South Etruria

Table 4.1 Locations of *figlinae* and lands owned/worked by *domini* and *officinators* as suggested by Figure 3.8

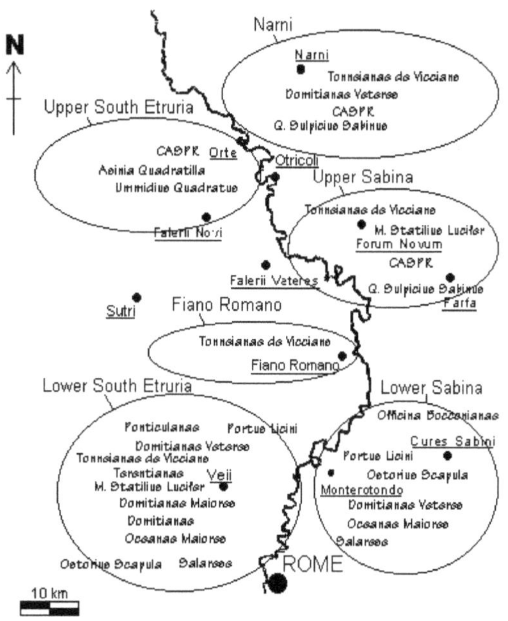

Figure 4.1 Locations of *figlinae* and lands owned/worked by *domini* and *officinators* as suggested by Figure 3.8

Let us consider the *figlinae* connected with the *Domitii* family. The chronology and development of the *figlinae* was discussed in some detail in 3.7.5. If we believe that a *figlina* is located at one point in space only, then the conclusions to draw are clear. The *Maiores* and *Minores* together comprise the territory of the original *Domitianae*. The other *figlinae* were in the neighbourhood, some to one side, some to the other. A further division of *Domitianae*, perhaps of the *Minores* section, occurred quite late and this new section probably encompasses the earliest part of the estate to have been exploited.

The *Domitianae, Domitianae Maiores,* and *Domitianae Veteres* all appear in stamps associated with the tested bricks, but when they are cluster-mapped, they seem to appear all over the valley. This does not however contradict Steinby's story but rather adds to it. If the *Veteres* and the original *Domitianae* are largely the same, then we should find bricks with these stamps cluster-mapped to the same area. A boundary stone naming Domitius Tullus and Domitius Lucanus was found near Bomarzo, north of Orte (Filippi, pers. comm), suggesting that this area was part of the original estate. Of the tested bricks of *Veteres*, SE 4 falls into Cluster B3 (Narni), SE 14 into Cluster D3 (lower Sabina), and SE 26 into Cluster C7 (lower South Etruria). The tested brick of *Domitianae*, SER 1, falls into Cluster C6 (lower South Etruria). We have then a *Domitianae* and a *Veteres* of very similar composition and source, and another *Veteres* from near a probable estate location. A post-Diocletianic stamp of the *officina Domitiana*, SE 37, also falls into cluster C7. There appears to be a long-lived nucleus associated with the holdings of the *Domitii* at C6/C7 in lower South Etruria, although various other sources are scattered about, primarily in the lower Sabina.

In Cluster D3, the associated stamps name the *Portus Licini* (SE 45), the *officinae Bocconianae* (SE 51), the *Figlinae Veteres* (SE 14), and also include the production of Aprilis, named as a slave of Domitia Lucilla (SE 7). The *officinae Bocconianae* is thought to be the late incarnation of the eponymous *figlinae.* Huotari connected the *figlinae* and *officinae* with a medieval *fundus Buccunianus* and the modern Bocchignano to the north of the Farfa River (Steinby 1978: 1508-9; cf. 3.1.4). DeLaine (1997: 90-1) connected the *Portus Licini* and *Figlinae Veteres* through *officinator* names with the *Figlinae Ponticulanae* and seemingly related toponyms in the landscape to an identified medieval site along the Farfa river as well. In D6, example SE 29 carries a stamp of the *Domitianae Maiores*. These two clusters (D3 and D6) are both from roughly the same area, which does not disagree with Steinby's connection of the two through the transference of Sex. Publicius Consors. Albertazzi *et al* (1994: 364-6) have shown that some bricks stamped with *Portus Licini* stamps are composed of material from the banks of the Aia, some way upstream from the Tiber itself. However, SE 45 and SE 5 are also bricks stamped with *Portus Licini* stamps, but they would seem to have been made from clay near Monterotondo Scalo and from roughly the same distance from Rome on the other side of the river in South Etruria. The lands belonging to the *Portus Licini* (or at least, contributing bricks to it) were scattered throughout the landscape. Given DeLaine's location of the *portus*, the port itself could be roughly equidistant from the three production loci at the confluence of the two rivers (but see below 5.3.2 for an alternative), here making *portus* both a warehouse and a port. In this case, there would have been a short overland journey to the port, or a transhipment from the smaller river to the larger one.

The locations of *figlinae* as suggested by the cluster map do not seem to disagree with what has been deduced from other forms of evidence. The only major discrepancy would seem to be for SE 1 and SE 21, examples carrying stamps of the *figlinae Oceanae Maiores*. The *figlinae* in their original *Oceanae* form are thought to be located in the area around Orte (Helen 1975:80-82), but these two tested examples from the *Maiores* section are sourced to further down the valley on both sides. In the Trajanic period the *Oceanae* were divided into two sub-sections, the *Maiores* and *Minores* (Steinby 1974: 69), a development which parallels the course of the *figlinae Domitianae*. In the case of the latter, the *Maiores* section does not use the same sources as the original *figlinae*, but the *Minores* does. In the nomenclature of *figlinae*, perhaps *Maiores* signifies the development of a new richer source than that in the boundaries of the original estate. No *Oceanae* or *Oceanae Minores* bricks were available to be tested for this study to corroborate this idea, but if the *Domitianae* are any guide, I might predict that *Oceanae Minores*, as a later development of *Oceanae*, would use the same sources as the original *Oceanae* (the heart of which should be near Orte, according to Helen 1975: 80-82).

4.2.2 Patterns of Land Ownership and Exploitation

In the Roman agronomists' discussions, clay is not strictly an agricultural product, but it is considered perfectly acceptable for landowners to exploit if it can be done profitably (cf. Varro I.2.22-3). It is not inappropriate therefore to consider the brick industry in terms of agriculture and the associated patterns of land ownership and exploitation. Perhaps rather than discussing the brick industry as if it was somehow separate from the other uses of the landscape, we could imagine the manufacture of brick and tile as a proto-'agribusiness' (to use a modern term). By this is meant a sector of agriculture exploited on an industrial, rather than on subsistence, level but not as in slave-run plantation agriculture (cf. below). It is hard to imagine a wealthy landowner's entire estate given over to the making of bricks, if only because *inter alia* an entire estate would not necessarily be composed of suitable clay resources. It is easy on the other hand to imagine the exploitation of suitable pockets of land on an industrial scale. In the archaeometric study conducted by Olcese (1993) discussed in 3.2.3, her methodology seems to be predicated on the assumption that all bricks carrying the same stamp were made from the same clay. This assumption is based on the wider, if unspoken, view in studies of Roman brick and tile that the *figlinae* named are in actual fact single contiguous estates, or *fundi*. It would seem to be not the case. *Figlinae* are fragmented, with scattered holdings throughout the hinterland. This realisation has quite different implications for our understanding of brick and tile industry-cum-'agribusiness'.

Scattered Holdings

First of all, what other evidence is there for fragmented estates? Most of the evidence is either epigraphic or literary, especially from the juridical sources. In the *tabulae alimentariae* from Veleia (*CIL* XIX.1455) and Ligures Baebiani (*CIL* XI.1147) in central Italy, which list the various properties in the area, their sizes, and the properties to which they adjoin, there is only one very large estate, that of the Antonii family (De Neeve 1984:167). In De Neeve's study of the tables, he demonstrates that the pattern of land ownership in these areas had to be quite fragmented. Even the Antonii property had been divided between four heirs. De Neeve found that there were three levels of land division, starting with large *fundi*, then complexes of adjoining yet economically independent *fundi* of different value and then isolated *fundi* which had very low value (De Neeve 1984: 168). The fact that fragmented estates were common throughout the Empire can be inferred from the Digest of Roman law. Book 8 of the Digest is concerned with rights of way and access (praedial servitudes, the rights of *iter*, *actus*, and *via*). The whole point of a right of way across someone else's property is to guarantee access to the isolated parts of your own property (De Neeve 1984: 167).

Fragmentation of property was not only a phenomenon of private land-ownership. Even properties owned by the imperial *fiscus* in North Africa could be fragmented, as is attested in six inscriptions from the Medjerda valley (ancient Bagradas) in Tunisia (Kehoe 1988: 19, 55-64). These inscriptions record how the land was to be exploited, and the various duties and obligations of the assorted people involved. In passing they reveal how the land was organised and how the ownership of diverse parcels of land could be transferred from one estate to another. For instance, a telling passage in one inscription reads:

> ... by permission of his [Hadrian's] providence the authority accrues to everyone to occupy even those parts which are in the leased out centuries of the estate of Blandus and Udens and in those parts which have been joined to the Tuzritan estate from the Lamian and Domitian estate, and are not being worked by the lessees...
>
> (Aïn-el-Djemala inscription 2.11-14, Kehoe 1988:59)

This reorganisation of estates' land is paralleled later on in Mauretania Caesariensis, where during the reign of Alexander Severus the imperial procurator reassigned land from an imperial estate to the people of Kastellum Turrense (for fragmented estates in Egypt cf. Rathbone, 1991). A similar phenomenon may be responsible with regard to *figlinae* which share the same clay sources over time (4.2.1). Ownership, and hence the name, changes, but it is always the same source. It is worth noting that in the inscription from Aïn-el-Djemala the 'Domitian' estate probably gained its name from its first owners, the brothers Cn. Domitius Lucanus and Cn. Domitius Tullus (Kehoe 1988:10, 208). If their estate in North Africa could be fragmented, there is no reason to suppose that their *figlinae* in Italy could not be similarly organised.

There is literary evidence for scattered landholdings being intertwined with those of other landowners as well. In Pliny's celebrated correspondence regarding his estate in the upper Tiber Valley community of Tifernum Tiberinum, there are hints that this is the case. This estate came into Pliny's hands from his adoptive father, but it was not a unitary body. De Neeve draws attention to two passages, 3.19 and 9.39. Pliny writes in the first passage that the neighbouring estate he wishes to buy is surrounded by his own fields. In the second, he mentions a temple that stands on his property, but this temple is clearly separate from his other holdings (De Neeve 1990:373). Horace's Sabine farm is another estate which is probably in at least two parts- one in which slaves work, and the other which is cultivated by five tenant farmers (Ep. 1.14.1-3; De Neeve 1984:72).

Finally, there is direct evidence related to the manufacture of brick and tile for fragmented land holdings. At Alastair Small's recent excavations at Vagnari on the Via Appia (between Venosa/Venusi and Gravina/Silvium), a stamped brick was found which archaeometric study demonstrated came from a local clay pit. The stamp was of an imperial slave called Gratus. Another identically stamped brick was found at a kiln site 11 km distant. The clay of the second brick was different from the first, and so it was concluded that the same Gratus was working at the two sites which both belonged to the imperial house (A. Small, pers. comm. 2002).

Estate Management

It is important to note that the form of the estate has implications for how it was managed. A contiguous estate with a central farm (what might be called a plantation) was tended by gangs of slave labour which could easily be overseen by the *vilicus* (Kehoe 1997: 3). On a fragmented estate, as in the North African inscriptions and Horace's Sabine farm, tenant farmers were preferred because in this way the various parcels of land could earn a return without the landlord making costly investments in security and supervision. How did farm tenancy work then? De Neeve makes a distinction between two types: tenancy proper and share-cropping. In both of these, there is a *locatio-conductio* (l.c.) agreement over the use of the land. There are a few varieties of *l.c.* contracts, and the position of the *locator* (he who 'places' something for use) and the *conductor* (he who carries it forward) changes depending on the variety. In hiring and leasing (known as *l.c. rei*), the *locator* makes the land or thing available -he is the landlord- whereas in the case of labour (*l.c. operarum*) the *locator* is the employee. There is a further peculiarity in *l.c. rei* contracts over the *conductor*'s right to use the object and the right to exploit it. In the first case he is merely hiring the object, but in the second he is leasing it. Tenant farming is covered by the second case (De Neeve 1984:4-5).

The difference between tenancy and share-cropping lies in the payment of rent, or the *merces*. If this is a fixed amount, (in money or in kind), then we are dealing with tenancy. If it is a percentage of the yield, again in money or in kind, then it is share-cropping (De Neeve 1984: 15-16). The agreement between the landlord and the farmer could in principle be indefinite, but in practice the most frequently cited period was of five years (cf. 5.3.4 on consular dates in brick stamps). This could be extended for another five years, or it could be extended on an ad-hoc basis from year to year. The long periods involved in land leasing also distinguishes these types of contracts from other types of *l.c.* (De Neeve 1984: 10). The relationship between a tenant farmer and the landlord was therefore on a legal basis, rather than being based in ties of patronship or debt bondsmanship (yet see below), and therefore he was an independent entrepreneur. His social standing was not necessarily low because the principal necessary in order to lease the land in the first place could be quite high (social standing being determined in the census of your holdings). The land owner did not have to provide any means of conducting the business, though if he did this only affected the amount of remuneration, not the legal relationship between the two individuals. Generally speaking, the tenant farmer was on his own from the point of view of risk, and the land owner's involvement was limited to making sure the tenant lived up to the terms of the agreement (which could include clauses on the upkeep and maintenance of the property and so on). Most importantly, profits resulting from the operation remained those of the tenant (De Neeve 1984: 15-16).

In share-cropping on the other hand the arrangement was more like a partnership between the farmer and the landlord. The landlord provided some of the means of the business, and recouped a percentage of the yield, depending on how much money the farmer put up in the first place. This system allowed poorer farmers access to better land, and reduced some of the risk, but at the expense of their autonomy and profits. De Neeve (1984:17) argues that there is little point in improving yields if that only means you lose more in payment to your landlord.

It is probably incorrect however to apply such cut-and-dried distinctions from legal texts to social reality, especially given the central role of patronage in Roman society (Wallace-Hadrill 1989: 65). Legally 'independent' tenants, by virtue of their lower social and economic status would not have been as free to act as De Neeve's study implies (Rathbone 1985: 330). Tenancy may have been a method by which the land-hungry farmer could be transformed into a *de facto* client by the landlord (Garnsey and Woolf 1989:160; Foxhall 1990:97). Ethnographic and anthropological studies of modern peasant-landlord relationships in the Third World have demonstrated that tenant farming is a method of land exploitation which actually increases productivity at the expense of the peasant's independence. By leasing out under-sized plots, the landlord forces the tenant to overproduce to pay the land rent and also to feed himself and his family from the same allotment. By using short term contracts (which in practice are rarely ever cancelled) the landlord uses the continual threat of eviction to intimidate the tenant (Foxhall 1990: 101-2). Economies of scale can thus be achieved using many small scale land units, a significant difference from modern economies (Foxhall 1990:100). Morley (1996:75-77) assumes wrongly that tenants and landlords alike aim for self-sufficiency, hence economies of scale can only be achieved through using large parcels of slave-farmed land. Increasing productivity by strangling the supply of available land means in practice that self-sufficiency is impossible; there would always be a lack of some item or another which might only be obtainable through trade or through reciprocal arrangements with the landowner or other tenants. Over time, the tenant farmer who entered into a legal agreement with the landlord could become, through coercion or necessity, more or less identical to a client. Clientelism, like tenancy, could be entered into on a voluntary basis (Garnsey and Woolf 1989:154), and there is no reason to suppose that an existing client could not become a tenant of his or her patron. Where land is concerned the distinction between clients and tenants, patrons and landlords could become rather blurred. (The importance of patronage as both a relationship and a system of relationships as per Johnson and Dandeker 1989, will be taken up in Chapter 6).

Letting Out the Brick Yards

Figlinae in the Tiber valley seem not to have been contiguous bodies, but rather fragmented. Such a situation was quite common in antiquity. The usual way of exploiting the individual parcels of land in a fragmented estate was to use *l.c.rei* to set up leasing agreements with tenant farmers or share-croppers. If a landlord had a suitable parcel of land containing brick clay, or inherited such a parcel with the kiln, drying sheds, and so on already established, *l.c. rei* could be used to exploit that plant and property or to establish the plant in the first place. It would be an effective way of attracting the skilled labour necessary to make brick. Brick makers, like any skilled workers, would find that demand for their services would wax and wane with the fortune of the wider economy. But given that demand in Rome was probably more or less constant (cf. 2.3.1), an independent brick maker may have been able to 'shop around' for a situation suited to his or her needs (there were the occasional female brick makers).

If we assume that this was the system used to lease out the parcels of land containing suitable clay outcrops for brick making, then certain aspects of brick stamps begin to make sense. Let us call this the 'tenancy hypothesis'. First of all, we should regard the *dominus* as the *locator*, and the *officinator* as *conductor*. This hypothesis opens up the (remote) possibility of *officinatores* of being men and women of some means, though we would not expect many of them to be terribly prominent (although it happened at least once that an *officinator* became a *dominus*, Setälä 1977: 202-3, Vismatius Successus). Steinby regards brick stamps as being an abbreviated *l.c.operarum* agreement, a labour contract specifically for making bricks. In her formulation this makes the *officinator* in essence the employee (or *locator*) of the *dominus* (*conductor*)(Steinby 1993: 139-144), the opposite to the case of *l.c. rei*. Problems with the *l.c. operarum* view were discussed in 2.2.2. If this view were correct, we would expect never to find *officinatores* to be named as *conductores*, yet certain stamped bricks do indeed name *conductores* in the usual formulaic position of the *officinator* (e.g. *CIL* XV.1 761, *CIL* XV.1 1477). As concluded in 2.2.3, stamps are not nor do they represent, *l.c. operarum* agreements.

Under the tenancy hypothesis, stamps on bricks could represent (or be a mechanism for) the *merces*, the payment for the leasing of the land in money or in kind, whether as a fixed amount or as a percentage. The clear indication of the property or estate from whence the brick came, and the name of the person who made the brick are the two most constant aspects of brick stamps, and would clearly be necessary for the land owner to know who had met their obligations and who had not. The addition of a consular date flows naturally from this observation, although consular dating in brick stamps has never been adequately explained. If we regard the *officinator* as a tenant, and assume that the landlord lives in Rome, then the stamps on bricks with consular dates would serve the very important function of indicating the payment of land rent *in kind* for a given year. In *CIL* XV.1, of nearly 2000 individual stamp types, only a quarter of them carried the consular date. If a consular dated stamp can be equated with paying the land rent, then an overall average of one quarter of an *officinator*'s output does not seem an unlikely land rent to pay. A certain amount of brick without a consular date may therefore represent production for the profit of the *officinator* himself. St. Marcius Lucifer appears in dated stamps mentioning the *figlinae Caepionianae*. His stamped brick tested in this study mention only his name in the stamp, and come from widely dispersed areas of the Tiber Valley. Under this hypothesis, this is no longer quite so mysterious, but rather represents production for profit, after his land rent in the *Caepionianae* estate was paid.

I cannot say whether we are truly dealing with a tenant farmer or a sharecropper, following De Neeve, because we do not know whether the payment was of a fixed amount or a percentage. But we might take the results of Helen's 1975 study where he tries to determine whether the *dominus* had an active role in production as an indicator. If the *dominus* did have an active role, then we might be dealing with a share-cropping system of land exploitation. If not, then we would have a tenant farmer system, and all the concomitant social and economic effects that that implies. Indeed, Helen finds that *dominus* did not have an active role. Nearly 80% of *officinatores* named in stamps in Helen's interpretation are independent of the *dominus* from a legal standpoint (23, 108-109). We might consider the remaining ca. 20% to have arrangements with their *domini* on a sharecropping pattern. The independent *officinator* in this view might correspond with the skilled tradesman, searching for a suitable location to set up shop, whereas the dependent *officinator* might represent a situation where the *dominus* is trying to establish another element in the overall estate portfolio.

These arguments of Helen's are based on the position of named individuals in the formulae of brick stamp texts, and whether they are freedmen or slaves. Yet from a social point of view there are many kinds of dependency, and the balance inactive *dominus* – independent tenant might have been reversed given the above arguments of Wallace-Hadrill (1989), Garnsey and Woolf (1989) and Foxhall (1990). Because the tenant *officinator* needed a *dominus* who already had set up the infrastructure, there was the potential for the *dominus* to exercise the continual threat of eviction to control his tenant (cf. discussion above, Foxhall 1990: 101-2). Also, if tenancies are handed down through families, there can come a point where a new *dominus* might raise the rent above the level the tenant family can afford. Using tenants would inculcate a certain conservativeness in tenants, to ensure that they can meet the rents, and not disturb their relationship with the *dominus* for fear of eviction. The putative dependency of sharecroppers, and the unlikeliness of the sharecropper voluntarily increasing yields (De Neeve 1984: 17) can also be reversed. De Neeve assumes that the *dominus* took the larger percentage of the yield, hence there was no incentive for the sharecropper to increase production. If, however, there was competition in brick manufacturing (if bricks were valuable, cf 4.5), the offer of a percentage rather than a fixed rent might have proven attractive to the *dominus*, if the *officinator* promised a high yield.

4.3 The Manufacture of Brick

4.3.1 Modes of production

What strategies can the tenant brick maker or the sharecropper use to make bricks effectively? Chapter 3 demonstrated that the term *figlinae* can encompass more than one clay source, and that one clay source can accommodate more than one *figlinae* (3.6). In order to determine whether the first or second aspect of the term (or indeed both) should be applied to any one stamped brick, it is necessary to consider the relationship between a stamp and the brick which carries it. Because there are multiple examples of individual stamps at the same site, or different sites, it becomes necessary to examine the interplay between all three variables concurrently: findspot, stamp, and fabric (the brick itself).

Each variable represents a dimension in which a particular stamped brick exists. There may be more than one brick at a findspot, so it is necessary to relate this brick to the others present for they may have all been part of the same building event. The fabric of a particular brick will be similar to all other bricks which were made from the same clay. The distribution of that particular fabric needs to be examined. The stamp impressed on the wet clay contains information relating that brick to all other bricks carrying the same stamp. The distribution of that particular stamp must be explored. The simplest situation will be a findspot where all of the stamped bricks present carry the same stamp, and are made from the same clay. Note that in 3.2.5 that particular situation was not expected. The expected situations are more complex, with the same

Combination	Findspot	Stamp	Fabric	Combination	Findspot	Stamp	Fabric
A	1	1	1	E	0	0	0
B	1	1	0	F	0	0	1
C	1	0	0	G	0	1	1
D	1	0	1	H	0	1	0

Combination	Findspot?	Stamp?	Fabric?
A	same	same	same
G	different	same	same
Implication:	common origin and distribution		
B	same	same	different
H	different	same	different
Implication:	geographically dispersed production		
D	same	different	same
F	different	different	same
Implication:	single source exploited by different *figlinae*		
C	same	different	different
Implication:	builders at a site had access to variety of sources		

Table 4.2 Logical combinations of findspot-stamp-fabric and the attendant implications for understanding brick production. 1 = same, 0 = different

stamp appearing on different fabrics, and different stamps appearing on bricks made from the same fabric. Table 4.2 shows the possible logical combinations of findspot, stamp, and fabric for any given set of bricks under consideration. There are only six meaningful combinations, which in actual fact are different aspects of only three ideas.

The table can be used to classify the relationships present at a site (recall Figure 1.5, which indicated the kinds of relationships in evidence). If a pair of stamped bricks are recovered from the same findspot, and their stamps and fabrics appear to be the same, then these bricks can be said to participate in Combination A. Combination A is the most basic situation. Combination G is in actual fact the same as A. Together they imply the distribution of materials from a single source. For simplicity let us call these two combinations Mode 1. Combinations B and H (Mode 2) imply the situation where the *figlinae* encompasses different areas of production, that is, the establishment of multiple kilns operated by a single entity. Combinations D and F (Mode 3) suggest the case where a single clay source is exploited to produce brick and tile for different *figlinae*. In this circumstance, where the bricks are at the same site might imply the sharing of transportation costs. As was seen in 3.2.5, that a *figlina* may exploit more than one source, and that one source may produce brick and tile belonging to different *figlinae*, is to be expected: Modes 2 and 3 are not necessarily mutually exclusive. Combination C does not imply a particular production mode, but when found at a site may indicate that the builders had access to multiple sources of material, hence we can think of stamped bricks found in this combination as evidence for 'consumer choice'. Table 4.3 lists the various tested stamped bricks and the combinations in which each brick participates, while Table 4.4 summarizes the counts by mode and period.

Mode 1 implies that the bricks had a common origin and travelled through the same distribution channels. This suggests that there was a tight connection between the consumer and the producer, for the consumer made (or had) the choice of one producer only. This tight connection might be one of social bonds of patron-client, or between clients of the same patron, or perhaps the producer and consumer is the same person if the movement of the bricks is from one part of the estate to the other. Mode 2 suggests that the brick maker had multiple kiln-sites throughout the region. These may have been let out as one 'package' by the *dominus*, and may have appealed to the brick maker for reasons of economy. By having the capability to produce bricks at a number of points, transportation costs may have been minimised, and there may have been the ability to exploit economies of scale. Such an individual would obviously have to be a person of means, and therefore some social standing.

Site	Catalogue	Period	Source Cluste	Combinations
	fnv 9		A3	
	fnv 13		A6	
	fnv 4		B6	
	fnv 6		B6	
	fnv 8		A4	
	fnv 5		B6	
	fnv 14		B6	
	fnv 15		D6	
1732	se 21	Antoninus Pius-Commodus	C8	H (se 1)
2672	se 1	Antoninus Pius-Commodus	D5	H (se 21)
	se 60	Nerva-Hadrian	D5	D (se 1) (B se 107?)
	fal 1		C2	
	fal 2		C3	
	fal 3		C6	
1758	se 63	Julio-Claudian	B2	A (se 65)
	se 65	Julio-Claudian	B2	A (se 63)
868059	ser 1	Severan	C6	D (ser 3)
	ser 2	Julio-Claudian	B2	F (se 85, se 108, se 63, se 65, se 20, se 172, se 155)
	ser 3	Severan	C6	D (ser 1)
1495	se 37	Late	C7	F (se 30, se 26, se 168, se 56)
	se 51	Late	D3	F (se 7, se 45, se 154, se 14)
1547	se 156	Julio-Claudian	A2	H (se 36)
	se 151	Julio-Claudian	C1	B (se 153, se 160, se 163, se 166, se 170, se 174)
	se 153	Julio-Claudian	C2	B (se 151, se 160, se 163, se 166, se 170, se 174)
	se 160	Julio-Claudian	C2	B (se 151, se 153, se 163, se 166, se 170, se 174)
	se 163	Julio-Claudian	D1	B (se 151, se 153, se 160, se 166, se 170, se 174)
	se 166	Julio-Claudian	C2	B (se 151, se 153, se 160, se 163, se 170, se 174)
	se 170	Julio-Claudian	C1	B (se 151, se 153, se 160, se 163, se 166, se 174)
	se 174	Julio-Claudian	C1	B (se 151, se 153, se 160, se 163, se 166, se 170)
	se 172	Julio-Claudian	B2	H (se 176), H (se 177)
	se 10	Nerva-Hadrian	C8	A (se 54)
	se 54	Nerva-Hadrian	C8	A (se 10)
	se 56	Severan	C7	F (se 30, se 37, se 26, se 168)
	se 22	A3	F (se 111, se 106, se 72, se 176)	
2650	se 36	Julio-Claudian	D5	H (se 156)
417	se 42	Antoninus Pius-Commodus	B1	A (se 47, se 53)
	se 47	Antoninus Pius-Commodus	B1	A (se 42, se 53)
	se 53	Antoninus Pius-Commodus	B1	A (se 42, se 47)
420	se 30	Julio-Claudian	C7	D (se 168)
	se 168	Julio-Claudian	C7	D (se 30)
	se 155	Julio-Claudian	B2	D (se 85)
	se 85	Flavian	B2	D (se 155)
452	se 50	Flavian	B7	H (se 48, se 13, se 18)
452	se 100	Flavian	C9	B (se 89, se 90)
	se 89	Flavian	C2	A (se 90)
	se 90	Flavian	C2	A (se 89)
	se 7	Nerva-Hadrian	D3	D (se 45)
	se 5	Severan	C8	B (se 45)
	se 45	Severan	D3	B (se 5)
454	se 52	Severan	C6	B (se 57, se 58)
	se 57	Severan	C8	A (se 58)
	se 58	Severan	C8	A (se 57)
2673	se 72	Nerva-Hadrian	A4	F (se 106)
	se 113	Nerva-Hadrian	C4	H (se 27)
	se 171	C6	F (se 52, ser 1, ser 3, se 28, se 8)	
2678	se 154	Flavian	D3	F (se 7, se 45, se 51, se 14)
	se 48	Flavian	B6	H (se 50, se 13, se 18)
40726	se 176	Julio-Claudian	A5	B (se 177), H (se 172)
	se 177	Julio-Claudian	B3	B (se 176), H (se 172)
2304	se 28	Flavian	C6	F (se 52, se 171, ser 1, se 8)
	se 152	Flavian	A2	F (se 116, se 104, se 156, se 18)
3592	se 104	Julio-Claudian	A2	G (se 116)
	se 13	Flavian	B3	H (se 48, se 18, se 50)
2322	se 14	Antoninus Pius-Commodus	D3	H (se 26, se 4)
	se 23	Antoninus Pius-Commodus	B4	?
	se 111		A3	?
2337	se 16	Severan	B6	F (se 48, se 16)
	se 26	Severan	C7	B (se 4)
	se 4	Severan	B3	B (se 26)
2357	se 18	Flavian	A2	H (se 48, se 50, se 13)
	se 20	Antoninus Pius-Commodus	B2	F (se85,ser2,se108,se63,se65,se172
2391	se 27	Nerva-Hadrian	D2	H (se 113)
	se 136		B3	?
	se 148		D2	A or D with se 27
2413	se 29	Severan	D6	?
	se 141		B3	?
2414	se 106	Julio-Claudian	A4	B (se 144, se 110, se 116, se 114)
	se 144	Julio-Claudian	B3	A (se 110, se 114)
	se 110	Julio-Claudian	B3	A (se 144, se 114)
	se 116	Julio-Claudian	A2	G (se 104)
	se 114	Julio-Claudian	B3	A (se 144, se 110)
	se 2	Nerva-Hadrian	A1	B (se 8)
	se 8	Nerva-Hadrian	C6	B (se 2)
	se 19		D2	?
3284	se 107	Julio-Claudian	D4	? (B se 60?)
	se 108	Julio-Claudian	B2	F (ser 2, se 63, se 65, se 20, se 172

Table 4.3 Different combinations for individual bricks (the other bricks making up any given combination are listed in parentheses). 'Site' refers to the site catalogue number in the Tiber Valley Project database, held at the BSR.

	Julio-Claudian	Flavian	Nerva-Hadrian	Antoninus Pius-Commodus	Severan	Late	Total
Mode 1 (A/G)	10	2	2	6	2	0	22
Mode 2 (B/H)	54	14	4	4	6	0	82
Mode 3 (D/F)	13	13	7	7	2	0	42
Combination C ("Consumer Choice")	17	7	6	2	1	2	35
Total	94	36	19	19	11	2	181

Table 4.4 Summary of production modes by period (numbers of all relationships in which the tested bricks participate)

With Mode 3 we can observe a different strategy to rationalise brick making by exploiting the same general clay source as other brick makers. In doing so there may have been the opportunity for the brick makers to share resources, be they kilns, labour, or transportation. (This supposes that the clay source underlies parcels of land owned by different patrons). If resources were shared, we could imagine that this co-operation was formalised in a partnership agreement. Lirb's study of the legal sources on rural *societates* (partnerships), found that this form of organisation was quite common and was often specifically aimed at increasing overall efficiency and profits (Lirb 1993: 280-2). Eleven *societates* are known in brick stamps (Helen, 1975: 115), although many more may have existed, if Lirb is correct in assuming that in a *societas* one partner could function as the front man for the *societas*' legal dealings. In such a situation, it would be impossible for us to know whether the named individual in a stamp was a single entrepreneur or was acting on behalf of a larger number (Lirb 1993: 285).

When we examine the number and kinds of relationships which exist between the tested bricks, we can see that any one brick will participate in multiple relationships. It will be related to others from the same clay, others again which have the same stamp, and still others found at the same site. Deciding which relationships are most important, hence identifying the mode of production is therefore difficult. By what rational can one prioritise one point of view over another? Such a rationale would have to depend on whether one thinks that find spot is most important, or the stamp is, or the clay is. Prioritising one aspect over the others means discarding potentially useful information. A solution therefore is to examine *all* the relationships which can be observed. Accordingly, Figure 4.2 plots the numbers of every relationship as a percentage of the total number of all relationships by time period.

Historic Trajectory of the Industry

It must be remembered at all times that our arguments are based on a limited (although representative) sample. That being said, the first thing to observe about this chart is the way that Modes 2 and 3 change over time. They seem to mirror each other. When the one is ascending the other is descending. The direction for both changes around the middle of the second century. Geographically-dispersed production (Mode 2) is the most common mode of production in the early first century but declines with the years while the exploitation of single sources by multiple brick-makers (Mode 3) increases. When, in the first half of the second century, Mode 3 reaches its height, so too does 'Combination C', and consumers are therefore able to obtain a wide variety of products. Mode 1, the seemingly monopolistic situation, is never overly important. It accounts for never more than ten percent of relationships in each period until the second half of the second century when it is almost as important as Mode 3.

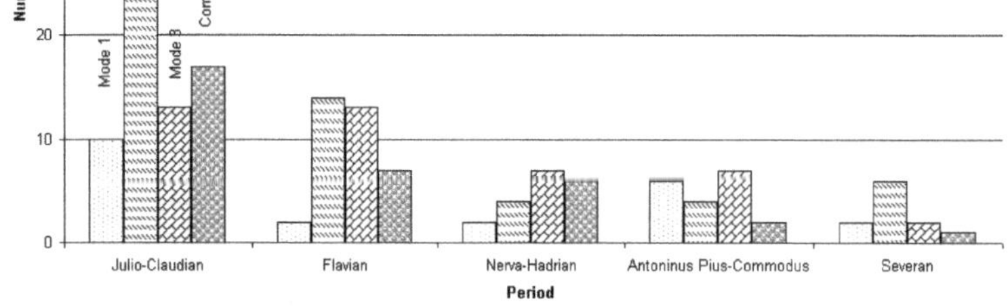

Figure 4.2 Production modes by period

The mirrored trends of Modes 2 and 3 suggest that these two modes are closely related. They might be two faces of the same process. What we are seeing is a pattern where relatively wealthy tenants with the necessary capital to operate multiple establishments on their own are reorganising for corporate action (a pattern reminiscent of the Ottawa Valley timber industry cf. 2.3.2). We could imagine that in the early days of brick-faced construction many different sources were leased into production as landowners discovered that they owned various suitable parcels of land. Some of these (presumably the largest?) would be shared resources, straddling property lines. As the industry develops, and the poorer tenants' resources (of capital or of economically-feasible clay deposits) are exhausted, there would be some necessary consolidation of production at these shared resources. This would not necessarily be a bad thing for those remaining because it would enable the pooling together of resources and distribution networks, making it more economical to both produce and ship bricks. There seems even to have been a dividend for builders in that a wide variety of bricks from different sources seems to have become available.

At the same time however there is always a small proportion of brick being supplied through Mode 1. We should be surprised at the relatively low preponderance of this mode, in all periods. The usual discussion of the evolution of the brick 'industry' places great store in the fact that the various estates mentioned in brick stamps tend to become the property of the imperial household as time goes on, especially by the reign of Marcus Aurelius (Helen 1975: 98-9; Setälä 1977: 239-40; Anderson 1991: 1). Combination C relationships outnumber Mode 1 from the beginning of the first century until the middle of the second. Mode 1 outnumbers Combination C thereafter, but it does not outnumber, ever, the exploitation of a single source by multiple productive entities (Mode 3). This would suggest that domination by the Imperial Household may not have been monopolistic in the fashion it was supposed to be. This is not to suggest that the Emperor was not an important landlord, or not that important in the production of brick and tile. The Emperor's 'monopoly' was in the ownership of land, *not* the production of bricks.

However, the ownership of land did affect the patterns of exploitation of clay for brick. A watershed seems to be reached by the mid second century, just when the Emperor's presence as a landowner becomes pronounced. By this point, 'Combination C' (consumer choice in brick) was no longer as significant as it had once been. It would seem as if whatever sort of market economy -as indicated here by the ability of consumers to obtain brick from a variety of sources- which once obtained in the early Empire (cf Temin 2001: 181 on the 1st century as a market economy) did so no longer. What caused this watershed? It may be that consolidation of land-ownership had the effect that formerly disparate parcels of land became united in a single estate (a process perhaps in evidence in North Africa, as attested in the Aïn-el-Dejemala inscription). There may have been proportionately fewer small parcels of land available for leasing, i.e. formerly parcels X, Y, Z were available whereas latterly there was simply one parcel, $W_{(x,y,z)}$. The conditions then that led to the formation of *societates* (the existence of which may be inferred from Mode 3) could well have been removed, leading to the resurgence in Mode 2 after the mid point in the century (Figure 4.2). If this is correct, and leaseholds were now bigger, the capital necessary to hold and exploit the lease should also have been bigger. Brick makers (tenants on the land) in the second half of the second century were probably wealthier than their forebears if they could successfully exploit what formerly had taken a *societas* to do. Alternatively, if the relative amount of capital available remained the same, then perhaps a change in the cost of transport or some other factor made the exploitation of these larger, geographically-dispersed parcels economic.

4.4 The Meaning of Stamps

The biggest obstacle in the interpretation of the brick industry is in knowing the meaning of stamps. Why were bricks stamped? What information does a stamp convey? Having determined the likely provenance of bricks' clay sources, I have arrived at a position which suggests that for many bricks, stamping might be tied to meeting the obligations of the land lease, principally the paying of the *merces* (4.2.2). However, that is not the entire story behind the phenomenon of brick stamping. Broise (2000: 113) has pointed out that anepigraphic stamps appear on bricks, together with epigraphic stamps, as early as the reign of Hadrian. As was argued in 2.2.2 anepigraphic stamps are probably equivalent to the slave-name stamps on the *bessales* bricks of the *figlinae Salarese* and *Quintanensia*. The earliest slave-name texts predate the earliest appearance of anepigraphic stamps (although not by much; Steinby 1974: 83 for example dates the earliest *Salarese* stamp to just before, 'anteriore', the year 123). After a few years of stamping the name of the slave who did the work, anepigraphic stamps might have been found to be simpler and quicker. Every brick would be marked by its maker in this scheme, so that it would be easily calculable how many were spoiled, how many were fired, and so on. Anepigraphic stamps in all likelihood serve the same function as the kiln dockets from La Graufesenque recording which potter had produced how many of what item (Parca 2001: 68). That is, anepigraphic stamps serve to differentiate production within the individual productive unit (Broise 2000: 115). Different anepigraphic stamps appear on *bipedalis* and *sesquipedalis* bricks all carrying the same epigraphic stamp (Broise 2000: 116).

If Broise is correct, then the usage of epigraphic stamps was probably aimed at an audience external to the *figlinae*, and was not therefore tied to the production process (*pace* Steinby 1993: 141). This is in fact corroborated by the data establishing the clay sources of the tested brick. The situation seems to be in fact that epigraphic stamps serve at least two clear purposes. In

	By river	Roman miles	Direction	Direct overland distance Roman miles
Aia	Furthest Nearest	27.81 0.68	downstream upstream	34.22 3.65
Monterotondo Scalo	Furthest Nearest	24.77 8.14	downstream upstream	22.05 4.99
Orte	Furthest Nearest	37.37 2.04	downstream upstream	53.72 1.87
Narni	Furthest Nearest	33.6 10.57	downstream upstream	32.73 7.46
Fiano Romano	Furthest Nearest	19.37 7.5	downstream upstream	----- -----
Valle Aurelia	Furthest Nearest	37.49 8.62	downstream upstream	29.43 6.16

Table 4.5 Distances in Roman Miles (1 = 1.48 Km) to the nearest and furthest sites which use the materials of particular source areas

the first situation, a particular stamp will be used for a particular source. The implication is straightforward. The use of stamps is to differentiate the produce from different sources.

In the second situation, a particular stamp will be used concurrently at different sources. The use of stamps does not differentiate different sources, just the output of different individuals. It does not matter where it comes from: suffice it to know that Lucifer made it. Or at the very least (remembering anepigraphic stamps), one of Lucifer's underlings made it. The importance of these sorts of stamps might be tied to distribution rather than production. The role of stamps in distribution was hinted in the discussion of how the Tiber might function as infrastructure (2.3.2). If we imagine for a moment that the stamp acts like a sort of 'shipping label' (cf. Helen, 1975: 24) then it does not matter that the bricks have different sources as they are destined for a particular warehouse or dock.

There is seemingly a hierarchy of stamp use. There is the simplest situation where one individual uses one stamp for all of his or her output, regardless of the source, the legal situation, or the owner of the land. Such a stamp would not be connected with the *merces*. Then a more complex situation exists where the landowner is recorded as well, and so while retaining its basic function the use of the stamp is nuanced. This type of stamp would be connected with the *merces*. Presumably the leasing of a different parcel of land, if additional to the original agreement, would necessitate a new contract. A new stamp would therefore be necessary when indicating payment of the *merces* and compliance within the terms of the agreement for this new parcel. This would account for those situations where an individual brick maker uses different stamps on bricks made from different clays.

The idea of stamps having a role in the distribution network is the basic stratum on which any other use would rest. It does not prohibit any other additional functions whereas any particular one of the other posited uses (as listed for instance in 2.2.2) does not fit the data, and would preclude each other. The homogeneity of stamp forms and elements also suggests one overriding imperative in the usage of stamps, despite our ability to see multiple purposes.

4.5 The value of brick

4.5.1 An Experiment in Transportation Costs

In the shadow of Rome in the Tiber valley however, the problem of distribution is more complex. The sites from which the stamped bricks were recovered are all more or less in the same part of South Etruria (1.4.2), but the sources from which the bricks were made are scattered from Rome to Narni. This observation allows me to perform a rough experiment to test by source whether overland or riverine distribution is more economical, in terms of time taken for transportation, when all of the sources have access to the same geographical market.

For each source (assumed for the sake of the experiment to be at the same location as the modern examples studied), the distance in a straight line was measured to the nearest and furthest site. This was taken to represent the amount of distance necessary to cover using oxen-drawn carts to transport the brick. The straight line distance was used because in some cases the road system is not clear, and in other cases which road to take is also in doubt. For estimating river travel distance, a straight-line distance was measured to the nearest navigable river (assuming, for the sake of the experiment, that the port would be at the closest point on the river to the kiln site), then the distance following the river to the closest point to the site, then the straight-line distance to the site. This would have necessitated transhipment at two points with two short cart journeys. Accordingly, the longest journey overall would be one overland from Orte of 54 Roman miles (1 Roman mile = 1.48 km). The shortest journey overall was also overland from Orte, a distance of 2 Roman miles. On average, overland journeys were of 20 miles, while riverine journeys were 18 miles in length. Table 4.5 charts the distances to make the shortest and longest journeys by land and by river (see Figure 4.3 for where these are).

To estimate the amount of time it would take to transport a quantity of bricks from the kiln to the building site in the Tiber Valley necessitates some assumptions. DeLaine (1997) developed a methodology for estimating the amounts of men and materials necessary for the construction of the Baths of Caracalla based on ethnographic comparison with 19th century quantity surveyors' handbooks. In a later publication

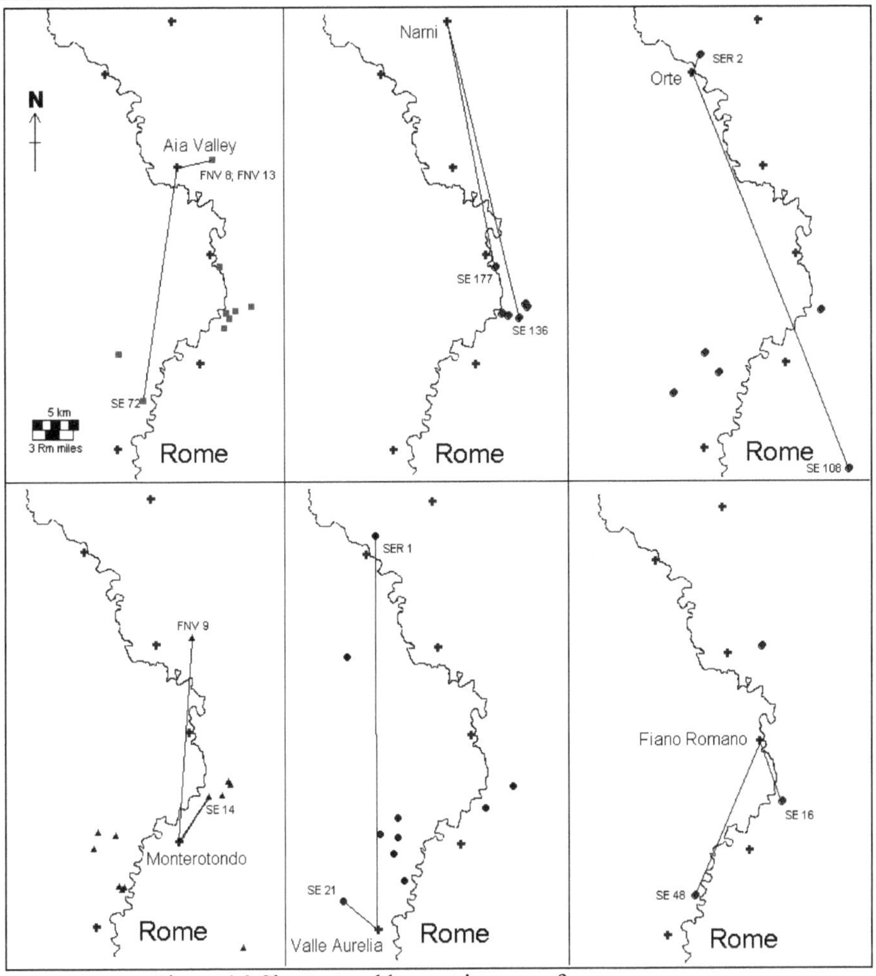

Figure 4.3 Shortest and longest journeys from source areas

(2001) she refined the methodology to compare the economics of different types of Roman construction in terms of amounts of time, material, and labour necessary to build the same amount of different types of wall (*opus incertum, opus reticulatum,* etc. 247-259). In this experiment, I follow the formulae in the 2001 publication for the calculation of transportation costs. The relative costs of transportation types are worked out from Diocletian's Edict on Maximum Prices (DeLaine 1997: 210-11). Roughly, this gives a ratio for sea : river (downstream) : river (upstream) : ox-cart of 1 : 4 : 8 : 42. By comparing this ratio with the daily rate of pay of a labourer (5 *modii* of grain/month, which works out as 36 *denarii* for labourers) the transport ratios can be expressed in terms of a labourer's daily pay. Thus to ship one tonne of material one Roman mile by ox-cart costs 1.44 times as much as the labourer's daily pay, 0.26 times by river upstream, 0.13 times by river downstream, and 0.035 times as much by sea (DeLaine 2001: 234). There are of course difficulties in extrapolating from the time of Diocletian backwards to earlier periods. DeLaine's work rightly stresses that although these figures are approximations they are nevertheless still the best available. The same caveats apply here.

In this experiment, the calculation of the time necessary to ship one tonne amounts resulted in final figures in decimal divisions of days. For ease of comparison I have calculated for four tonnes instead which gives results in whole days (four tonnes represents a shipment of 300 *bipedales*. I consider in more depth the problem of load sizes in 5.3.2, but one tonne represents roughly 75 *bipedales*). However, because the relationships are constant, to work back to one tonne involves nothing more complicated than dividing the final figures by four. In the following example I work out the time involved for the longest and shortest journeys using the river and overland only for bricks using Aia clays. Table 4.6 tabulates the results from all sources.

From the Aia source, time taken for **longest** downstream journey for 4 tonnes of bricks (*bipedales*). 1 *bipedalis* = 58 cm x 58 cm x 4 cm = 0.013456 m^3. 300 *bipedales* = 4 tonnes
mle = man labour equivalent

- Brick kiln to river (where port is assumed to be), with oxen, Roman miles = 2.84
- 1.44/mle/tonne mile x distance x 0.013456 tonne/*bipedalis* x 300 = 16.5 mle
- Distance to closest point on river to findspot (where port is assumed to be), by river, Roman miles = 22.13
- 0.13 mile down river/tonne mile x distance x 0.013456 tonne/*bipedalis* x 300 = 11.6 mle
- From river to findspot, with oxen, Roman miles = 2.84
- 1.44/mle/tonne mile x distance x .013456 x 300 = 16.5
- Total mle for four tonnes of *bipedales* = 45 mle or
- In 12 hour days = 4 days

- To the same findspot, completely overland is a distance of Roman miles = 34.22
- 1.44/mle/tonne mile x distance x 0.013456 tonne/*bipedalis* x 300 = 199 mle or
- In 12 hour days = 17 days

From the Aia source, time taken for **shortest** downstream journey for 4 tonnes of bricks (*bipedalis*). 1 *bipedalis* = 58 cm x 58 cm x 4 cm = 0.013456 m^3. 300 *bipedales* = 4 tonnes
mle = man labour equivalent

- Brick kiln to river (where port is assumed to be), with oxen, Roman miles = 0.68
- 1.44/mle/tonne mile x distance x 0.013456 tonne/*bipedalis* x 300 = 4.0 mle
- Distance to closest point on river to findspot (where port is assumed to be), by river, Roman miles = 3.65
- 0.13 mile down river/tonne mile x distance x 0.013456 tonne/*bipedalis* x 300 = 3.83 mle
- From river to findspot, with oxen, Roman miles = 0.68
- 1.44/mle/tonne mile x distance x .013456 x 300 =3.95 mle
- Total mle for four tonnes of *bipedales* = 11.74 mle or
- In 12 hour days = 1 day

- To the same findspot, completely overland is a distance of Roman miles 3.65
- 1.44/mle/tonne mile x distance x 0.013456 tonne/*bipedalis* x 300 = 21.21 mle or
- In 12 hour days = 2 days

As would be expected, river transportation is more economical in terms of time than land transportation in all cases (Table 4.6). For the longest journeys, if those journeys had to be made by ox-carts, they would take twice as long from the Valle Aurelia to eight times as long from Orte. However, ox-carts would be feasible for short distances, being only 1.3 times longer from Orte to 3.3 times longer from Narni. These results are unsurprising. However, what is unexpected is the overall uniformity in travel times by river. Keeping in mind that the destinations for the bricks made from these sources are all more or less in the same place, we can see that the different (geographically scattered) sources are therefore competing against each other. In a sense, what Table 4.6 indicates is the competitiveness or efficiency of the different sources, in terms of how quickly they can get their products to market. They are all capable of shipping to the same places in roughly the same amount of time, even though they are in some cases quite distant from the site. To the nearest site which uses their bricks, the range of travel times is from 2/3 of a day to two days; to the furthest sites the range of journey times is about three to six days. If we discount Valle Aurelia, the furthest journey times are only three to four days.

The conclusion to draw from this experiment is that, so long as a brick maker can get access to river transport, all sources are more or less identical in cost-of-travel and therefore these costs are not a significant factor in explaining why one type of brick should be used at a site over another. However, these observations are dependent on brick producers having free access to the river, which is another question entirely. If the brick maker had to rely on ox-carts, for whatever reason, then the production location suddenly would have become significant; it is conceivable that a shift to a different mode of transportation and the concomitant costs could have upset what previously had been a successful brick making enterprise (5.2.1 on the changing fortunes of brick makers).

4.5.2 Profits

If the cost to make and to transport brick is high, the final price of the brick will be high and increase with distance from the point of manufacture. The profit margin will shrink with distance, and there will come a point where consumers, if they are able, will not pay the price. If the estimates in the previous section are of the right order of magnitude, this was not a concern in the Tiber Valley (as long as the river was used for transport). Nevertheless, cost ratios, as deduced from the Diocletian's Price Edict, are usually taken to mean in the context of brick that brick was a low-value commodity, not likely to be used very far from where it was made. Greene (1986: 40) writes that 'it is obvious that no low-value bulky

Source Area	River, days to furthest	Land, days to furthest	Furthest by land is how many times slower?
Monterotondo Scalo	3.25	10.66	3.28
Orte	3.2	26	8.13
Narni	2.75	15.86	5.77
Aia	3.71	16.57	4.47
Fiano Romano	4.33	---	---
Valle Aurelia	6.66	14.25	2.14
Source Area	**River, days to nearest**	**Land, days to nearest**	**Nearest by land is how many times slower?**
Monterotondo Scalo	2	2.43	1.22
Orte	0.66	1	1.51
Narni	1	3.6	3.6
Aia	1	1.75	1.75
Fiano Romano	1	---	---
Valle Aurelia	1.8	3	1.66

Table 4.6 Journey times from source areas

cargoes could have been traded profitably overland for any significant distances'. The catch here is 'significant distances'. If one can prove that a certain distance was probably profitable, then that obviously is not a 'significant distance'. The appearance overseas of Tiber Valley brick is taken to be evidence of transport as ballast, implying that brick could never be a cargo in its own right (Aubert 1994: 240, yet cf. Thébert 2000: 374-8, and below). Despite the fact that the Price Edict gives prices for shipping *upstream* as well as downstream, brick is thought never to have been traded upstream in the Tiber Valley (Filippi, pers. comm.; cf. Steinby 1981: 239). This last assumption carries with it the interesting effect that the *figlina* named in any particular brick stamp must necessarily be located further upstream than the highest site where the brick was found. That idea can safely be dispensed observing that several of the tested bricks were probably made downstream from their find-spots, implying that however much they were worth, it was certainly enough to absorb the cost of being shipped upstream. Cost of production and transportation, as Laurence has noted, while important is not the be-all-and-end-all of economics (Laurence 1999: 99).

The value of brick can be inferred from the different production modes discussed in the previous sections. Because all three modes are in evidence for every period under discussion, different bricks had different values. Mode 1 accords best with the idea that brick was too bulky and too low-value to get anywhere very far. A single producer working from a single source could produce only a set amount of bricks, and would find it easiest, given the small scale of his operation, to ship to the local area. This producer's bricks would be the most expensive to consumers, because the unit cost of each brick would be higher. In Mode 2, producers have found a way to cut the costs of transportation by exploiting a variety of sources which could be closer to the ultimate destination. They would also be in a position to meet demand effectively by shifting workers and production to whichever site was closest. In Mode 3, by exploiting a source with other producers and presumably sharing kilns and so on, the cost to any one producer would be divided by the number of other producers involved. The cost of transportation would also be reduced. Even if the ultimate price to consumers remained the same amongst all three modes, the profit margins would differ, allowing some to prosper while others would founder.

4.5.3 Industrial Siting

Another way of reducing the overall cost of production would be to share workers and plant between different industries. Since no brick-kiln sites have ever been excavated in the Tiber Valley (at the time of writing), allowing us to examine just how production was organised, it falls to other methods to explore the possibility of ancient 'industrial parks' (for lack of a better term). The key here is pollution. In general, the chemical make-up of the bricks ought to correspond well with the mineralogy. The amount of quartz present should be correlated with the amount of silica, feldspar

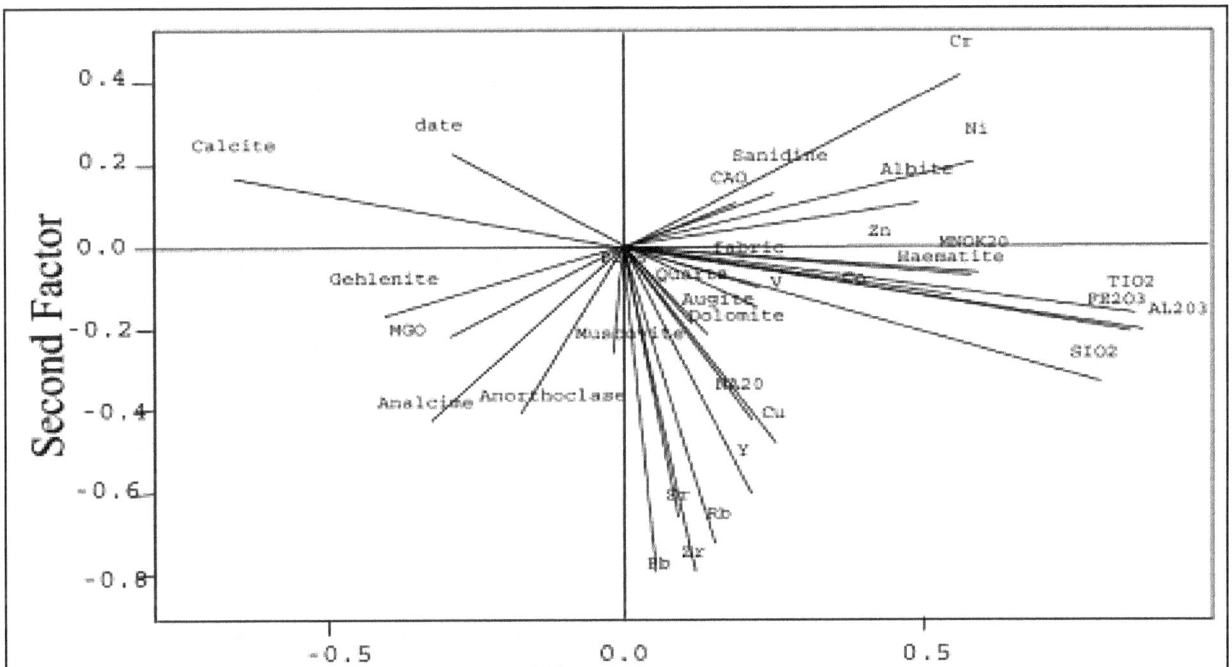

Figure 4.4 Plot of factor analysis of minerals and chemistry. Lead (Pb) and copper (Cu) are not associated with any of the minerals present.

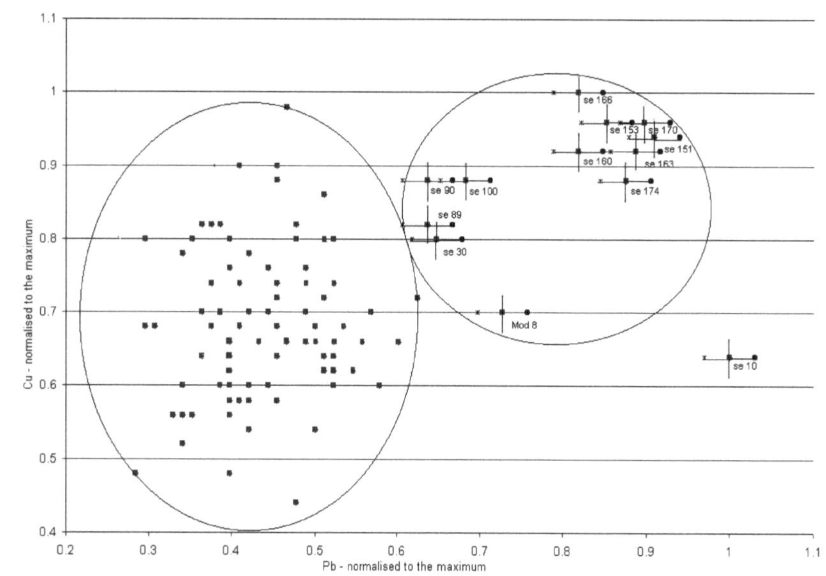

Figure 4.5 Lead (Pb) versus copper (Cu) in tested SES bricks (confidence intervals at 95% for mean: Pb = +/- 0.025; Cu = +/- 0.030).

with sodium and aluminium, haematite with iron, and so on. Where it does not, or where the amounts present account for much more than the minerals present, we should look for other explanations, including human agency. Elements (e.g. lead and copper) which cannot be correlated with naturally occurring minerals might be present as a result of some other human activity unrelated to the making of bricks. Figure 4.4 is the plot of a factor analysis on the minerals and chemistry of the tested bricks. The only major discrepencies are lead and copper, which do not seem to be correlated with any of the minerals present. Traces of copper are a usual by-product of lead working. Figure 4.5 is a scatter-plot of lead versus copper for each sample, normalised to the maximum lead and copper levels present in the tested bricks. Although the overall levels are low, there appears to be two distinct clusters. One, to the bottom left of the scatter-plot, contains the majority of the tested bricks. The other cluster is towards the top right of the diagram. This small cluster contains the examples from bricks carrying the stamps of only two individuals. Because this small cluster is apparently differentiated from the large cluster containing all of the other examples, the smaller cluster may indicate bricks which were made in an area of particular lead concentration while the large cluster might be interpreted as a background level of ambient lead pollution.

When ore is smelted for its lead, the lead reduces to metal quite easily at low temperatures, and so tends to be quite pure. Much higher temperatures are required to extract silver from lead ore, and so other metals present (lead, copper, bismuth, etc.) are also reduced, becoming concentrated in the leftover material (litharge). The litharge can be re-smelted to extract its lead. The resultant lead can therefore contain significant levels of copper (Craddock 1995:210-211). Lead production in antiquity (about 80,000 metric tonnes/year at its height) was on a level comparable to that during the Industrial Revolution; pollution levels were about four times greater than natural as discernible in the Greenland ice sheets. It is estimated that about 5%, or ~4000 metric tonnes/year entered the atmosphere through the mining and smelting process (Hong, Candelone, Patterson, and Boutron. 1994: 1841-2).

There is a natural connection between lead manufacturing and the brick industry, as many people known in the brick industry (primarily land owners) are also known from lead-pipe stamps (Bruun 1991:67,153, 156 239-42; Setälä 1977: 32-3, 69-70, 149-50). Manufacturers of lead pipes also made a variety of other objects (e.g. vessels, coffins) in lead. There are over 450 *plumbarii* named in lead-pipe stamps, making lead workers the single largest group of known artisans in the City of Rome itself, according to Bruun (1991: 380). He concludes that the manufacturers mentioned in stamps are probably the owners of the workshops, and that occasionally individuals might own more than one workshop, at a level above that of the manufacturer, a method of organisation not dissimilar to that envisioned for the brick industry (cf. Steinby 1982; 1993).

All the bricks in the small cluster in the scatter-plot were made by only two people: Ostorius Scapula and C. Iulius Neicephorus. The one-name form of these stamps indicates, according to Helen (1975: 130), that these individuals were probably both the landowners and the brick producers. An outlier (SE 10, Anteros) with high lead and low copper might not be an indication of manufacture but rather of particularly high pollution. The Ostorius Scapula bricks (seven in all) were recovered at the same site, along with a number of other stamped bricks which did not show any particular concentration of lead (SE 156, SE 172, SE 54, SE 56, SE 22; SE 10 being the exception) If the lead contamination happened at the site where these bricks were recovered, one would have expected some of the other bricks to have been contaminated as well. Similarly, at the site where the C. Iulius Neicephorus bricks were found, the other stamped bricks recovered did not exhibit high levels of lead (SE 5, SE 7, SE 45, and SE 50). The contamination therefore happened at some place other than the building site.

However, the cluster mapping suggested that C. Iulius Neicephorus and Ostorius Scapula *both* exploited the *same* clay sources. They were both using at least one source in common (Cluster C2 in Figure 3.8), while the other sources exploited by these two individuals were fairly close to one another (Clusters C1 and C11 in Figure 3.8). The other stamped bricks which showed high levels of lead were also using clays geographically close to those used by Ostorius Scapula and C. Iulius Neicephorus (SE 10, SE 30). The likely source for all these clays is South Etruria but not too far from Rome, probably near to the Tiber. Given the relative ease with which lead is worked (Craddock 1995:205) coupled with the emission of lead aerosol particles and the amount in the atmosphere (Hong *et al.* 1994:1841-3), the pathway for the contamination of the bricks might be part of the brick manufacturing process. Before clay can be made into bricks, it must be physically broken down. This was done in the winter time by simply leaving the clay in the open to be weathered by rain and frost action (DeLaine 1997:114). Lead may have been deposited on the clay pile with every rain shower. If this is correct, it may suggest then that the lead-working occurred in the winter time. If the two industries are in close proximity, as seems to be suggested by the chemistry of Ostorius Scapula's and C. Iulius Neicephorus' bricks, then I might make a further hypothesis. Brick-making ceases during the winter to allow fresh clay to be dug and weathered; these operations require comparatively fewer workers than during the summer manufacturing season. A rational use of the excess brick workers would have been to switch them over to lead-working during the winter season.

In the 'contaminated' cluster, Ostorius Scapula may therefore have been the prime manufacturer of lead, or in very close vicinity to someone else's lead works, with the other bricks being contaminated through proximity. Ostorius Scapula is not yet known in *fistulae* stamps, nor is C. Iulius Neicephorus (Bruun, pers. comm. 2001). There may be, however, a connection between Ostorius Scapula and lead. Lead mining began in Britain shortly after the Roman conquest (Healy 1978: 61-2), which was the same time that Ostorius Scapula was governor. The smelted lead was made into pigs to be shipped throughout the Empire. More sensitive methods of analysis (perhaps lead isotope analysis) might be able to establish where the lead came from in Ostorius' bricks. The presence of copper might indicate that the lead had been recycled from silver production, in which case perhaps Spain could be the origin. British lead, though optimistically stamped *EX ARG* when turned into ingots, is quite low in silver (Craddock 1995: 214).

Whatever the source, it would seem that there is at least one case of lead working and brick manufacture in reasonable proximity, perhaps even by the same person. Bruun suggested as much for brick and *fistulae* production in general, but added that there was no hard evidence other than the appearance of names in both types of stamps (1991: 242). We might expect individuals named in both brick and *fistula* stamps (such as Articuleius Pactus, Memmius Rufus, and Q. Canusius Praenestinus, Bruun 1991: 239-42; Aelius Felix, Annia Cornificia Faustina, Regina Claudia Capitolina, M. Cocceius Nerva, Domitia Lucilla the elder and the younger, C. Fulvius Plautianus, Petronius Septimianus, Petronius Mamertinus, Q. Servilius Pudens, and Ummidia Quadratilla, De Kleijn 2001: 261-307) to have centralised plant for what were probably fairly noxious industries. This would have also lowered the ultimate costs of production and certainly of distribution by allowing the one product to 'piggyback' with the other, increasing the profitability of both. It is significant that in the *CIL*, Ostorius Scapula has the second most recorded number of examples of one particular stamp type, *CIL* XV.1 1393 - some 89 examples all in the City of Rome - which indicates the success of his strategy. Other industries for which there could be the possibility of 'doubling-up' in this fashion might be brick and lime production (kilns) or brick and forestry (the production of fuel through coppicing).

4.5.4 Intrinsic Value

Brick might even have an intrinsic value, which goes beyond the labour input. The mineral assemblages in the bricks seem to be the result of technical choices made by the brick maker. The reasons behind the choice of these particular minerals may be what give Tiber Valley brick inherent value. The brick samples produced a suite of 11 minerals common to every brick and tile: quartz, augite, haematite, gehlenite, calcite, analcime, muscovite, dolomite, anorthoclase, sanidine, and albite. Confusingly, these minerals could come from various combinations of the three basic types of rock - igneous, metamorphic, and sedimentary.

The mineralogical data were therefore examined to determine if there were particular minerals which seemed to be related. Strong positive correlations were found between six minerals in particular, augite, haematite, analcime, muscovite, dolomite, and anorthoclase. The samples were then tested for similarities on the basis of their mineralogy using Ward's method of cluster analysis (the dendorgram result is illustrated in Figure 3.5). These six correlated minerals accounted for between one- and two- thirds of the minerals present in each cluster (Table 4.7). These minerals have widely different parent rocks and are unlikely to have occurred together naturally in a source clay. For instance, in the raw clays tested from the Aia valley in the Sabina (MOD 7) and from Orte in South

cluster 1	%	cluster 2	%	cluster 3	%
augite	30	augite	27	calcite	27
albite	23	albite	13	quartz	23
calcite	23	quartz	13	anorthoclase	10
dolomite	8	anorthoclase	12	augite	9
quartz	5	dolomite	8	albite	8
sanidine	4	calcite	7	gehlenite	8
analcime	2	sanidine	6	haematite	4
haematite	2	gehlenite	4	dolomite	3
muscovite	2	haematite	4	muscovite	3
gehlenite	1	analcime	3	analcime	3
		muscovite	3	sanidine	3
Correlated minerals make this percentage of the total:	44		57		32

Table 4.7 Minerals and their proportions in descending order within the three largest groups of bricks depicted in the dendrogram in Figure 3.5. Strongly correlated (>95%) minerals are shaded.

Etruria (MOD 9), there is no augite or other pyroxenes present. In the finished bricks from these yards, however, there are significant levels of augite (MOD 5, MOD 6, MOD 1). We might therefore infer human agency: we might interpret these minerals, especially the volcanic ones, as having been particularly sought out by the brick maker.

Close study of the available geological maps indicates that in volcanic South Etruria, clays are present at the base of Monte Soratte, a limestone massif. On the sedimentary Sabina side, basaltic lavas are present in large pockets around the Eretum area, and above the confluence of the Nera and the Tiber in Amelia. In 3.1.4 it was suggested that the use of *'fornace'* and its derivatives as place names in the landscape occurred at the interfaces between different geological zones. It may be the case, given the mineralogy and this occurrence of place-names, that the brick makers selected fairly uniform clays which by and large were the decomposed products of basaltic lavas and basic magmatic rocks, hence the source of the augite and other correlated minerals. The brick makers were exploiting the interfaces between different geological zones.

Clays formed in these interfaces may have had certain qualities that marked them out as desirable for the particular needs of the brick industry. The presence of basaltic sands would be important during the early stages of drying, according to Tony Mugridge, a long-time manufacturer of hand-crafted bricks and owner of Tuckies Brickyard in Shropshire. After the mould is removed from the brick, there would be a danger of the brick collapsing under its own mass as it is moved from the shaping table to the drying area. The basaltic sands would increase the strength of the unfired brick, and so prevent this from happening (Mugridge, pers. comm.). When we remember that prior to being sawn into triangles a *bipedalis* brick was two Roman feet square and about two inches thick, the importance of not collapsing can be appreciated.

This distinctive property of Tiber valley brick might have ramifications for our understanding of Roman architecture and should be explored, although that is not currently possible within the constraints of this study. Being able to make such massive bricks may have allowed the development of particular architectural forms (cf. DeLaine 1995 on the importance of geological interfaces in the Tiber Valley for the development of Roman concrete architecture). The appearance of Tiber valley bricks at sites around the western Mediterranean might also have a functional aspect especially in areas where volcanic material is not available to temper the local bricks. It might have not been possible to make bricks to the required dimensions given the materials at hand. (In North Africa, many Italian bricks were used in the construction of bath-complexes, especially hypocaust floors, Thébert 2000: 348. Perhaps there was also an ideological aspect, where in the minds of the locals, an 'Italian' building needed 'Italian' materials). The addition of volcanic materials as temper is what would give brick its inherent value.

4.5.5 A Word on Overseas Trade in Brick

Steinby (1981: 244-5) argues that brick, if found overseas, only represents a 'secondary commerce'. That is, when ships arrived at Rome and offloaded their cargo, they would take anything at all to put in their holds for the return journey, to recoup some of the cost of returning to their home ports. In this view, a small amount of brick would be taken onboard to function as ballast. The low value of brick, combined with local production throughout the Mediterranean, would prevent any real trade in this commodity. On the other hand, citing a wreck off Cap Dramont in the south of France, Steinby argues that the high number of mortaria found, combined with amphorae and other goods, represent a 'primary commerce'. These ceramic mortaria are in her opinion trade goods in their own right (1981: 244-5).

Her view seems to rest on an unfounded viewpoint that because brick has low monetary value, it therefore cannot be an item for any real long-distance trade (although it is hard to see why mortaria would not face similar problems to those she envisions for brick). In this section on the value of brick, I have indicated the ways in which the producers of Tiber valley brick may have reduced their production costs, hence increasing their profit margins. The parallel siting of other industries alongside brick kilns may also have been a strategy for reducing overall production costs. As long as a producer could get access to the river infrastructure, cost of transportation from any of our probable source areas to building sites was roughly equal in terms of labour. Finally I have argued that Tiber valley bricks abroad may have had an inherent advantage over local production, by virtue of their volcanic content, and hence they were valuable. Whatever the monetary price of a batch of bricks, it seems as if producers did take steps to ensure that they kept the cost of production low, enabling a higher profit. It seems a bit obvious, but perhaps it needs to be stated baldly: profit is possible in the brick industry because bricks had value (otherwise there would be no point in making them in any great quantity).

If profit can be made, could a long-distance trade be sustained? In Parker's 1992 study of shipwrecks in the Mediterranean, and in European rivers, he lists a catalogue of over 1200 individual wrecks, dating from the Bronze Age to the 13th century. The majority however are Roman. In a sample of 98 wrecks, he found that cargoes of *brick alone* made up 14% of the sample. He argues that in general cargoes were largely bulk or compound, and 'tramps' carrying a wide range of cargoes were few (1992: 20-1; which contradicts Hordern and Purcell's concept of 'cabotage', 2000:123). As for the question of ballast, of the wrecks he studied he found that 'paying ballast' (what he terms 'goods not worth carrying on their own', which fits Steinby's ideas regarding brick) could only be identified but rarely. Where it was possible to identify such cargo, it tended to be on top or at the end of the main cargo, although heavier goods were placed at the bottom of the hold (1992:28). One such might be wreck 123, Cabrera A off the coast of Spain where a cargo of amphorae was covered by a layer of tiles. However, this seems to be the only such wreck; known tile or brick cargoes proper are far more numerous. One in particular is wreck 543, in the Kerme Gulf off Turkey, which had a cargo of ca. 5000 tiles, along with some amphorae and coarse pottery. Unfortunately, whether any of the bricks are stamped or not is usually not mentioned in the catalogue, which is understandable given the condition most of these wrecks are in. Most have been looted and sport divers and fisherman can hardly be expected to study each brick on the bottom for a stamp and report their findings. No brick cargo ship has seemingly been excavated or published by archaeologists, either. There are a few exceptions, however, regarding the recording of stamps. Only one stamp type was found at the Capo Carbonara C (wreck 221) although the cargo was a complete load of tiles. The stamp is that of M. PROCILI MELEAGR (*CIL* XV.1 S. 363). Another wreck carrying nothing but tiles is Punta Scario A (wreck 961), but again there was only one stamp type found (reading TI CL FELIC EX OFFICIN).

What is apparent in the catalogue of shipwrecks and their cargoes is the frequency with which all-tile or brick cargoes are noted, for all periods. Long before the Tiber valley industry got underway, Classical Greeks were shipping Laconian roof tiles (wreck 808, Petrokaravo). So too were the Etruscans shipping roof tiles (wreck 894, Porto Venere). The latest cargo of tiles noted is a 13th century wreck off Turkey (wreck 1138, Tekmezar Burnu). A proper study of shipwrecks carrying brick and tile would be welcome, and could settle the 'long distance brick trade' debate. But for the time being it seems as if, rather than there being an incidental trade in brick and tile to suit occasional ballasting needs, brick and tile was a cargo which was worth shipping as an entire load. Sometimes, if the two stamps from the Capo Carbonara C and the Punto Scario A wrecks are any indication, the entire load could be procured from a single producer.

4.6 Chapter Summary

The geography of the brick industry is rather fragmented. Certain *figlinae* exploit sources at many different points in the Tiber Valley, while other *figlinae* are limited to one area. Fragmented landholding was a common pattern in antiquity. To exploit his or her scattered holdings effectively, the landowner could use *l.c. rei* agreements, where the land would be let out to a tenant farmer, who in return paid an annual land rent. In this chapter, I have argued that this pattern of agricultural exploitation explains the patterns of clay use in brick, and the information recorded in the stamps.

The manufacture of brick should not be understood as an industry divorced from other uses of the hinterland, but rather, as Varro implied, as an aspect of agriculture in its broadest sense. As an 'agribusiness', brick may have been exploited on an industrial scale, but this in part is due to the tenant's obligation to produce enough to pay the rent, and to make a profit to feed his own family and dependants. Yet, these rents may not have been onerously high. Because of the more or less continual demand for brick in Rome, the skilled brick-maker might have been actively 'courted' by the landlord who wished to develop this particular resource; one way of securing the deal would be to offer generous terms. That is a supposition, but what details of this industry-cum-agribusiness as can be determined, can be understood through an examination of the nexus of relationships represented by an assemblage of stamped bricks.

First and foremost, understanding the nexus allows us to sort out the different production modes for brick. By plotting the number and type of these modes over time, the broad patterns of brick making becomes clear. In the early days of this commodity, many different sources of clay were exploited throughout the valley as landowners discovered the suitability of their various parcels of land. Landowner's estates were fragmented, but clays could be intensively exploited using the system already employed in farming, of tenants and share-croppers. The tenants and share-croppers paid for the use and exploitation of these sources largely in kind. There is some evidence for parallel siting of production of brick with lead working. The likely net effect of this siting would be to lower production and distribution costs through the sharing of labour and infrastructure. The composition of the tested bricks also suggests that certain materials were actively sought out by brick makers for addition as temper. This volcanic material allowed bricks to be made which were otherwise larger than they might have been. It is not the case that a brick is a brick is a brick, and so the scarcity of Tiber valley brick gave it its intrinsic value. This value accounts for why Tiber Valley brick should be found so far afield, or transported upstream from its point of manufacture, or why the biggest names in Roman society should have participated in this branch of agricultural production. The next chapter continues the re-evaluation of the brick industry, especially in terms of distribution.

Chapter 5: Bricks to Rome, Bricks to the Valley

5.1 Introduction

In this chapter, I continue this re-evaluation of the Tiber Valley brick industry. Because brick has both inherent and added value (4.5), depending on the production mode, distribution patterns can be used to study how the bricks got, or were prevented from getting, to market (5.2.1). The interconnections between sites using stamped brick in the Tiber Valley are considered in 5.2.2 using a computerised dynamic settlement model, and seem to suggest a regular system for distributing brick. This reinforces the argument that it is not so much fixed physical costs which determine distribution, but rather, social interference (a theme which will be taken up in detail in Chapter 6).

In 5.3 some of these interpretations, including the findings in 4.3 and 4.4, are put to the (statistical) test. Different elements in the stamps are found to be associated with particular production modes or market orientations (5.3.1), and a model of the possible distribution logistics is developed (5.3.2). Consular dated stamps are the subject of 5.3.4, where it is found that there is a roughly 5 year cyclical pattern in the rate of stamping (which bolsters the discussion of land exploitation patterns in 4.2.2). Finally, the chapter concludes with a consideration of the unstamped material from Forum Novum and Falerii Novi (5.4). It is found that it is possible to date unstamped material on the basis of its geochemistry. It is also possible to fit sites at which only unstamped bricks were found into the wider patterns discussed in 4.2 to 5.3.

5.2 Consuming Brick

5.2.1 The Marketing of Brick

If brick has value, both inherent and added through the various modes of production which are deduced from the relationships between site/stamp/fabric, then how those bricks got to market was likely of great interest to the *dominus* as well as the *officinator*. Once the landowner received the payment of the land rent (a certain amount of bricks), selling those bricks in turn put the landowner into direct competition with the brick makers. Simply because land owners might not make the bricks themselves does not mean that they could not take actions to ensure that they won the competition. In 2.3.2 it was suggested that the 19th century Ottawa Valley timber industry could be a useful parallel for understanding how a riverine economy could function. Like brick, timber can be costly to produce, and for different reasons, it is also tied primarily to rivers for transport. We need to recall briefly the conclusions from that parallel:

- low start-up costs, combined with the inability to predict production or demand, create repeated cycles of over-supply and price collapse
- uncertainty and crises drive the tendency towards consolidation in the hands of the larger players
- the cheaper the price of the product, and the further the distance from point of sale, the more improvements necessary to bring that product to market
- large operators band together for the improvement of the river, giving them the concomitant right to charge others for the use of these improvements
- stamps on the timber are used for ownership, indication of destination, indication of origin, calculation of volume shipped, taxation, tolls
- stamps develop informally in response to the difficulties of shipping on the river; their codification in law is at the behest of the large operators, likely to the large operators' advantage

The Tiber is not a very large river, in terms of the actual area available for shipping. Wheat, oil, wine, fruit and vegetables, wood, and stone were all shipped down the river. This picture of the middle Tiber, from Orte to Rome, is of a very busy stream. In such crowded conditions, the parallel suggests that the stamps on brick could serve like the timber stamps in the easy identification of cargoes and when compared against shipping contracts the destination of those cargoes, primarily docks or warehouses. On the Tiber, the year AD 123 is perhaps the most frequent date occurring in brick stamps. It has been suggested that some sort of order was issued from on high that year (Bloch 1959: 237), which if true (but cf. 5.3.3 below) seems analogous to the situation on the Ottawa where, in an already mature industry where stamps had been in use for some time, the large timber operators requested a formal stamping law by Parliament. It was the largest operators who felt the need for a new law that regularised a customary practice according to *their* rules. By having all timbers stamped and the marks registered, unauthorised production could be curbed and the proper calculation of tolls and taxation effected. On the Tiber, the *Domitii familia* were the largest producers in AD 123; in that year the urban prefect was Annius Verus, the father-in-law of Domitia Lucilla, *domina* (or

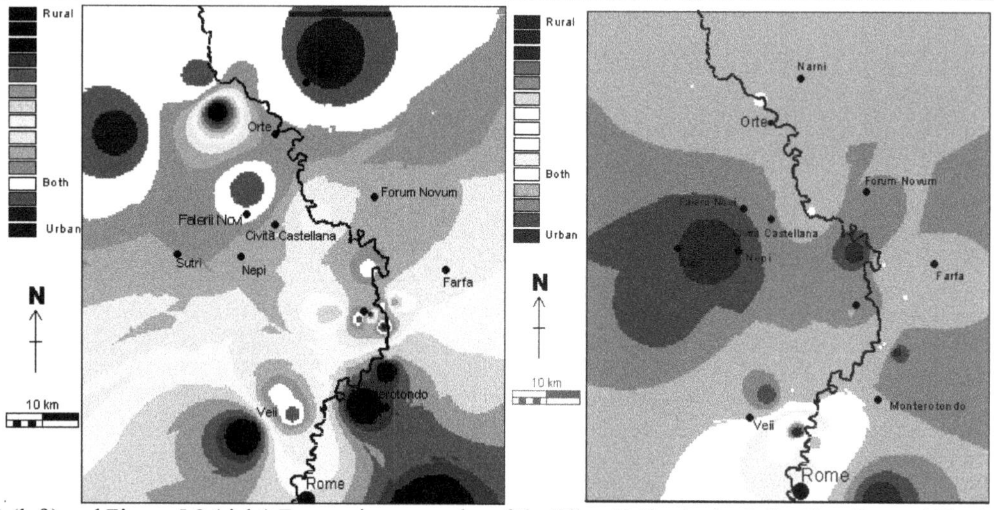

Figure 5.1 (left) and Figure 5.2 (right) Economic geography of the Tiber Valley in the Julio-Claudian and Flavian periods, respectively, based on the index values constructed for brick stamp types by period found at sites throughout the Valley.

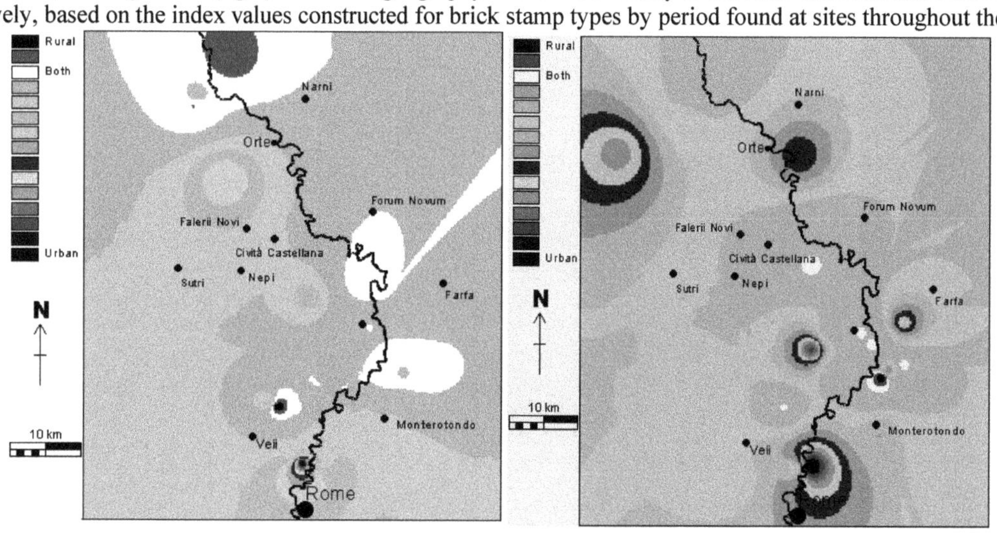

Figure 5.3 (left) and Figure 5.4 (right) Economic geography of the Tiber Valley in the Nerva - Hadrian and Antoninus Pius - Commodus periods, respectively, based on the index values constructed for brick stamp types by period found at sites throughout the Valley.

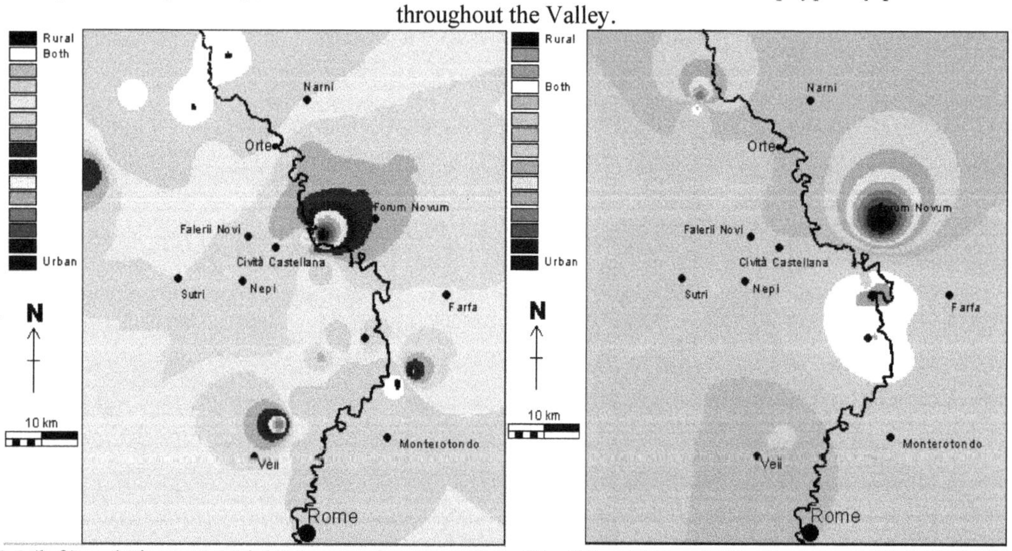

Figure 5.5 (left) and Figure 5.6 (right) Economic geography of the Tiber Valley in the Severan and Diocletianic periods, respectively, based on the index values constructed for brick stamp types by period found at sites throughout the Valley

mistress) of the estate. Here I will propose a third layer to the meanings of brick stamps. Because certain modes of production allowed for higher profit margins than others, stamps may be to do with controlling shipping and access to the Tiber and its associated infrastructure. Controlling the infrastructure for getting the brick to market would nullify the advantages of Mode 2 (geographically dispersed) production, while at the same time making Mode 3 (single source exploited by different *figlinae*) more economical by turning certain of these points into nodes in the wider distribution system. Mode 1 production would still be rather limited in scope, and would only be found in long distance trade if the brick producer had particularly close ties with a well-placed *dominus* (perhaps in a share-cropping arrangement).

To use this hypothesis about control of market access, let us assume that the major market for brick was the metropolis, Rome. If an assemblage of stamped bricks at a given site consists of the kind of stamp types which are usually found only in Rome, then for this site one could argue that the person paying for the construction had access to the distribution networks usually centred on Rome. This site can be imagined as having a high degree of integration with the city, a little piece of the city in the surrounding countryside. Conversely, if these stamp types are usually found not in Rome, then that site may be thought of as being excluded from the urban distribution networks, in effect a backwater.

The occurrence of stamped brick types in the Valley compared to the city can be used to investigate this differential access to the river infrastructure. I developed a simple index by tabulating for each stamp type the number of examples found in the Valley (the Valley frequency), and the number found in Rome (the Rome frequency) as recorded in *CIL* XV.1. The Valley frequency was divided by the Rome frequency, and the results were standardised so that a number greater than one demonstrated a production run oriented towards the Rome market, while a number less than -1 indicated production geared towards the Valley market. Results which ranged between -1 and +1 were taken to mean production aimed at both markets. Production runs oriented towards Rome were interpreted as having had access to the infrastructure which made it possible to market in Rome, whereas Valley production runs were interpreted as rather the opposite. The index is constructed from 523 examples of stamped bricks found in the Tiber Valley (Filippi and Stanco catalogue; this was the draft version of Filippi and Stanco 2005 that they made available to me in advance of publication); there are over 2000 examples of the same types recorded in Rome; cf Appendix E).

The series of maps in Figures 5.1 – 5.6 link the distribution of each stamp type found in the Tiber Valley to their degree of participation in the rural or urban networks. These maps give us a sense of the degree to which various parts of the hinterland were integrated with the City from the point of view of consumption, or even ideology (the desire to use the same materials as used in Rome). Functionalist economists would argue that integration with Rome depends on how close an area is to Rome, but that is clearly not the case. It is much more complex, and changes over time. During the Julio-Claudian period, there are only a few sites which have access to materials for which the primary market was Rome (Figure 5.1). There is quite a bit of variability and flux between the Julio-Claudian and Flavian periods, with most places in the Valley having access to urban distribution networks (but with quite an opposite pattern nearest the city) (Figure 5.2), and over the next hundred years (Figures 5.3, 5.4) the situation settles to a state where during the Severans (Figure 5.5) only one place uses the same materials as in Rome. In late antiquity (Figure 5.6), there is suddenly much more variety than once there was- it is almost a return to conditions prevailing in the Julio-Claudian period. Note firstly the gradual blurring of the rural/urban areas, and secondly the close proximity to Rome of areas which are completed excluded from urban networks. Finally, there is the opposite situation as well, where places quite distant from Rome are tightly bound within the urban networks. Forum Novum is quite stable, neither predominantly rural nor urban until late antiquity, when it suddenly is using the same material as in Rome. This pattern might be explained with reference to the early medieval period, with the establishment of Forum Novum as the papal cathedral centre for the Sabines (on the history of Forum Novum cf. Gaffney *et al.* 2001: 59-60).

Stamped bricks made by at least thirty different slaves of the *Domitii* appear in the Tiber valley, but their production alternates between Rome and the Valley at different stages in their careers. For example, the slave Trophimus Agathobuli first appears in the Tiber valley stamps in about AD 93/4, with Cn. Domitius Tullus as *dominus*. This stamp (*CIL* XV.1 1002) appears in a Rome zone. As time passed, Tullus died and his daughter Domitia Lucilla inherited his estate. The next stamp of Trophimus Agathobuli (*CIL* XV.1 263) dates to shortly before AD 115, with Lucilla as Domina, and the production run with this stamp demonstrates an equal weighting between Roman and Tiber valley production. After his manumission in AD 115, Trophimus Agathobuli appears again, but this time his production (stamp *CIL* XV.1 1108) is decidedly tilted towards rural production. A slave belonging to Trophimus Agathobuli also appears at this time, and his production (stamps *CIL* XV.1 1118a-b) is similarly positioned (on the history of the slaves of the *Domitii*, see Steinby 1974, 47-58).

For this example at least it is as if access to the major market is controlled by the important landowners (eventually the *Domitii* produced an Emperor, Marcus

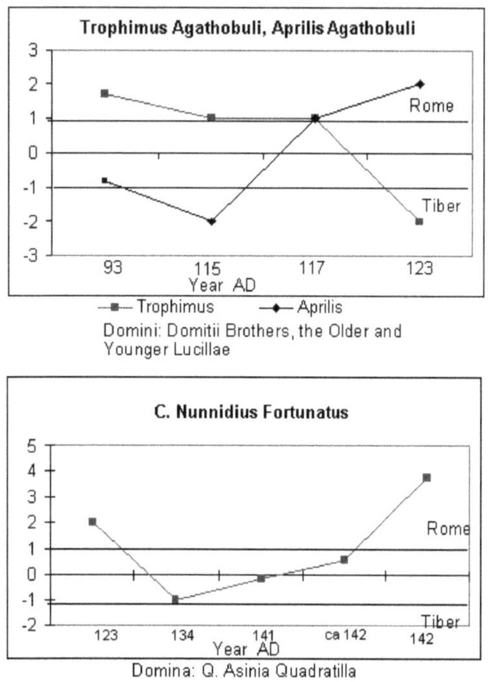

Figure 5.7 Three career histories. For each stamp type a particular brick maker used, an index value can be calculated, determining the relative degree of access to the urban or rural markets (the same index on which Figures 5.1 – 5.6 are based). Career histories can then be plotted and compared

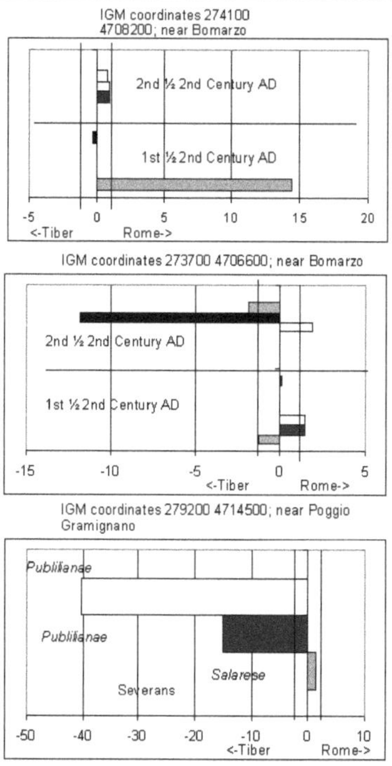

Figure 5.8 Three kiln histories. As for individuals, so too for production sites. All of the stamped brick recovered from a production site (each type represented here by a horizontal bar) can be used to create an index value by period, indicating to which market producers using a particular kiln had access.

Aurelius). While Trophimus Agathobuli was their slave, they were enriched by his output and so those bricks went to the market where they could command the best prices i.e. Rome. Alternatively, while a slave, he had access to the transportation network and warehousing facilities that made it economic to market in Rome. Then, as a slave of a new mistress, for whatever reason, the main market was slowly denied to him. Finally, as a freedman leasing the land to make bricks, while occasionally getting product to Rome, he and his slaves were largely unable to get major market access, his outfit being too small or lacking the resources to transport a very heavy, very bulky material the distance, or perhaps locked out by the dominance of the few and the powerful (cf. Wallace-Hadrill 1989: 73). Figure 5.7 plots the course of the careers of two other individuals based on the stamp indicators. It is noteworthy that Trophimus' contemporary, Aprilis Agathobuli, has a career that runs almost the opposite of Trophimus. The career of Cn. Nunnidius Fortunatus ends at a high level in the index, suggesting a fair degree of success at accessing the distribution networks, a success we could lay at the feet of Asinia Quadratilla, his *domina* who was also the granddaughter of a patron of Ostia (Setälä 1977: 72).

Questions of market access apply also to the *figlinae* as a whole. Recently, a production site of the *gens Domitii* has been located near Bomarzo. The identification of the two kilns as belonging to the *Domitii* is partly based on the finds of brick stamp wasters, but also on the recovery of a boundary stone explicitly naming the brothers *Domitii* (Filippi, pers. comm.). The index constructed for the stamps from the one kiln indicate a production split between Rome and the Tiber valley except in the Nerva to Hadrian period, when its production shifted dramatically towards Rome. The other kiln was similar in that its production was split between Rome and the Tiber valley, but in the Antonine period, production became centred to a large degree on the valley (Figure 5.8).

Across the river, at Poggio Gramignano, production of bricks from the *Figlinae Salarese* and *Publilianae* has been identified by the association of brick fabrics with wasters (but not stamped wasters) (Martin 1999: 374; Monacchi 1999: 382,9). Dating to the Severan period, production from *Figlinae Salerese* is slightly tilted towards Rome, while *Publilianae* is very much directed at the Tiber Valley market. The *Publilianae* stamps refer to production by women. Aemiliae Severa is mentioned as *domina*, and Iunia Antonia is called *negotiatrix*. To judge by the scale indicated here, they were a very successful outfit. In this case then we have one kiln used by different *figlinae* to serve different markets while over in Etruria we have two kilns from one *praedia* serving different markets.

5.2.2 Interconnections in the Tiber Valley

This index of market access is one way we can examine the interrelationships between sites using stamped brick in the Tiber valley. I expand on these observations below in 5.3.4 where I examine the other assemblages of materials from sites which have similar rankings in the index. However, given that some areas seemingly have access to the same materials as were used at Rome, and also used brick which had been shipped upstream, we ought to examine the *geographic* interrelationships first. If we examine the interrelationships or interconnections between sites using stamped brick, we should be able to work out the patterns of trade which enabled the use of this material at these sites. Traditionally, this would be the point where Thiessen Polygons or Central Place Theory would be introduced into the discussion (cf. Morley 170-174). But neither of these methods properly take into account the similarities or differences in size, importance, or interaction between sites. T.E. Rihll and A.G. Wilson developed a computerised model which expressly takes these considerations into account (Rihll and Wilson, 1991: 60, 62). Their work grew from similar models created and used by geographers at Leeds University, and represented an attempt to use the model to study a particular historical context (Rihll and Wilson, 1991: 59).

The model works by examining each site's position relative to every other site in turn, based solely on its x- and y- coordinates; no other information is entered. The model consists of a system of differential equations which when run, iterate over and over again until the system balances out, indicating a solution. It works out the relative site importance and the likely interactions between the sites based on their spatial positioning alone. This can then be graphed, indicating the size (or importance) of sites and the patterning of the interactions between them. The model's equations were adapted and turned into a computer programme for this study by Mark Wakefield of the Mathematics Department at this University. How the model functions mathematically is covered in detail by Rihll and Wilson (1991). Essentially, it is an *entropy-maximising* model (69-70), which means it is a probabilistic model that tries to find the most likely overall state of the system while making the fewest assumptions. As far as can be determined, this is the first application of the model since Rihll and Wilson published it, a fact no doubt related to its use of complex mathematics to settle on a solution.

Rihll and Wilson's model was used to study the formation of the Classical Greek *poleis* based on the pattern of settlements in the earlier Archaic period. Their principal question was:

> ...did location *vis-à-vis* other settlements, rather than location in a particular type of environment, have a significant effect on settlement growth and socio-political centralisation? (60)

Here we may ask what location implies for the interconnections between sites which use stamped brick. We might expect that, given the traditional concerns about the bulk of brick, the difficulty in transporting it, the differences between South Etruria and the Sabina, the role of the river as a barrier rather than a conveyance up- and down-stream, that interconnections should be minimal. However, the market access index (5.2.1) implies that the situation is far more complex, with some areas being as 'urban' as the great metropolis itself. That is, rather than every site in the Tiber valley being focussed on Rome (or every hierarchy of sites), how do sites interact with *each other* in the Tiber Valley?

Rihll and Wilson's model incorporates two basic hypotheses regarding site interaction. The first is that interaction is proportional to the size of the site where the interaction originates, and the importance and distance from that site to all other sites considered. The second is that the importance of a site is proportional to the amount of interaction it attracts from the other sites (63). Distances are assumed to be the shortest straight-line distance, which would seem to underestimate the importance of the complications of landscape. However, since there generally are some geographic reasons for where settlements are built (there are no settlements in the middle of lakes or rivers for instance), physical relief is automatically considered in the pattern of settlements themselves (64-65). Rihll and Wilson build a third hypothesis into their model, which allows us to escape the need to know the size of sites we are dealing with. The third hypothesis is that the site size is proportional to its importance (69-70). This allows us to run the model from an initially egalitarian state; all we need to know are the geographic co-ordinates of the sites. Essentially, using this model specifically allows us to explore the effects of geography, stripped of all other considerations..

The sites modelled included the major towns in the Tiber Valley (including Rome) and the sites from which stamped bricks were recovered. Selective inclusion of sites allows us to model site interactions related directly to the consumption of stamped brick. The sites considered are the same ones which were used to create the 'market access' index. Figures 5.9 – 5.13 plot the results against a map of the Tiber Valley. The first thing to observe are the 'centres of gravity' which appear, and the fact that they are interconnected (there is nothing inherent in the model which suggests that sites should be interconnected; this rarely happened with Rihll and Wilson's data on Greek pre-classical settlement for instance). Secondly, trans-Tiber ties are clearly

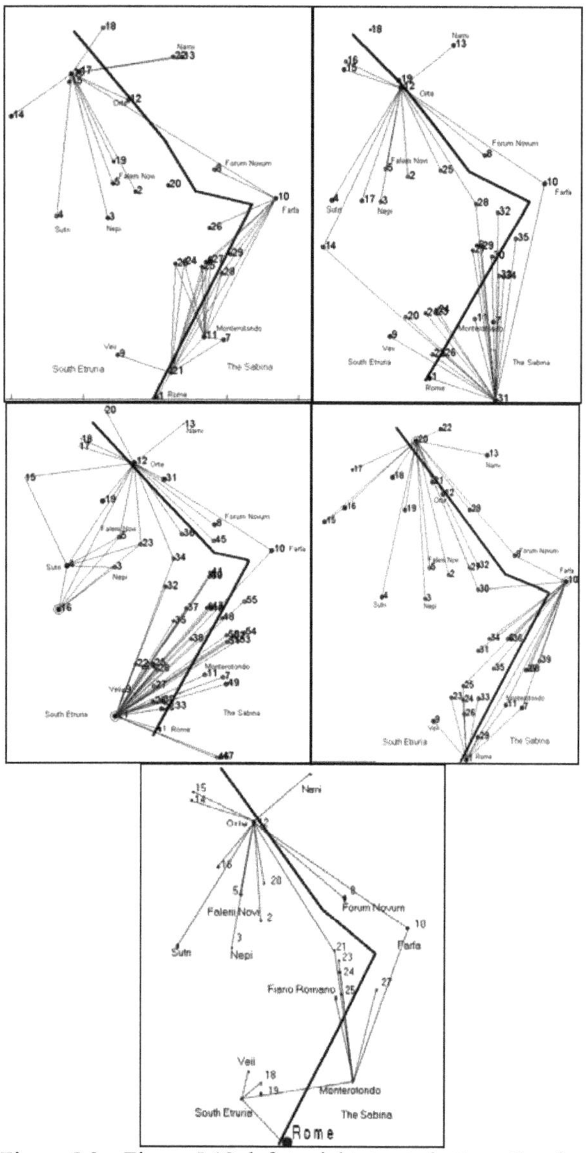

Figure 5.9 – Figure 5.12, left to right, top to bottom. Gravity model output of interactions between numbered sites using stamped brick in the Tiber Valley. Thick line represents schematised course of the Tiber. From left, Julio-Claudian period, Flavian period, Nerva-Commodus period, Severan period.

important. In every period, the Tiber does not seem to be a barrier between sites in South Etruria and in the Sabina. The greatest intra-regional difference seems to be more of north and south rather than east and west. Note the apparent differences in the usage of the river and the road system between centres of gravity. In the Julio-Claudian period (Figure 5.9), the interactions between settlements and gravity centres follow the river. There are three major centres in this network, based on Orte, Monterotondo, and just outside Rome. In the next period (Figure 5.10), the river is again important (with perhaps the Anio as well), while the usage of the Cassia-Clodia road system seems to be implied. There are only two major centres now. The

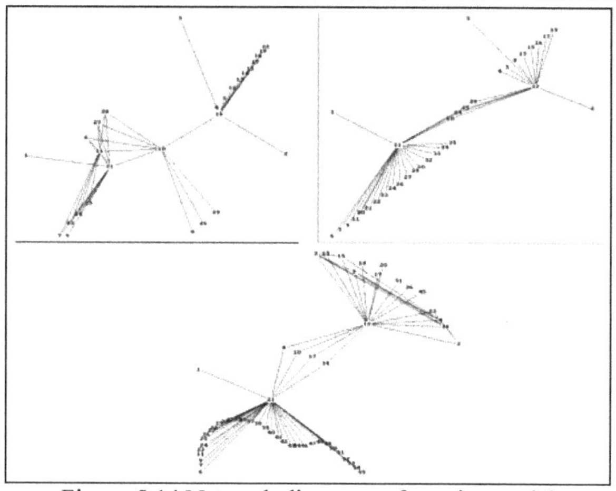

Figure 5.14 Network diagrams of gravity model interactions depicted in figures 5.9 – 5.11. Clockwise from top left: Julio-Claudian period, Flavian period, Nerva-Commodus period

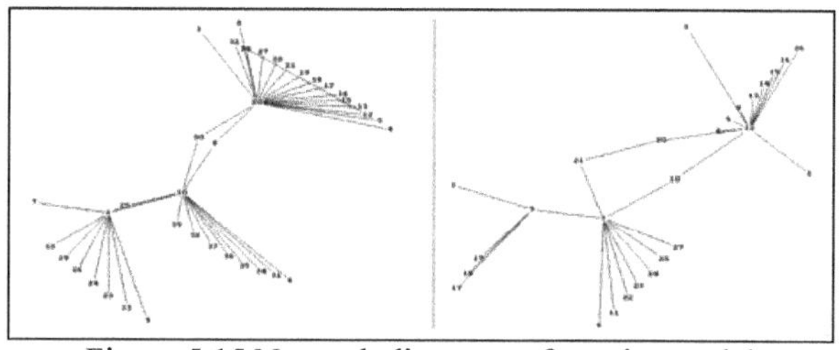

Figure 5.15 Network diagrams of gravity model interactions depicted in figures 5.12-5.13. From left: Severan period, Diocletianic period

period from Nerva to Marcus Aurelius (Figure 5.11) is more complex yet again, with a very complicated use of the rivers and the road system and two major, and a number of minor, centres. In the Severan age (Figure 5.12), the pattern seems to revert back to that of the Julio-Claudian period, a pattern which by late antiquity (Figure 5.13) is firmly re-established.

Why should there be such a difference in the usage of the communications networks? To answer this question we should imagine these maps not as pictures from a stratospheric viewpoint but rather as space to be moved through, as an itinerary of what comes next, and what river versus road travel implies. Road travel is far more constraining than river travel. When travelling on a road, one has to go *through* places. Roads control and reconfigure movement (cf. Laurence 1999; 2001a; 2001b). Often, this means being funnelled into a town, across a political boundary, into a new tax regime. A river on the other hand is often the border itself. A river's course is determined by nature, it has width, and it allows one to *bypass* places. It is therefore safer. In those periods where the river seems to be the only conduit used, does that equate to greater political insecurity? Or alternatively, using the river might allow one to bypass controls, and so is it in effect an attempt to subvert a higher authority? It is not a question of the river being easier to use to ship bricks. If this was the case, the road system would not be so readily apparent in this model. Rather it would seem to confirm that access to the river system was not universal.

Figures 5.14 and 5.15 show the same information as Figures 5.9 – 5.13, but in pure network terms, of nodes and connections (the geography has now been stripped out). The characteristic path length of these diagrams (the distance or number of links it takes to get from any two nodes, on average) is always about four. Despite the widely differing geographic distribution of sites in each period, the number of connections between them always remains the same in each period. This is not an artefact of the model used, because in Rihll and Wilson's results there were many 'islands' or settlement clusters which did not connect with the others (1991: 76-86). Also in these pure network diagrams there are the 'joiner' sites which connect the two (or more) centres of gravity. Given that *tegularia* and other redistribution sites for brick are known to have existed, it is tempting to interpret these pure network diagrams as representing a four step distribution process of site to warehouse to transhipment point to warehouse to site. Short cuts exist in the earliest and latest periods, which may suggest that the distribution infrastructure was being subverted at first and was slipping later. In Chapter 6, the relationships between the bricks in the archaeometric sample (cf. 4.3.1) are modelled as a network. These networks indicate the same strong cross-valley ties, suggesting that Rihll and Wilson's settlement model does capture essential aspects of the intercommunications of sites in the Tiber valley.

5.3 Purpose of Stamps

5.3.1 Associations

By examining the different modes of production as evidenced by the nexus of relationships centred on a given stamped brick, I have built up an interpretation of the meaning and usage of stamps. The (abridged) argument so far: tenancy and sharecropping can be used as models to understand how landowners may have arranged for the effective exploitation of their land. Brick makers, as skilled tradesmen, may have sought out particular arrangements with landowners so that they could pursue profitably their trade. The land rent (the *merces*) was paid in money or in kind; stamps might represent a mechanism to ensure that the rent had been paid in a given year. This meaning of stamps is probably a secondary development, while the basic purpose of *epigraphic* stamps is likely connected to distribution. Brick has inherent value, and so it was worth the while to invest in, to make, and to sell brick. Transportation costs from the different clay bodies were largely uniform so long as access to the river could be guaranteed; profit margins were therefore more dependent on the mode of production than on how the bricks were distributed. The different people involved in the industry jockeyed for position in the market. The pattern of interconnections between sites which had access to stamped brick suggests that there was a dedicated system for distribution; access to this system varied. Some were more successful than others in gaining access and were able to market their products to the metropolis. At the building sites in the countryside, some builders were able to build with the same materials as were used in Rome.

I now have quite a lot of evidence on the production and distribution of brick and tile, which puts me into a position to test certain hypotheses about the function of stamps, rather than merely to make guesses. I have determined that bricks can be made in three different modes of production; I have determined that the individuals named in the stamps have differential access to the urban and rural markets. If despite the argument so far the information in the stamps is indeed related to production, I should find statistically significant associations between the different modes and the different stamp types or elements (shape, presence/absence of *signa*, presence/absence of consular dating). If the information in the stamps is connected to distribution (which I argue is the basic layer of meaning) then any significant associations should be between the market orientations or Combination C ('consumer choice') and the stamps or stamp elements. The test used was the standard chi-squared test of association. This technique is used to test whether the occurrence of a particular trait is independent of the occurrence of another (Shennan 1997:100).

There were no significant associations found between the production modes and any other elements, whether stamp shape, market orientation, or the presence/absence of *signa*. The top part of Table 5.1 shows the chi-squared values and other statistics for the unsuccessful tests. The alpha statistic indicates the probability of an association (given in percentages in the parentheses). Cramer's V indicates the strength of the relationship (the closer to 1, the stronger) while Yule's Q, where appropriate to calculate, indicates whether the presence of one attribute implies the presence or absence of the other (109-118). The failure of these tests to find a statistically significant association suggests that the epigraphic stamps are not connected with the modes of production. The audience for whom the stamps were intended should be therefore outside the productive unit. It is when the stamp elements are tested against the market orientations and the 'consumer choice' option that I begin to find significant associations. At the bottom of Table 5.1 are recorded the successful tests.

It is the presence or absence of *signa* and the shape of the stamp which seem to be associated with distribution rather than production. If a *signum* is present in a stamp, there is a good probability (> 95%) that this stamped

Categories	chisquare value	Degrees of freedom	Probability of an association	Cramer's V	Yule's Q	Successful (probability >90%)?
mode v consular dating	too many expected counts less than 5; chisquare approximation probably invalid					no
mode v orientation	3.997	4	50-75%	0.03		no
mode v signa	0.404	2	<5%			no
orientation v consumer choice	2.256	2	50-75%			no
mode 1 v shape2	0.015	1	10%			no
mode 1 v signa	0.004	1	5-10%			no
mode 1 v consular dating	too many expected counts less than 5; 1 cell with expected count less than 1; chisquare approximation probably invalid					no
mode 2 v signa	0.665	1	50-75%			no
mode 2 v consular dating	too many expected counts less than 5; chisquare approximation probably invalid					no
mode 3 v shape2	1.8	1	75-90%			no
mode 3 v signa	0.126	1	10-25%			no
mode 3 v consular dating	1.733	1	75-90%			no
mode 1 v orientation	1.354	1	75-90%			no
mode 2 v orientation	2.059	1	75-90%			no
mode 3 v orientation	0.005	1	<5			no
mode 2 v consumer choice	1.516	1	75-90%			no
mode 3 v consumer choice	2.586	1	75-90%			no
consumer choice v consular dating	0.681	1	50-75%			no
signa v consular dating	3.635	1	90-95%	0.05		yes
orientation v shape2	7.637	1	99-99.5%	0.11	0.63	yes
orientation v signa	8.218	2	97.5-99%	0.12		yes
mode 2 v shape2	5.412	1	98%	0.1		yes
signa v orientation	7.775	1	99%- 99.5%	0.12		yes
mode 1 v consumer choice	12.594	1	99.90%			yes
consumer choice v signa	4.069	1	95-97.5%			yes
consumer choice v shape 2	6.08	1	97.5-99%	0.09	-0.56	yes

Table 5.1 Chi-squared tests of association between presence and absence of the various modes, urban/rural market orientations, and stamp elements. Shape 1 = rectangular, shape 2 = orbicular

brick was part of the urban marketing networks, but if it is absent, then it was probably part of the rural networks. Interestingly, those bricks to which consumers had greater access (as indicated by 'Combination C') are also more likely *not* to have a *signum,* suggesting that in the countryside one could do as one wished. In the city, where presumably the greatest profits were to be had, access to materials (which is the flip-side to the brick maker's access to transportation infrastructure) was controlled. Consular dating co-varies positively with the presence or absence of *signa*. If the *signum* is present, odds are that a consular date will be too, while if it is absent, so too will be the consular dating. Finally, the shape of the stamp seems to be associated with the different networks- rectangular with the rural and orbicular with urban. The shape of stamps has long been used as an indicator for the date of the stamp (Dressel 1891: 9; Bloch 1947: 23; Steinby 1974: 19-20), and so it might be argued that this association has nothing to do with market orientation but rather with chronology. Yet contrary to that expectation there exist many instances where the same individuals appear in different shaped stamps, as was noted in 2.1.4. If it were true that changing stamp shape represents nothing more than the chronological evolution of stamp types, then market orientations should occur independently of the stamp shape. This is not, however, what happens. The indicators for whether a particular stamped brick was distributed through the urban networks or the rural ones are summarised in Table 5.2.

Urban bricks will have...	while rural bricks will have...
...a *signum*	...no *signum*
...*or* a consular date	...no consular date
...*and* be orbicular in shape	...and be rectangular
	...and maybe even be "consumer choice" (Combination C) ones

Table 5.2 Distribution indicators

5.3.2 Logistics and *Signa*

In the tenancy hypothesis the literate parts of certain stamps are related to the *merces*, the need to demonstrate that the obligations and payment under the land leasing agreement have been met for the year. The need of *officinatores* to get their payment in kind to their landlords, resident in Rome, should necessitate some sort of infrastructure. This need may be what gave rise to the development of *signa* in stamps. In 2.3.2 where I considered how the Tiber might work as a piece of infrastructure, I argued that in stamps of the *Portus Licini*, the different *signa* might refer to particular docks within the port complex. This might be extended to all stamps which use *signa*, for in 5.3.1 above I noted an apparent congruence between stamp types with *signa,* and a marketing orientation towards the city. The logistics of getting the brick to the warehouses of Rome may be part of the explanation for the development of *signa* in the second century AD brick stamps. Any particular batch of bricks would be made up so that the stamped brick lies on top, easily visible and its *signum* indicating at which dock or wharf that batch is to be unloaded.

Could a *signum* indicate a place, whether a dock or warehouse? Roger Ling (1990) considered the problem of a stranger in town, trying to find his or her way to an unfamiliar house. He points out that in literature (*inter alia* the plays of Terence, the poems of Ovid, and the Acts of the Apostles) addresses as such do not exist. Rather, *ad hoc* directions serve the purpose instead, relying on prominent landmarks (identifying a location in this way is a practice which continues in modern Italy, despite street names and house numbers!) At a town like Pompeii, the different fountains with their different emblems would offer convenient identifiable landmarks, while inns and apartment buildings could be identified by a painted sign (the Elephant inn, the inn of the Cimbrian Shield) or by name (the *praedia* of Julia Felix) (210-1). These emblems and paintings are not all that different in conception than *signa*, and were useful ways of marking out a location and reference points in an era where literacy was not universal.

In this reading, a brick stamp works on two levels. One, it refers to the exploitation of clay sources, a sort of accounting function. Two, it refers to the Tiber, and the complexities of getting the finished bricks from the kilns to the warehouses. A two-level conception of the meaning of stamps raises a number of questions, not least of which are:

- did the people handling the bricks read the stamps in this way?
- is there any real proof that the docks were organised in this fashion, that they had individual names and symbols?

The Docks of Rome and London

I will take the second question first. There are indications in the topography of Rome that certain areas along the Tiber were in fact given over to the unloading of specific goods. Mocchegiani Carpano (1984: 39) gives as an example the area called '*ad ciconia nixas*' on the left bank of the Tiber in the Campus Martius which was linked to the *Portus Vinarius*, given over to the unloading of wine amphorae. Coarelli (2000: 375-378) argues that a certain portion of the Severan marble plan of Rome actually indicates a storage area for brick along the slopes of the little Aventine. The piece in question seems to indicate a large open structure, and carries the label 'NAVALEMFER' (reproduced in Carettoni *et al.* 1960, plate XV, fragment no.2), which is usually taken to indicate one of the lower *navalia* or ship yards (*navale inferius*). However, in Coarelli's view, this label is very similar to the text on a stamped brick from Pannonia (*CIL* III, 11382) where '*navalia*' seems to be used in the same sense as *portus* in the *Portus Licini* stamps. He then suggests that this strange use of 'ship yard' is a colloquial expression equating brick yards, with their long rows of brick, with ship yards and the furrows made in the ground by the pulling up of ships onto dry land in the boat sheds. Coarelli notes that there is a second piece of the marble plan (reproduced in Carettoni *et al.* 1960, plate XLI, fragment no.201) which seems to indicate an open area with rows of sheds. He then re-interprets the label on the marble plan to read *Navale M(arci) Fer[ocis]*, a man known to history as Cn. Pompeius Ferox Licinianus (consul suffect in AD 98), and connects the two pieces of the marble plan together. He concludes that here, on the marble plan, is the *Portus Licini* of brick stamps. Unfortunately, the fragments of the plan do not indicate how these structures communicate with the river. The structures Coarelli identifies could well be to do with the storage of brick and tile, which is an important identification. (The arguments connecting these structures with the Portus Licini, however, are beside the point and probably reflect a desire to attach a famous name to the structure).

It seems reasonable to argue that the docks along the Tiber could be indicated by a particular name (in a way similar to how Ling 1990 suggests urban space could be organised) and that these docks could be given over to the trade in particular goods, as Mocchegiani Carpano (1984: 39) argues. Warehousing connected with brick has always been suspected (cf. Steinby 1974: 74 - 74) and Coarelli (2000: 375-8) may have discovered where at least one such complex was located. The example of the Port of London in the 17th to 19th centuries (Weinreb and Hibbert 1995: 235-237) may be taken as an indication of the dynamics of how all these pieces fit together.

In the tidal reaches of the Thames, ocean-going vessels had to anchor in mid-stream, to be unloaded by lighters, which then took the goods to a variety of different docks. Piracy and smuggling took their toll as goods did not always reach their intended destination, and duties were lost. Queen Elizabeth I passed a law requiring that all goods were to be unloaded at a limited number of 'legal quays', ensuring that customs duties could be collected. As trade grew in proportion to the Empire, the various quays, wharves, and docks began to specialise in certain trade items. Over the 18th century limited docking facilities, congestion, inadequate warehousing, and theft began to have a serious effect on the efficiency and profitability of trade in the port. By the beginning of the 19th century, to address these problems private companies had obtained charters to open new docking and warehousing facilities, including the East India docks, the West India docks, St Katharine's docks, and the Surrey Quays. These charters enabled them to obtain long-term monopolies in certain goods. Rum and hardwood were unloaded at the West India docks, softwood was unloaded at the Surrey Quays, sugar and rubber were unloaded at St. Katherine's. The private docks took a percentage of the value of all the cargoes which passed through, making docking and warehousing a particularly lucrative business. However, not just the dock complexes themselves were named, but also individual docks and even the staircases which lead from the embankment to the water's edge - names like 'Goat Stairs'; 'Three Cranes Wharf'; Limehouse Dock'; 'Elephant Stairs'; 'Puddle Dock' (Place names in Rotherhithe, 2002).

With regard to the brick trade, the scale of the trade in the second century and the continual demand for building materials in Rome, a city of about one million people in the first century AD (Morley 1996: 38-9) ought to have contributed significantly to congestion on the Tiber. Congestion was a significant problem for 18th century London, when that city had a population nearing (if not surpassing) a million people (Weinreb and Hibbert 1995: 631). The Thames in London is a much wider river than the Tiber in Rome; if congestion on the Thames was a problem in the 18th century, it could well have been significant on the Tiber in the 2nd century (allowing for the fact that the sea-going vessels docked and were off-loaded into barges at Portus and Ostia). Similar physical constraints in one time and place may have given rise to similar solutions in another. In Antiquity, at Ostia and at Arles (France) taxes and duties were paid on goods when they were unloaded (Mocchegiani Carpano 1984: 52-3); this was also the case at Rome itself as is indicated by a partial inscription found near San Silvestro (*CIL* VI 1785, 31931). Mocchegiani Carpano suggests that the curator of the Tiber may have acted to a degree as a sort of 'harbour master' (39); if so, he may have played a role in collecting these duties, but more importantly (from the point of view of the dock owner) in deciding which docks would have the right to unload what product in the first place.

It may have been that, like in Georgian and Victorian London, warehouse owners could be the people who also owned the docks at which the goods for their warehouses were unloaded; alternatively, the docks might be owned by a person connected to the warehouse owner by ties of patronage. It is not all that far-fetched to suppose that a warehouse owner in Rome might have specialised in providing storage space for brick. Warehouse contracts, specifying the type of good and the space in which it would be stored, are known from Puteoli (Rickman, 1980: 236-8). In our scenario, the *Portus Licini* (with its four different stamp *signa*) might well indeed be identified with those structures identified by Coarelli (2000: 375-8), but the docks serving the complex may have been operated by four distinct individuals. Because the docks specialised in particular goods (perhaps brick needs special ramps, pulleys, or cranes to unload?) brick makers might have long-standing arrangements for their bricks to go to these particular docks. The *signa* could therefore be incorporated into a stamp because the destination in Rome for each consignment would already be known.

Perhaps there is some more archaeological evidence for a brick-related installation along the river, if these arguments are correct. Near the Campus Martius, the 'Tor di Nona' is a 96m long mole which jutted into the Tiber and had a temple at its end (Quilici 1986: 202). The temple had capitals decorated with the skin of a wild beast, either a panther or a lion (Delaine, pers. comm.). Panther skins are an attribute of Bacchus, while lion skins are an attribute of Hercules. Both Hercules and Bacchus appear in *signa* (Hercules: *CIL* XV.1 156, 214-6, 241, 324-5, 686, 715, 768, 772, 1247, 1497 *Figlinae: Domitianae, Favorianae, Genianae, Marcianae, Voconianae*; Bacchus: *CIL* XV.1 126, 382 *Figlinae: Caninianae, Oceanae Minores*). The construction of this mole is Augustan in date, and was later incorporated into the Hadrianic embankment (Quilici 1986: 202). The *figlinae* date from the mid 1st century through to the reign of Commodus (Steinby 1974: 29, 34-5, 37, 41-45, 61-6, 69-71). In 2.3.2 I argued that temples might have functioned as landmarks for the navigation of the Tiber, and that *figlinae* might take their names from these landmarks. If we imagine that the docks would also take their names from landmarks or other prominent features, then it is a small matter to connect the *signa* of Hercules or Bacchus with the temple at the end of this mole. Perhaps the Tor di Nona is one of the putative specialised docks, dealing exclusively with the produce of these *figlinae*.

In order for the consignment to get to these docks, not every brick would need to be stamped. Only a few would need to be stamped to ensure that at least one clean stamp would survive the firing. This particular brick could then be laid on the top of the batch. Steinby (1981: 245) suggests that between as many as one in two or three, to one in 10 or 15 bricks were stamped (the figure of one in two or three probably is in reference to *bessales*, which earlier I argued were similar to anepigraphic stamps in conception, cf. 2.2.2). In AD 123, 16 different stamp types use only seven *signa* (wolf: *CIL* XV.1 28a,b and probably 266a; corn measure: *CIL* XV.1 648a,b, 1157; pine nut: *CIL* XV.1 1034, 1429; fish: *CIL* XV.1 847; Mercury: *CIL* XV.1 121; sistrum: *CIL* XV.1 248; 'X' : *CIL* XV.1 1020, 1029a,b,c). If AD 123 was the single busiest year in the brick industry (cf. 5.3.4), seven dedicated docks to the brick industry does not seem unreasonable (seven *signa* in AD 123, the most heavily stamped year ever in the industry also points to the unlikeliness of *signa* being used as heraldic devices for producers).

A river boat wreck carrying a load of bricks (of which four were stamped), was found in the Stella river in north western Italy (Zaccaria and Gomezel 2000: 301; Parker 1992 as wreck 777, although the stamps were unknown to him). The bricks were stacked neatly one on top of the other (Zaccaria and Gomezel 2000: 301). Information regarding how many stacks there were, and where exactly those four bricks were placed in the stacks, was not indicated, but the fact of there only being four is interesting. One *bipedalis* brick measures roughly 58 x 58 x 4 cms. At a Roman tilery excavated in Sussex, some tiles were marked with the numerals 'CCXX' (220), which the excavators took to represent a tally mark identifying the number of bricks in a batch (Rudling 1986: 211). To load one batch of 220 bricks (assuming for the sake of argument that batch sizes were roughly uniform throughout the Empire, and that 220 was a typical size) onto a boat would present difficulties. Stacked in a single pile, this sum would reach nearly 9 metres in height. However, stacked on edge in rows, 220 bricks could be laid out in only four rows, each eight Roman feet long (1 Rm ft = ca. 29 cm). Each row would contain 58 bricks. If the first brick in each row was the stamped brick, it would suggest a stamping rate of about 1 in 60, which is four times greater than Steinby's highest rate (1 in 15; 1981: 245). Of course we do not know (for it is not recorded) how many bricks in total were on the Stella wreck. We are now in the realm of speculation, but if we accept Steinby's stamping rate, there were only 60 bricks on the Stella wreck (or a quarter of a complete batch) arranged in four rows each two Roman feet long. It is tenuous but this reconstruction of kiln batches, row lengths, and stamping rates all are multiples of, or divide cleanly into 12; this figure may represent a 'rule-of-thumb' in the brick industry (cf. DeLaine 1998: 65 on rules-of-thumb in construction).

Literacy of the Working Man

Which brings us to the first question: did the men who worked on the river and the docks and in the warehouses read the stamps in this fashion? It is in fact an impossible question to answer. However, if we ask if they *could* read the stamps, that is a different matter. For the system to work, as I have imagined it, the working men would only have to be able to distinguish between the different stamps, and know what to do when they encountered a particular stamp type. This is what UNESCO calls 'functional literacy' (UNESCO 1988: 8; Hanson 1994: 162), and it is the basis for Hanson's (1994) study of the problem of ancient illiteracy. The thrust of her argument is that numeracy, reading, and writing are different skills and that any analysis of ancient illiteracy should take account of the fact that these skills were distributed in the population in varying amounts. Functional literacy means that an individual might have sufficient skill to perform his or her job, yet not qualify as 'literate' in the usual sense. In the brick industry, recognising *signa* and stamp combinations would be a matter of memorisation; when new variants appeared, only one person would have to be able to decipher the stamp, and then the others could learn to recognise the new variant. The industry could probably have sustained quite low levels of literacy and still functioned effectively.

In the *CIL* there are 46 distinct types of *signa* (listed in Table 5.3). These 46 *signa* appear on 428 different stamp types, totalling 3003 examples, or 29% of all examples in *CIL*, and 21% of all types. Forty-six individual *signa* does not seem like an overwhelming amount of types to remember, although it would probably take some time to learn them all. Of course, not all of these *signa* would be in use at any one time, and those that were would be in use for greater or lesser periods of time. Also, it seems unlikely then that a *signa* could be a heraldic device identifying the brick makers themselves (cf. 2.2.2), because there appears to be too many individuals (355 individual *officinatores*, according to Helen 1975: 23) and not enough symbols. However, it is conceivable that the 46 *signa* are related to 46 individual families, but these families are not necessarily the landowners from which the bricks were made, again because the symbols shift through stamps from one particular land-owning family to another. It is all rather circumstantial, but if *signa* were indeed connected with identifying the appropriate docks a particular batch was destined to be unloaded, then it is possible that the *signa* might be connected to the iconography of the dock-owners. Where clear connections between *signa* and the *dominus* named in stamps occurs, such as for Rutilius Lupus and the *lupus signum*, then perhaps we have an indication that not only did Rutilius have interests in brick manufacturing, but also in the docking facilities or in the warehouses.

Signum	# of types	total # of examples of those types	Signum	# of types	total # of examples of those types
Nut	88	609	Silvanus	1	45
Mercury	27	222	Dove	3	37
Canine	19	178	Stag	6	34
Crown	45	173	Sistrum	8	33
Victoria	19	161	Ram/Goat	4	31
Eagle	9	117	Dolphin	3	28
Vase	13	102	Capricorn	3	22
Cattle	23	100	Snake	4	17
Minerva/Roma	11	85	Castro	2	17
Cock	12	84	Africa	1	13
Fish	3	75	Boat	7	12
X'	8	72	Elephant	2	11
Hercules	12	71	Modius	6	11
Trident	11	71	Bacchus	1	10
Fortuna	7	64	Vulcan	3	10
Caduceus	16	63	Jupiter	1	10
Mars	5	59	Isis	1	10
Hilaritas	5	59	Bonus Eventus	1	9
Boar	10	57	Bear	1	7
Lion	8	54	Lizard	1	6
Scorpion	2	48	Panther	1	6
Horse	9	46	Spider	1	4
Sun	4	46	Diana	1	4

Table 5.3 *Signa* in numbers of stamp types. 40 types of *signa* (from the pine nut to Isis) account for 422 stamp types for which ten examples or more are known, making a total of 2967 examples. The remaining six types account for another 36 examples.

In that case other bricks using canine *signa* may have passed through his docks, no doubt paying well for the privilege. (In Rutilius Lupus' stamp *CIL* XV.1 18 dating to AD 110, the first stamp to use consular dating, the *signum* is a club, an attribute of Hercules. Perhaps he did not yet own any docks at this point, and this batch went instead to the Tor di Nona?) Finally, the rest of the information in stamps on this understanding may have been intended for the (truly literate) manager of the warehouse. Stocks would be dated with consular dates, and whoever chose to inspect the warehouse would be able to tell exactly where the different lots came from, and when.

It was argued in 5.3.1 that bricks having stamps without *signa* were not destined for the city. Because the shipment of these did not therefore have to contend with the congestion and confusion of Rome, their stamps were accordingly less complicated. Their distribution may have been overland, using ox-carts. Possibly they could have been shipped by river, but if my interpretation is correct they were not destined for the docks which usually handled the brick trade. It is possible that the docks identified by *signa* may have been connected with wholesalers, which would account for why *signa* are associated with urban consumption (the rural market may have been too small for it to be worth the while for the dealer in brick to make the effort).

5.3.3 Study Sites

The maps considered earlier (Figures 5.1 – 5.6) divided up the landscape into conceptual zones, where the builders had access to these urban and rural materials. It has been argued that the usage of consular dating is connected with the payment of the *merces*, the land rent. Consular dated bricks were associated with 'urban' networks in 5.3.1. 'Rural' bricks, which are not associated with consular dating, should therefore represent production for the *officinator*'s own profit. It might be that in identifying different market orientations I have mapped the difference between zones of rents and profits. In the countryside, sites which use 'urban' bricks might have a closer relationship with the *dominus* mentioned in the stamp than with the *officinator*, and vice versa. DeLaine (2002) identifies a number of different procurement routes for brick in building projects of various sizes at Ostia. In some buildings, a connection between the person who commissioned the building and the land lords from whose estates the bricks came can be demonstrated. This is especially clear with regard to the Forum Baths, built by Gavius Maximus who was Praetorian Prefect under Antonius Pius. Over 90% of the stamped brick came from either the imperial family or Asinia Quadratilla and Flavius Aper (who were possibly married to each other) (DeLaine, 2002). Given that most 'urban' bricks by definition were used in the City, their appearance in the countryside marks those sites being as remarkable in the countryside as the Forum Baths are in Ostia.

In the Tiber Valley Project database there are at present over 5000 sites recorded, noting everything from pottery recovered, to visibility, to bibliography of past work at that site (Patterson and Millett 1998). This database probably represents the most comprehensive source of archaeological data in Italy, on a par with the better Sites and Monuments Records in Britain. What is immediately striking then is the paucity of stamped brick recovered or noted at a site, in comparison with ordinary brick and tile. Stamped brick is noted at a mere 2% of sites, while unstamped brick is found at virtually every site. Most of the stamps selected for archaeometric analysis came from sites where more than one stamp was recovered. This represents 22 sites; a further 28 sites from which single stamped bricks were recovered were selected at random for study in this section. These study sites represent a proportion of just under 1% of the entire number of sites in the database, but nearly half of the sites from which stamped bricks were recovered. The sample has a probability of being representative of wider trends in the Valley (as recorded in the database) of 95%, nine times out of ten. Given the standards of field survey at that time which provides the bulk of the material in the Tiber Valley database (see 1.4.1), this would seem to be an acceptable level of confidence.

When the other materials recovered from these sites, are considered it should be possible to make certain predictions: sites using the 'rural' bricks/tiles will have much more coarseware than fine-ware, decorative marble will not be in evidence at sites using rural brick, and so on. Is there an association between the type of stamped brick in evidence and any traditional indicators of site status/function? Chi-square tests were used to examine the relationship between bricks and various pottery forms. These were coarseware, Italian *terra sigillata*, African red slip, *vernice nera* (black painted ware), and African cooking wares. There were only exceedingly low probabilities for any associations between brick and the different kinds of pottery. Generally, only probabilities of 95% or above are accepted as statistically significant for chi-square tests. The only chi-square test which found evidence of an association at all was that between architectural marbles (veneer, columns, etc.) and brick ($\chi^2 = 8.003$, df =1, V = 0.16, Q = 0.73). The test was significant at the 99% level (the probability of an association), but the strength of this association (V) was somewhat weak. The Yule's Q statistic found that the presence of urban-oriented stamped bricks implies the presence of architectural marbles nearly every single time. No other building material (travertine, selce, tufa) was associated with stamped bricks.

What does this mean? Firstly, of two traditional indicators for a site's status (namely, marble and fineware pottery), only the one related to the building of the site was associated with stamped brick. If a site has a stamped brick at it, of the type associated with urban networks, chances are very good that there will also be architectural marble used at the site. It is not however correct to say that if stamped bricks of the rural kind are found then there is no chance of finding architectural marbles. At sites with rural stamped brick in the sample, the presence/absence of marble was equally split. What this suggests is that there was a hierarchy in the Tiber Valley. At some sites, those who could build with the same materials as were used by builders at Rome (marble, certain types of brick), did. These materials might have been made available through the networks of patronage, as at Ostia which enabled Gavius Maximus to build his splendid baths. At other sites, costs were cut by using local, 'rural' materials (this however would represent a boon for local suppliers) for the parts of the construction which would be hidden anyway when the walls were clad in marble. The builder of such a building would be correspondingly further down the social ladder, having to obtain his materials on the open market (again, a situation attested at Ostia, cf. DeLaine 2002). Finally there were those sites which did not bother with marble at all, and used the local brick.

Foxhall (1990: 109-111) suggested that sites where the buildings seemed to be of good quality but the pottery was low status might be equated with occupation by a tenant farmer. In our case, it is the type of building material itself which might indicate when we are dealing with simple farm buildings. For a number of sites the South Etruria Survey field surveyors felt confident enough to label the site a 'farm' or a *villa rustica.* In testing the association between 'urban'/'rural' brick and whether the interpretation of the site is villa or not (using the chi-squared test), I find that there is a 90% chance of an association between 'urban' and 'villa' ($\chi^2 = 3.125$, V = 0.0625, Q = 0.48). One might have expected, had the South Etruria Survey simply applied 'villa' as a default category for a site, that there would not be any evidence at all for an association. That there should be such an association, however weak, lends credence to the idea that in identifying 'market orientations' for particular stamp types, a difference in the patterns of usage which corresponds with the type of site has been identified.

5.3.4 The Problem of the Year AD 123

'Urban' stamps are associated with consular dates. The overwhelming prevalence in stamp types of the date 123, when Paetinus and Apronianus were consuls, has suggested to most commentators that there was an order from on high in that year for all stamps to be marked with the date (Bloch 1959: 237). This impression is based on the seeming 'fact' that every brick was stamped in that year. The problem of the year AD 123 is that any explanation of stamping practice must account for this extraordinary year. In absolute terms, the year 123 does seem exceptional. There are 255 types from this year (which necessarily carry the consular date); the next closest year is not until AD 134 when there are 45 types. In *CIL* XV.1 there are recorded over a thousand examples of all types from AD 123, but in AD 134 there is only a quarter of that number.

It is important to note the difference between the number of types from a certain year, and the number of *examples* of those types. Which is more remarkable, more important: a year in which there are 30 different types recorded, but only one example of each, or a year

Year AD	# of types	total # of examples	total # of examples/type
110	1	12	12
114	2	8	4
115	6	45	8
116	3	67	22
117	4	98	25
120	1	6	6
121	1	4	4
122	2	8	4
123	255	1198	5
124	16	44	3
125	13	38	3
126	20	85	4
127	8	77	10
128	3	18	6
129	4	35	9
130	3	32	11
131	6	17	3
132	2	3	2
133	5	20	4
134	45	257	6
135	8	70	9
136	2	3	2
137	5	26	5
138	7	68	10
139	4	19	5
140	5	8	2
141	4	37	9
142	2	41	21
144	4	5	1
145	2	13	7
146	1	1	1
147	1	1	1
148	2	19	10
150	6	25	4
151	2	2	1
152	3	9	3
154	5	41	8
155	5	6	1
156	1		0
157	1	3	3
159	1	26	26
161	2	9	5
162	1	2	2
164	1	22	22
203	3	9	3

Table 5.4 Stamping rates of consular dated stamps, including both *litteris cavis* and normal stamps.

in which there are only 10 different types, yet 50 examples of each? This is the problem we are dealing with: how to account for the differences in stamping rates to obtain a truer picture of what is the overall practice in the industry. One approach is to compare the year 123 to other years known in brick stamps by the number of examples of each stamp type, on the basis of the *CIL* XV.1 catalogue. For each year I can work out what was the average rate for that year, and then compare the years against each other. It may well be that in some years there was a 'boom', and more bricks were produced, but the interest at the moment is not in absolute numbers. Rather, it is relative numbers which matter. By considering each year on its own terms I determine whether there was a change in stamping practice from one year to the next.

Part of the problem is that stamps on *bessales* are included in the count in that famous year 123. These should be ignored for the time being, because as argued earlier (2.2.2), they are probably concerned with internal arrangements of the productive units, and so are not at all the same as the remaining stamp types which use consular dates. This removes 95 types accounting for 485 examples from consideration, leaving 160 types accounting for 713 examples. The year 123 still appears significant. But how many examples are there per type? For *bessales*, only five (+/- .5, at 95% confidence). For normal stamps, the number is 4.5 (+/-.1, at 95% confidence). On this reckoning, even though the purposes of *bessales* stamps and normal stamps are different, the rate is more or less the same. Five examples per type does not seem particularly abnormal, when the remaining 44 years for which there are consular dated stamps are considered. Table 5.4 tabulates the figures, while Figure 5.16 plots the number of examples per type per year.

What is striking about Figure 5.16 is the periodicity of the graph. It has a fairly regular saw-tooth effect. There are peaks at the years 110, 115, 120, 123, 127, 130, 135, 138, 142, 145, 148, 154, 159, and 164. There is then a gap until 203 (which is not plotted on this graph). On average, the length of time between peaks is four years, although the most common period (the mode) is five years. When the information is included from consular dated *bessales* stamps, the pattern is similar, suggesting that the same periodicity was also at play. Stamps from the *Domitii familia* also are plotted on the graph; although the average period in *Domitii* stamps is slightly higher, at 5.5 years, the same peaks and troughs are observed.

It is a striking pattern, and immediately brings to mind the discussion in 4.2.2 on the typical length of tenancy agreements. The slight staggering around the five year mark might be connected with landlords trying to exercise control over their tenants by creating instability over the terms of their leases (cf. Foxhall 1990: 101-2). The saw-tooth in the graph also suggests a high turnover of *officinatores*. This situation is not unlike what the parallel with the Ottawa Valley timber industry suggested, where low start up costs and the difficulty of predicting supply or demand created repeated cycles of oversupply and price collapse. When the price is rising, more land is exploited, but when the price collapses, the shock means that the *officinator* is driven out of business (or can no longer afford the land rent). The decline in absolute numbers of examples and types over time from the high in 123 to when consular dates cease to be used is evocative of the consolidation described for the Ottawa Valley timber industry which results from the shocks and shake-outs of that sort. It also suggests a transformation of land organisation away from free tenancy. The year 123 itself however does not appear to be fundamentally any different than other years in terms of number of examples per type. There is no need therefore to posit some form of government intervention in AD 123 concerning stamping rates or the organisation of production; whatever was going on was happening within the bounds of normal practice (cf. 6.2 on the characteristics of a complex system).

Government intervention as an explanation seems rather too statist. It is of course entirely possible that a government official could have decided that output ought to be raised that year hence more bricks would be stamped. This is a slightly different explanation from the traditional one, which envisions a more direct intervention in the industry ("all bricks must be stamped"). However, the government intervention theory relies on the exceptional nature of the year 123, but in at least one important regard stamping in that year was the same as in other years.

Given the ethnographic parallel with the timber industry, and the recognition that brick production is more akin to extractive, heavy industries (cf. Darvill and McWhirr 1984: 241), a more colourful interpretation might be a land rush: California 1849; the Klondike in the 1880s; the Ottawa Valley in the 1830s; the Tiber valley in the 100s. The year 123, if it does not represent a government intervention, might represent a real increase in the amount of land exploited for brick production: a real boom in land rents. It would be a situation analogous to the 'rush' which happens in the early days in certain extractive industries, as in timber and in gold, when everyone is trying to open up as much land as possible. Demand for brick in the year prior to AD 123 should have been enormous (or at least the expectation of a sudden increase in demand was) to foster such a run on the land; this run could have been precipitated for instance by the announcement of future plans for public/private building programmes, such as the rebuilding of Ostia in the early second century AD (cf. DeLaine 2002 on scope and scale of building activity in Ostia; Bloch 1959: 236 on Rutilius Lupus storing brick in anticipation of the Emperor's return

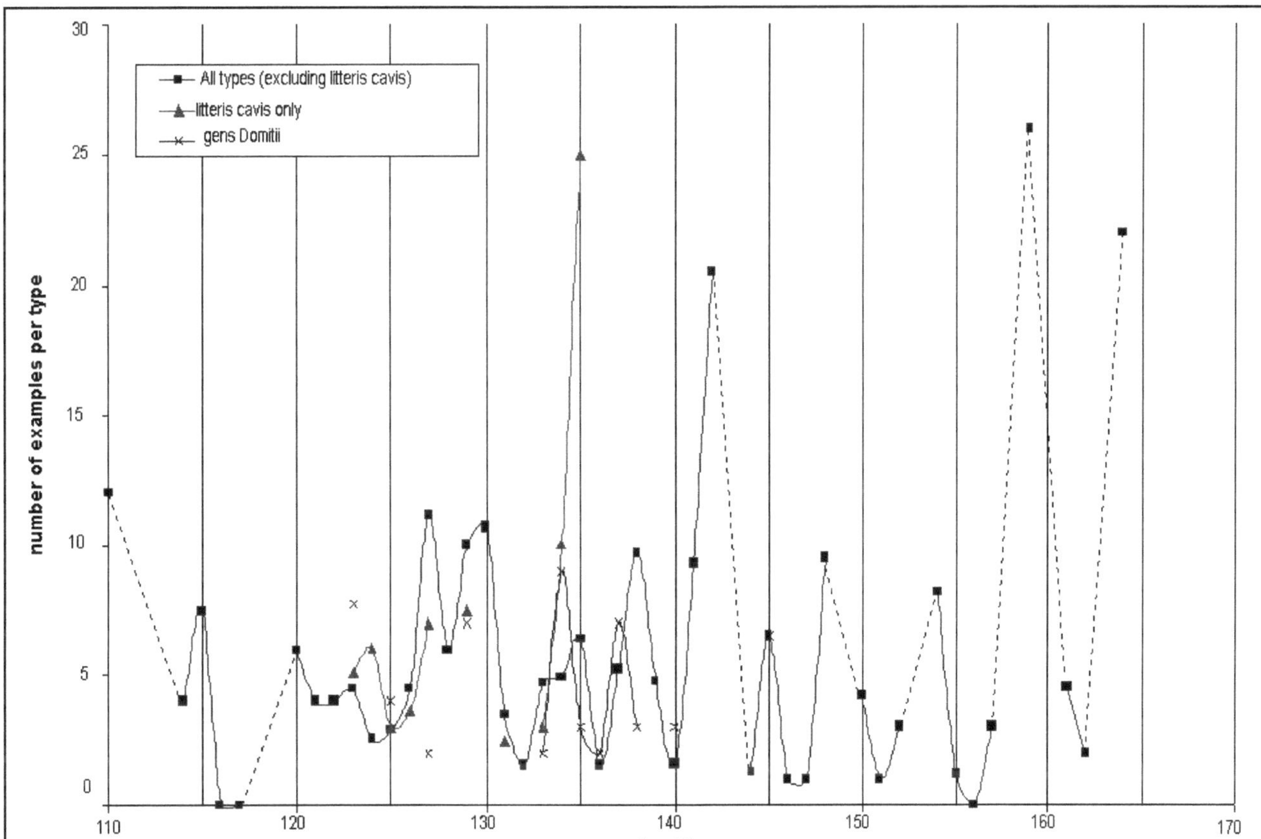

Figure 5.16 Stamping rates for consular dated stamps, number of examples per type per year. Broken line indicates likely continuation of trend, for years where no examples are yet known.

after AD 117 and the resumption of building activities). Regardless of why there were so many stamped bricks in 123, the stamping rate that year was no different than what was normal throughout the 44 years in which brick stamps carried a consular date.

5.4 Unstamped Bricks

We should return to the observation of the relative paucity of stamped bricks in the Tiber Valley. If, under the tenant-farmer hypothesis, a stamp is regarded as being bound up in someway to the paying of the *merces*, then presumably the builder procured the brick from the *dominus* named, his agent, or a third party (perhaps a wholesaler who bought the bricks from the *dominus*). This implies two things. First, the construction of any site with a stamped brick represents a flow of money to the *dominus* concerned. He has a parcel of otherwise unproductive land which he leases out for the production of bricks. He receives payment in a certain proportion or fixed amount of brick which as we have seen has inherent value, and which is stamped and shipped to the city. He sells this brick on to the builder at an enormous profit ratio since there was no cost to him to produce it. Secondly, *at sites where no stamped bricks were used, the brick and tile present signify an officinator's production for profit.* This means the majority of sites in the Tiber Valley. Under the tenancy hypothesis, after the *merces* was paid, the rest of the *officinator*'s production could be sold for his own profit. These bricks would not necessarily be stamped, unless they were to be shipped some distance or sold direct to a wholesaler in the city.

With unstamped bricks, the fact of there not being a stamp reduces the number of aspects we can study to two: the source of clay used in the brick, and the combinations of these sources at the site. Yet this may be enough to enable the dating and phasing of a site, and to understand a site's position in the wider industry. The unstamped brick and tile from the villa at Forum Novum, excavated by the British School at Rome, can be taken as a case study to explore these ideas. Given the exploratory nature of this analysis, the results should be approached with caution.

5.4.1 Shipping Within the Estate

If landowners own plots of land up and down the valley, the seemingly uneconomic shipment of brick and tile upstream may in fact be quite rational. In Cluster B6 (see Figure 3.8) there are some of the unstamped bricks from the Forum Novum villa. This cluster is provenanced to the area of Fiano Romano, some 15 km downstream in a straight line, or about 40 km following the most likely riverine routes. Only two of the Forum Novum unstamped bricks tested came from the local neighbourhood (FNV 8 in Cluster A4; FNV 13 in Cluster A6). Of the two stamped bricks in Cluster B6, the one most similar to the Forum Novum material is from the *Tonneianae de Viccians*. Although this establishment had many often contemporaneous owners, this stamp type belongs to members of the Tonneii family, who presumably gave their name to the brickyard. Given the form of the stamp, the Tonneii may have been producing the brick themselves, rather than via a tenant. By the reign of Claudius, the Tonneii are no longer found in brickstamps (cf. Steinby 1974: 94-100 on the *figlinae*). According to its excavators, the unfinished villa at Forum Novum was abandoned by the middle of the 1st century (Gaffney *et al.* 2001: 70). It may be premature to connect the Tonneii to this grandiose unfinished villa, but assume for the moment that we may on the basis of the coincidence of the family history and the building history, both extinguished at roughly the same time. If the Tonneii were building this villa, the apparent non-rational movement of brick in such a case could be seen as a transfer of materials between different properties belonging to the same family. Owning the infrastructure to make the brick, it would make more sense for them to use it rather than to purchase materials made closer to the site. A similar argument may also explain the movement upriver of SER 1, the brick which carries the *figlinae Domitianae* stamp. It was found at Seripola across the river from Orte. This river port is not very distant from Bomarzo, the locality of a likely estate of the *Domitii*. The brick is sourced to somewhere in South Etruria not far from Rome. While one cannot say for certain where the brick was destined, it does seem to be a similar situation as what we imagined at the Forum Novum villa. There is a great deal of uncertainty involved, but it might suggest that the movement of materials from one landholding to another would be a normal economic activity.

5.4.2 Dating and Phasing

In the clustermap in Figure 3.8, the stamped bricks seemed to be clustered together according to date. The MDA in 3.7 agrees with this observation. It should be feasible to assign unstamped bricks to these chronological groups on the basis of their composition alone. This section represents an attempt to open another line of inquiry in the study of unstamped brick, and its conclusions should be regarded as a tentative first step in this direction. Given the ubiquity of this material, the possibility of extracting useful information is exciting.

A Control Group for Undated Stamps

Another multi-discriminant analysis was used. The examples from stamped bricks were considered first of all, to create a dated control group (75 examples) against which the unstamped material could be compared. Suffice to say, as in 3.7, the MDA implied a gradual shifting of the work areas within particular clay sources, which is reflected in the composition of the

Example	Period	Example	Period
FAL 1	Flavian	FNV 8	Julio-Claudian
FAL 2	Julio-Claudian	FNV 9	Nerva-Hadrian
FAL 3	Nerva-Hadrian	FNV 13	Nerva-Hadrian
FNV 4	Severan	FNV 14	Flavian
FNV 5	Late	FNV 15	Julio-Claudian
FNV 6	Severan	---	---

Table 5.5 Assigned datings for unstamped material, via MDA comparisons with dated stamped material

bricks. Much subtler variations in geochemistry and mineralogy can be connected to particular periods.

The Undated Stamps

With stamped bricks there is an overall chance of correctly dating them of just slightly over 9 in 10, based on their composition. The problem now is to place the samples from unstamped bricks into this framework. The samples from each unstamped brick in turn were tested one at a time against the control group or 'dated base' - first FAL 1 was assigned to the Julio-Claudian period and tested, then to the Flavian period, and so on. The 'best' assignment, and hence closest to being chronologically true, was the one with the overall highest probability of being correctly dated, and where the dated base group was correctly assigned to its proper chronological periods.

After the individual samples had been tested and assigned, they were considered *en masse* with the dated base. This final result had an 85% probability of correctly discriminating chronological period from the geochemistry and mineralogy. Table 5.5 gives the examples from unstamped bricks, with their assigned periods.

The unstamped bricks were tested for the express purpose of provenancing. They were collected from surface survey at Falerii Novi (FAL) and from the villa site at Forum Novum (FNV). The Falerii examples are not associated with any particular archaeological context and so cannot be independently dated. However the phasing and chronology at the Forum Novum villa site suggested by the excavators (Gaffney *et al.* 2001:70) does not disagree with the datings suggested here for the unstamped materials; three of the examples date to the first century, corresponding with the initial construction and occupation of the villa; two date to the first half of the second century corresponding with the last of the villa's initial occupation period; and three date to the third century or later, when a different part of the villa site was occupied.

Visual Dating?

A closer examination of the mineralogy further suggests that there might be the possibility of discerning the chronological period from a visual examination of the fabric of a brick itself. However, the connection between the mineralogy and the visual appearance of a brick is not without problems, and so the mineralogical – chronological correlation in a larger sample might not prove as significant. For the tested bricks under consideration here, the dating of the examples was highly correlated with the amount of calcite present in an individual example. The mineralogy results for each example were quantified as a ratio relative to quartz, since quartz, being one of the most abundant minerals in the Earth's crust is therefore likely to be present in similar relative amounts in any example. Figure 5.17 presents the average amount of calcite present relative to quartz by period. Flavian bricks for instance will have roughly the same amount of calcite as quartz, but Antoninus Pius-Commodus bricks will have hardly any calcite at all compared to quartz. This may be due to where the clays were being extracted in the first place. If shallow alluvial clays were being exploited along the riverside, there might be proportionately higher calcite in those clays, due perhaps to the erosion of limestone in the Sabine hills, or the presence of organic matter (snail shells, etc) in the floodplains. Clays in South Etruria are largely overlaid by volcanic ejecta and are only accessible in stream cuts in the gullies and ravines, hence the lesser likelihood of calcite. This may also suggest that over time the easiest sources to exploit

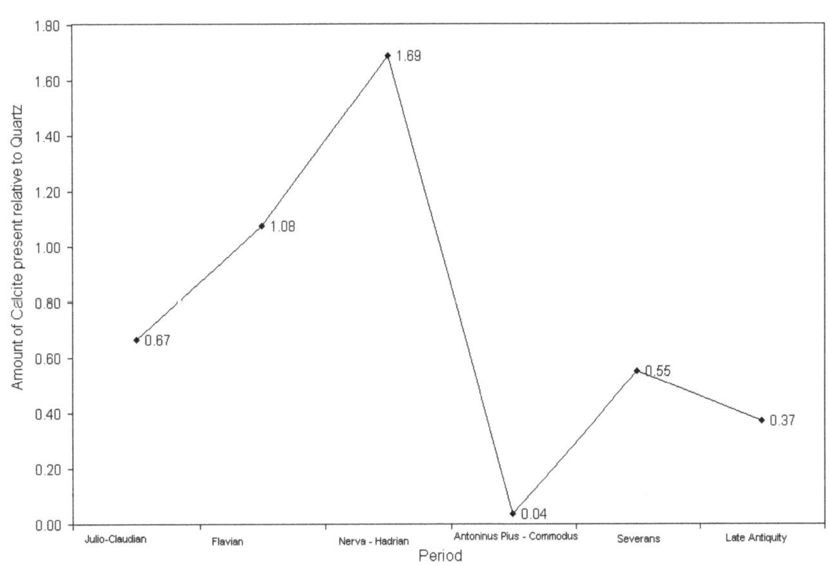

Figure 5.17 Calcite to quartz ratio in tested SES bricks over time

Petrofabric	Principal components	Corresponding cluster	Likely source area	Date(s)
1	high Ca, pyroxene, low feldspars	C3	lower South Etruria	Julio-Claudian
2	high quartz	B2 or C8	Orte Scalo or lower South Etruria	Julio-Claudian, Severan or later
3	high sanidine, no calcite	C10	lower South Etruria	Severan
4	high mica	C9	lower South Etruria	mid-late second century
5	sanidine, albite	B1	upper South Etruria	mid-second century

Table 5.6 Potential correspondence of 'petrofabrics' from the Mola di Monte Gelato with the cluster map in Figure 3.8

were exhausted, and deeper sources of clay had to be found. Certainly, the results of the MDA in 3.7 point to the gradual shifting of production from one clay body to another.

Correlating Fabrics

The relative amounts of minerals to each other can be used as a rough guide for correlating other studies' descriptions of ceramic fabrics with those in this study, especially when the description is based on petrographic thin-sections. At the Mola di Monte Gelato in South Etruria, a major occupation site was excavated by the British School at Rome during the 1980's (Potter and King 1997). By comparing the description of the petrofabrics of the bricks (Freestone 1997: 235-6) with the relative amounts of minerals of those tested in this study, one can connect the villa to the wider picture of distribution in the industry, and also work out if there is agreement between the site's phasing and the bricks used.

This site along the Treia River was occupied from the first century onwards, becoming a medieval *domusculta* in its final phases. It was originally a villa of some opulence. The occupants of the villa seem to be quite connected with the metropolis and with Veii. The excavators connect this initial phase with a certain C. Valerius Faustus who was an *Augustalis* at Veii. The closer centres of Falerii Novi and Nepi do not seem to figure in life at the villa. Then, in the second century, the villa seems to lose its opulence to a degree, becoming much more agricultural in focus. The excavators connect this with Trajan's decree that all senators must own land in Italy. Whereas formerly the villa was the retreat of a member of the *nouveau riche*, by the second century it had become somebody's investment property. At the end of the century there is sudden and dramatic evidence for the abandonment of the villa, which the excavators would like to connect to the proscription of the owner during a Severan purge (Potter and King 1997: 45-46).

Given this story, we can expect that the source of clays used for the bricks might come from the Veii area for the original build, while in later periods more local materials might be used. Also, given the very urban focus of the early villa, the materials used might be most similar to those used predominantly in Rome. Table 5.6 illustrates the correspondence between the petrofabrics and the sourced bricks from this study.

In cluster C3, there is only one tested example, FAL 2 from Falerii Novi. Yet it seems to correspond best with Petrofabric 1, which comprises nearly 90% of all the ceramic building material recovered from the Mola di Monte Gelato. The date of FAL 2 was assigned through discriminant analysis against the dated material tested in this study. This date agrees with the initial construction of the villa. What is even more interesting, the source of the material used in FAL 2 while not pinned exactly to a particular locality, seems to have been located in the lower part of South Etruria. The investigator who performed the analysis of the petrofabrics at Monte Gelato decided that the likely source for the Monte Gelato material was 'local' (a term which can be infinitely elastic). The material in Quadrant B in the cluster map (Figure 3.8) was sourced to Fiano Romano, Orte, and Narni, while the material in Quadrant C could only be sourced to the area above Rome. The Mola di Monte Gelato is further north and inland than Fiano Romano, but the material used there in the majority of the initial build seems to have come from much further south. This accords well with how the villa's occupants looked to Rome and Veii in their public lives. Petrofabric 2 accounted for nearly ten percent of the remaining material. It could be sourced, via comparisons, to either Cluster C8 at the bottom of the valley or Cluster B2 at the top. However, the material from the bottom of the valley dates to the Severan period or later while the other material dates again to the Julio-Claudian period. If we believe that Cluster C8 is most similar to this petrofabric, it might indicate that there was a small rebuilding programme quite late in the villa's life, after its destruction at the end of the second century. Indeed, there is a late Roman settlement at the site. Alternatively, if we accept that Cluster B2 is most similar, then during the initial construction the builders lowered costs by obtaining a certain amount of materials close to hand.

If it had been possible to test the material from Monte Gelato for this study, it would have been possible to be more certain about these options. However what this comparison and the discussion of Forum Novum demonstrates is that it may be possible to work out from the building material alone the phasing of a site, and when incorporated into a larger framework, perhaps the kind of site as well.

5.5 Chapter Summary

In this interpretation, monopolies on *production* were never important. Stamps were related primarily to the problems of distribution and this is especially apparent with the patterns of use of *signa*. The presence of *signa* in stamps was connected with the problems of getting material to the right docks in Rome. Binomial stamps and the use of consular dating point to another layer of usage. This second layer probably represents a mechanism to prove that the *merces* was paid for a given year and consequently all obligations had been met. The use of consular dated stamps is related to the payment of the land rent, the *merces*. In the entire corpus of brick stamps recorded in *CIL* XV.1, there is a roughly five-year cycle in the rate of stamping for consular dated stamps, which corresponds with the typical period for a land-lease agreement. Some of these sources would have been better or worse for exploitation, in terms of quality of the clay, or amount of the resource, or access to markets. Over time there was a certain amount of consolidation as the less economically viable sources were abandoned and/or tenants (or even landowners, for political reasons or death/inheritance) were forced out of the game. This is a development which parallels, for example, the Ottawa Valley timber industry. The consolidation would have the additional effect of making it cheaper and easier for those still in production to make brick, by allowing them to pool resources and distribution costs, or gain access to distribution infrastructure

If bricks are more valuable than the usual 'low cost/low value' commodity dismissed in studies of the Roman economy, and it can be inferred from the various modes of production (4.3) that some would have been cheaper to make than others, thus increasing the net profit, then it can reasonably be concluded that controlling this market would give someone an advantage. Indeed there is evidence to suggest that some *domini* (or their agents) took active steps to control the trade. This control manifested itself in the countryside by limiting some areas to materials which were produced locally while others could obtain the same materials as were used in the City itself. I found that the type of brick (whether 'rural' or 'urban' in the index constructed in 5.2.1) corresponded well with the types of building materials at the site, where the presence of 'urban' brick for instance was associated with veneer marbles. This variability in access to physical materials (and by extension, social realities) means that maps which delimit a boundary between the *urbs* and its hinterland based on cost-ratios, (cf. Morley 1997 map 1), implying economic and social differentiation, are useless. However, using the current methodology which is formulated from the point of view of production, the obvious questions (e.g. Who has overall control of the commodity? Who benefited most? How does this advance our understanding of Roman society?) cannot be answered. A different methodology must now be employed.

It would seem that the presence or absence of certain elements in stamps on bricks is associated with how the bricks are distributed, and not with how they are made. Hence it can be argued that the epigraphic information in the stamps (the names of individuals, *figlinae*) relates to the control of distribution. The key to understanding this control lies again in the nexus of relationships represented by an assemblage of stamped bricks, and the relationship between one set of relationships and another. It is the *pattern* of these relationships themselves which must be investigated to understand how social power is manifested in the brick industry. The next chapter, which uses the well established methodologies of social networks analysis and 'small-worlds' research, is concerned with how this power and control is manifested, and how it changes over time. If these last two chapters have been primarily concerned with the tenant, the next deals with the patron.

Chapter 6: Dynamic Social Networks in the Brick Industry

6.1 Introduction

One result of the archaeometry, discussed in Chapter 4, was a drawing out of the nature of the social relationships in the production of brick and tile. The archaeometric relationships suggested that land was exploited through the agency of land tenancy; tenancy puts the focus squarely on patronage. Patronage as a system allowed the Roman state to function; it allowed it to adapt and expand to incorporate new peoples (Wallace-Hadrill 1989: 75-6; in this guise it could be equated with romanisation cf. Millett 1990: 7-8). If patronage can be considered as a *complex* system (cf. 6.2.2), then one can explore how all the various interrelationships in evidence interact, and study the new phenomenon which emerges. In this chapter, I consider the patterns of patronage in land exploitation and the social and geographical interrelationships petrified in brick (the relationships deduced in 4.3.1, which were based on the cluster map in 3.5). In their totality, these patterns of patronage and of brick manufacturing can be represented as networks, where individual people or bricks form the nodes and social relationships, however deduced (epigraphically; archaeometrically) form the links.

Structure of this Chapter

In 6.2 I explain how the brick industry may be conceived as a social network, and explore how the shape of this network changes over time. The shape or pattern of a social network has implications for the choices available to any one person, if only because the shape implies that different people have access to different pathways over which information may reach them hence influencing their actions. I examine a case where the flow of information through this social network can be documented: the uptake of the practice of including consular dates in brick stamps. If a new idea may be imagined as being infectious like a disease, epidemiological models for how disease spreads in networks can provide an alternative explanation for the emergence of consular dating. It is a property of the dynamics of the social network itself, of how people interact, and we do not need to imagine a government intervention legislating stamping practice (which reinforces the argument in 5.3.4).

In 6.3 I argue that the changing shape of the brick-industry-as-network represents a very special class of network called a 'small-world'. In a small-world most individuals gather together into many distinct clusters, but a few long-distance links between otherwise distant clusters tie the whole population together (Buchanan 2002: 208). Small-worlds come in at least two distinct varieties, 'aristocratic' and 'egalitarian'. 'Aristocratic' networks are characterised by super-connected 'hubs,' individuals with many long-distance links, while in 'egalitarian' networks everybody has more or less the same number of links with long-distance links randomly distributed in the population; there are no hubs (119). I explain how small-worlds develop, how patronage may have influenced their growth in the brick industry, and how the realities of the brick fields may have limited this growth. The economic ramifications implied by the appearance of the different small-world types in the brick industry are explored both at the level of patronage and at the level of manufacturing.

In 6.4 I examine the relative positioning of every individual person in these patronage and manufacturing networks over time to estimate their potential power to influence these networks. Whittaker (1995: 22; cf. 2.4.1) argued that the study of urbanism in antiquity was really only a proxy for studying the operations of power in society. A network approach provides a secure theoretical framework for studying the relative levels of power for a given set of individuals. I argue that the hereditary succession of Emperors by adoption rather than by blood in the 2nd century AD may be a property of the social networks at the time: it is related to how power in that particular network is a function of 'who-you-know' rather than a result of being at the apex of a broad pyramid of dependants.

In 6.5 I consider the problem of what happens when somebody dies, creating holes in the network. Small-worlds are incredibly robust to the random removal of nodes; in human terms, this removal of nodes may translate most dramatically as the death of an individual. However, if it is the most powerful individuals, the ones who act as hubs in the network, who should die, then the consequences can be more severe. The elimination of hubs can cause a network to fail catastrophically. I examine the proscriptions and deaths surrounding the dynastic change-over from the Antonines to the Severans, and consider the role of proscriptions in general in the development of the brick industry. I argue that in the cases of Commodus and Septimius Severus, proscription can be viewed as a reaction against the structure of society at the time, a return to power based on dependants rather than interlocking circles of 'who-you-know'. In network terms, it is a shift from an 'aristocratic' network to an 'egalitarian' network (noting of course that these terms refer to the shape, the skeleton, of the network, and not necessarily to a literal aristocracy or new brotherhood of equals).

The network approach allows us to connect what we find archaeologically to wider historical patterns in a way that is theoretically grounded in the physics of dynamic networks (a field which has exploded since

1998 and the research of Watts and Strogatz). Indeed, it allows the archaeology to drive the writing of history because the social structure of Roman society can be drawn out from the archaeology, independently of our knowledge of Roman history.

A Word on Patronage

Patronage, at its most fundamental, can be understood as an asymmetric relationship between two people of unequal social status; there is a reciprocal, or two-way, exchange of goods and services (n.b. a slightly different meaning to Erdkamp's 2001 formulation of 'reciprocal exchange' cf. 2.4.1); and the relationship is of some duration and is personal, not transient as it would be for purely commercial reasons (Saller 1989: 49). To this can be added the idea that patronage is a relationship which may be entered into voluntarily; a client may withdraw his allegiance, and may have more than one patron (Garnsey and Woolf 1989: 154; Drummond 1989: 101). However, according to Dandeker and Johnson (1989: 225) this understanding of patronage is not really the empirical issue. For them the quest to identify a patronage relationship, or the reciprocal exchange of goods and so on, is a red herring. Rather, they suggest that patronage should be understood as a structure (a system) where these relationships mesh together into a network of relationships. The issue is to identify whether or not patronage-as-a-system plays a central role in organising the state, society, and the economy - 'in what kind of society does 'who are your friends' have decisive importance?' (222-3). As for reciprocity, it is not one-off exchanges that matter, but rather the vertical and diagonal ties between all levels of society which legitimise and control the social order that matter (Wallace-Hadrill 1989: 77; Dandeker and Johnson 1989: 225)

These networks, of patronage, and of manufacturing existed not only in 'social space' but also in the real, physical world (a consideration explicitly taken into account in the drawing out of the manufacturing networks, cf. 6.3.3). They intersected with real places, whether the villa of a *dominus*, the local *municipium* where his bricks were used to build a basilica, or the local docks where those bricks were unloaded. It is in this fashion that these networks bind the hinterland with Rome.

6.2 The Brick Industry as a Dynamic Social Network

6.2.1 Networks in the Brick Industry and the Transmission of Ideas

Rutilius Lupus and Consular Dating

Rutilius Lupus was *praefectus Aegypti* (prefect of Egypt) in the years 113 – 117. In the Tiber Valley, he owned the *figlinae Brutianae* (Steinby 1974: 27-9), the first *figlinae* in the area which used consular dating in stamps. He is thought to have instigated the use of stamps so that he could keep track of his estates' production while on duty in Egypt (Bloch 1959: 235-6). Yet, after only a decade, the use of consular dating exploded in the brick industry, with hundreds of examples carrying the date 123. I argued in 5.3.4 that the rate of stamping with consular dated stamps was on average no different in 123 than in most other years, and so a government-legislated requirement to use consular dated stamps in 123 was unlikely. However, there has been little consideration of how the idea of using consular dates spread in the brick industry to the point where, even if government intervention did account for 123, this specific form should be chosen.

The first consular-dated stamps in Italy appeared in the areas around Parma, Piacenza, and Velleia in the mid 1st century BC (235 n.40). Significantly, one of the families which used consular dates in their stamps at that time was the Naevii. The *Figlinae Naevianae*, which is related to that family (Steinby 1974: 67-8), produced bricks in the Tiber Valley, shipping them even to Ostia. The *figlinae* eventually fell into the hands of Rutilius Lupus. In the high reaches of the Tiber valley (the Città di Castello and Todi areas), consular dating was used by a T. Papirius T. l Synhister in AD 12 and by an Ampliatus in AD 93 (Bloch 1959: 235 n.40). The idea of using consular dates was already in the Tiber valley in areas not too far removed from where some of Rutilius Lupus' properties were (he also owned the *Figlinae Narnienses,* Steinby 1974: 68).

Before he was prefect of Egypt, Rutilius Lupus was in charge of the *praefectura annonae,* the imperial food supply, between AD 103 and AD 111. In the discharge of his duties, Rutilius Lupus would have become quite familiar with the logistics of Portus and Ostia, and the local government. It is easy to image how his experience administring the food supply would have suggested the idea of using consular-dated stamps to keep track of items moving in and out of the warehouses under his care, even if the idea of using it on bricks had not yet occurred to him. His ownership of one of the pioneering *figlinae* in this regard, and that *figlinae*'s already established trade with Ostia might also be routes by which the idea came to him.

Why it occurred to Rutilius Lupus to stamp the date on his stamps is not the most important question. How the idea spread is more interesting, for the answer shows how information passed around the brick industry, and the important position certain individuals had which allowed them to influence the industry as a whole. It demonstrates how aspects of the brick industry are actually properties of the structure of the network itself.

Each time a different landowner required that the consular date should be included in a stamp we can imagine that that landowner had become 'infected' with the dating 'virus'. In this way, brick stamps with consular dates allow us to track the progress of the idea throughout the industry. *Officinatores* sometimes

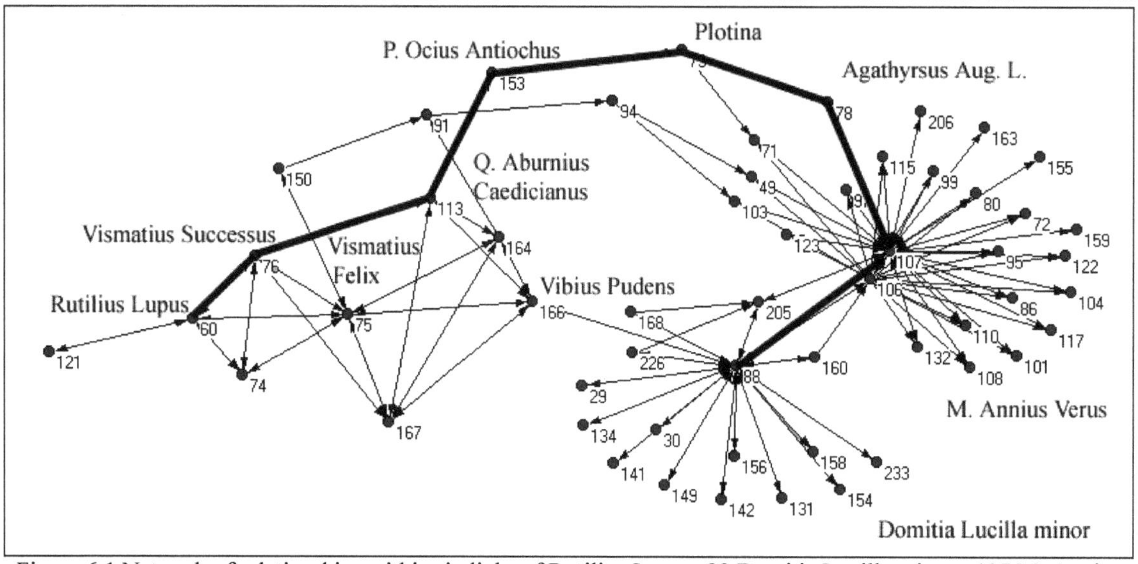

Figure 6.1 Network of relationships within six links of Rutilius Lupus. 88 Domitia Lucilla minor ; 107 M. Annius Verus. Note Vismatius Felix's circle, of which Rutilius Lupus actually appears to be of minor importance (at least, in network terms).

appear in stamps of two or more *domini*, an observation which allows us to connect the *domini* by virtue of their common contact with the *officinator*. It is also the case that sometimes *officinatores* are members of the same family, and so too are certain *domini* related to each other. Certain *figlinae* had more than one concurrent owner, and so the landlords can also be connected through their communal ownership of these *figlinae*. I will go into more detail about how we can discern the networks of relationships which tie all of these individuals together into 'The Brick Industry' in 6.3, but for the time being, let us assume that any such connections represent communication between these individuals. Laid out in full, we have a network of individuals who are related to each other because they spoke to each other regarding the production of brick. Let us call this a 'patronage network'. The prosopographies of every *dominus* known in brick production are studied in Setälä 1977. This work, alongside Helen's list of *officinatores* who appear in stamps of more than one dominus (1975: 139-150), provided the data for drawing out the relationships in the social network of brick production. Figure 6.1 is a detail from this network, centred on Rutilius Lupus and extending as far as six distinct links from him.

Diseases and Epidemics

How then did the consular dating 'disease' spread on this network? The mathematics of epidemiology are conceptually rather simple. If more than one person on average catches the disease from the initial carrier, the disease will multiply and spread. If the figure is less than one person on average, the disease will decrease and eventual die out. It is also (remotely) possible that only one other person will catch the disease, and so the disease will linger in the population (Buchanan 2002: 163). Epidemiologists have found that for whatever the disease, however it is transmitted, and even given various levels of immunity in the population, there is always a point where the disease can suddenly explode into an epidemic. This is the 'tipping point', the transition from one phase to another, an idea in this sense similar to that in nuclear physics of 'critical mass' (163). With regard to consular dating, during the initial decade in which it was used, only the *Brutianae* were 'infected'. The *figlinae Naevianae*, also owned by Rutilius Lupus, became infected in 121 and 122 (*CIL* XV.1 344 and 345). As the season opened for brick production in 123 the tipping point was surpassed and the entire industry seems to have become infected with the idea of consular dating. Why this happened is not related to individuals' susceptibility to new ideas but rather is related to the structure of the network.

(There is an interesting 'prelude' before we reach the tipping point in 123. The next consular dated brick stamp is from 113 (*CIL* XV.1 2157), where a certain C. Allienus Proclus is *dominus*. This man is otherwise unknown to history, and we cannot connect him to Rutilius Lupus in our network. However the form of his stamp bears a striking similarity to Rutilius Lupus' first consular dated stamp, *CIL* XV.1 18. Rutilius Lupus' stamp is circular in shape, with three straight lines of text across the stamp. A *signum* also appears (a club, usually an attribute of Hercules; might the absence of the wolf *signum* indicate that Rutilius Lupus did not have his own dock/warehouse facility at this time, using instead the docks at Tor di Nona, cf 5.3.2 ?). Allienus Proclus' stamp from three years later also has three lines of text, displaying a *signum*, but is orbicular in shape. No other stamps at this time had three lines of text in them. His stamped bricks are well represented at Ostia (LSO 1131, Steinby 1977: 329). Subsequently, Rutilius Lupus' stamps from 114 to 117 (*CIL* XV.1 19

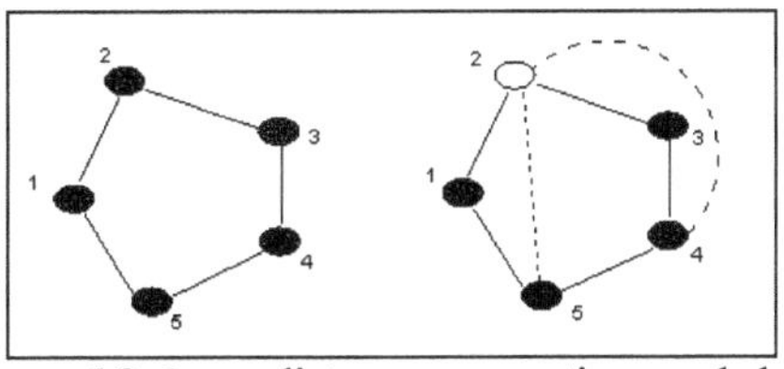

Figure 6.2 Long distance connections and short cuts in a network. In this very simple network of only five individuals (left), everybody has only two relationships ('connections', 'links', 'ties') and they share equal positioning. After two 'long distance' connections are established which by-pass previous links (right), person 2 is now in a position of power, able to cut 1 and 3 out of the loop altogether.

– 25) have the same shape as Allienus Proclus' stamp. There seems to have been a back-and-forth dynamic between the two men connected to Ostia, mutually 'infecting' each other, influencing the shape and format of brick stamps for the next century).

Zanette (2001) modelled the spread of rumours on an epidemiological model of the spread of disease. He ran the simulations using a wide variety of parameters related to virulence and immunity, on a number of different types of networks. The different networks correspond to the structure of relationships between individuals in a population. On a regular, lattice-work network (where everybody has the same number of connections to every body else), the percentage of people hearing the rumuor (becoming 'infected') always remained close to zero. The disease created its own obstacles: people whose immune systems mounted an effective defence, others who died without passing the disease on, and so on. In terms of ideas, these obstacles are people who reject the idea outright, or take it on but never transmit it further.

However, the introduction of a few, long-distance connections between different parts of the regular network (Figure 6.2) changed the dynamics of the spread of the disease. In fact, he found that when roughly 20% of connections were between otherwise unconnected parts of the network, the propogation of the rumour ('the disease') varied significantly. It either died straight-away or came to infect as much as 33% of the network within a short period of time. He concluded that the 'epidemic' succeeds or fails on the basis of the pattern of interactions between individuals; the tipping point is related to the structure of the network (2001:3).

The Structure of Rutilius Lupus' Relationships

In the network centred on Rutilius Lupus (people who can be linked to him directly, his friends or relations, and those who worked in the same *figlinae* owned by him, and their friends and relations), it is obvious that we are not dealing with a 'regular' network. A 'regular' network is one where everybody has the same number of connections. In fact, there are three distinct hubs in Rutilius Lupus' local network: Vismatius Felix, an *officinator* of Rutilius Lupus', Annius Verus, and Domitia Lucilla (Figure 6.1). These 'hubs' are individuals who have many more links or connections than the majority of people. There is only one 'bridge' between the Vismatius Felix cluster and the other two which is known to have existed before 123, the connection linking P. Occius Antiochus and Plotina (wife of the Emperor Trajan). The structure of this network seems to be much the same as the kind that Zanette (2001: 3) found which lowered the tipping point for an epidemic: a network with a few long distance connections. That is, while most people are strongly connected to a few mutually-connected others, there are individuals who have connections with otherwise unconnected clusters of friends. In the brick industry, this might be where an *officinator* produced bricks in two geographically distant *figlinae*. He would be the long distance connection between otherwise unconnected estates. The links between Rutilius Lupus and Annius Verus, that is the bridge formed by the Occius Antiochus and Plotina connection, may have been the very route through which consular dating became a normal part of brick stamps, for once Annius Verus was 'infected', it was but short steps to the rest of the industry (in fact, when the entire network of the brick industry from the 1st century to the reign of Caracalla is plotted, cf. Figure 6.4 below, Domitia Lucilla, the daughter-in-law of Annius Verus is the single most well connected individual, able to be reached in only three steps – information would only have to pass via three people – by nearly 50% of all people in the industry).

Research is showing that for diseases which spread through a network characterised by hubs and long distance connections, *there is no tipping point* (Buchanan 2002: 180; Pastor-Sartorros and Vespignani 2001). That is, the structure of these networks always creates the right conditions for the progress of the disease to be tipped into the epidemic regime. A single infected individual never infects, on average, less than one other individual because of the pattern of connections. This suggests that, with regard to consular dating, once the idea was introduced, *it was always going to spread.* At first glance, the pattern of connections in Figure 6.1 does not seem to be dense enough for this to take place, but it is important to remember that we drew this network from a very select environment, the relationships which could be deduced from brick stamps. However, if every single relationship which existed could be drawn in, it is likely that the pattern produced would not differ significantly from this pattern: of hubs and long-distance connections. This similarity of structure, at whatever scale we examine, is a hall-mark of complex systems. Another is the power-law distribution of each individual's connections: most of the individuals in Figure 6.1 have only one or two links with others, while two individuals (Annius Verus and Domitia Lucilla) have over a dozen. (Parts of the discussion in this section appear in Graham 2006).

6.2.2 Complex Systems and Social Networks

A complex system is different from a complicated system. If a system, be it a microwave oven or the Space Shuttle, can be described or understood in terms of its individual parts, then it is simply complicated. On the other hand, in complex systems the whole cannot be understood in terms of the components, and the interrelationships between the different parts are not fixed but subject to change out of the dynamics of the system itself (Cilliers 1998:viii-ix). Social systems, economies, ecologies are all complex systems. The spread of consular dating suggests that the brick industry can be considered a complex system as well. The pattern of interaction between the components in a complex system is crucial to understanding the system as a whole. But how does this structure come into being in the first place? What did individual Romans do to create this structure? What is the significance of considering the brick industry as a complex system?

As we have seen, different network shapes, different patterns of relationships, have very different implications for the larger-scale phenomena which emerge from the interactions which make up the network. Network shape also has implications for how interpersonal relationships can be mediated, and hence the range of possible actions open to individuals taking part in these networks. The study of network shapes helps us to understand the significance of these relationships. Networks have meaning.

A Philosophy of Networks

How is meaning generated? In the brain (the prime example of a complex network), millions of neurones are connected to each other via synapses. Neurones fire when they have received a certain level of incoming signals from other neurones. These signals are mediated by the synapses, and can therefore determine whether the receiving neurone becomes excited or inhibited, and the degree to which this occurs. Initially, such a network should be relatively undifferentiated, but as it is exposed to new experiences or new information, causing neurones to fire, structure begins to develop:

> Consider three neurones, A,B,C. Each time both A and B are active simultaneously, the strength of their interconnection should be increased slightly, but when they are not active, it should decay slowly. In this way, if A and B are often active together, [the connection strength] will grow, but if A and B are only associated spuriously and A and C more regularly [that connection instead] will grow. In this way, a network will develop internal structure, based only on the local information available at each neurone. This development of structure can also be called 'learning'. (Cilliers 1998: 28; cf. 6.3.2 on preferential attachment)

An event in an individual's life can be likened to the firing of a neurone, and the relationships which mediate this event, the synapses. What causes the neurone to fire is decided within the neurone itself, not the network: a person has many choices regarding action. Action or an event can cause new relationships to form, or old ones to wither away. But to understand the meaning of any given event, the entire pattern needs to be considered. If the pattern changes the meaning must also change. Note that any particular neurone can participate in numerous configurations of connected neurones. This suggests that events can have multiple meanings. If the pattern, the structure, has meaning, then its development (what Cilliers who is a philosopher of science calls 'learning') should equate with history.

When two people come face to face, for whatever reason, a relationship exists. When those two people continue to meet, the depth of that relationship increases. *Structure* is the network pattern of these relationships between people. Of course, these relationships can be of all different kinds, and they do not necessarily need to be positive. Two people who have a falling out can have a relationship of mutual animosity or antagonism.

The shape of the structure has implications for the choices people make. Someone with many different circles of friends is potentially in a better position because a wider range of information is available to inform his or her choices (Granovetter 2022). Conversely, someone whose friends all happen to be friends of each other is more circumscribed because what one knows all the others are likely to know as well (Figure 6.3; in Rutilius Lupus' immediate circle, only Vismatius Felix was well connected to other groups)

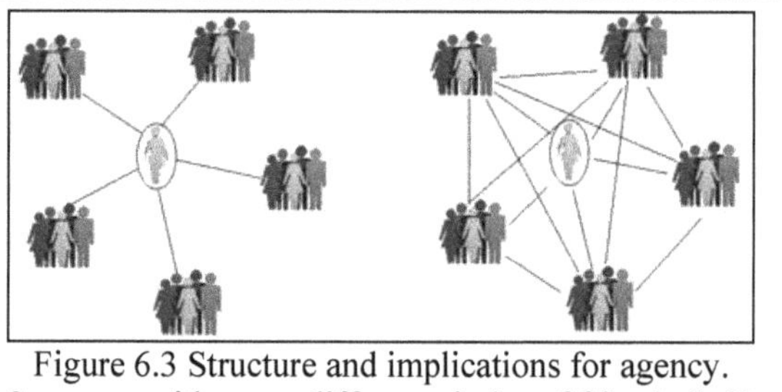

Figure 6.3 Structure and implications for agency. Someone with many different circles of friends (left) has access to different sources of information. Someone whose friends all happen to be friends of each other (right) all have access to the same information, leaving no-one with a comparative advantage.

Here structure is described in terms of networks of friends and information exchange, but networks can be discerned in many different spheres of action. We can study these networks from the point of view of one person in particular, following the connections through (as was done for Rutilius Lupus), or attempt to study all the relationships between a defined set of people (Hanneman and Riddle 2005). Certain patterns might lend themselves well to the transmission of ideas and information (power) while individuals within it may well be excluded, being poorly connected to the wider network (for examples in a modern context, cf. Emirbayer and Goodwin 1994: 1420-1421 on the uptake of new political causes in 1960s America, and on the spread of civil unrest in Paris 1968).

6.2.3 The Actual Shape of the Brick Industry: Networks over time

A Marriage of Sociology and Mathematical Physics

There are two strands to what I have been discussing so far. In the first, I have been looking at the global properties which result from the interactions of individuals in a network (as in disease spreading, epidemics, and the transmission of ideas, 6.2.1). In the second strand, I am looking at local properties, of one individual's position in the network and what this implies for him or her (6.2.2). Much of this chapter involves shifting perspective back and forth from the local to the global. Whenever I am concerned with analysing the local level, I use the sociologist's tools of social network analysis, following especially Hanneman and Riddle's on-line text book (2005) and using the UCINET analysis package (Borgatti, Evrett, and Freeman 1996). When I am concerned with the global level, there are some statistical routes to follow (principally those developed by the mathematician Watts 1999), but I will also refer to the results of simulations as guides to interpreting my own data (the study of dynamic networks is so new that there is not as yet a generally available 'tool-kit' for their study; each researcher programmes his or her own).

To accomplish this marriage of sociology and mathematical physics to understand the shape of the brick industry (global) and the opportunities for individuals (local), the core study group has to be identified. There are a total of 1325 known individuals in the brick industry from the 1st century to the 3rd (Helen 1975: 23). Of these there is a group of 234 people who I can knit together into a network, on the basis of family ties, of co-ownership of *figlinae*, and of transference between *figlinae*. Of course, these are just the relationships which can be observed today; there were probably relationships which have not left any trace (similar to the hypothesized contacts between Rutilius Lupus and Allienus Proclus). There is a theoretical maximum of all the possible unique relationships within this core group of 54 522 (# of individuals x # of individuals - 1).

Relationships

The theoretical maximum of the total possible number of relationships supposes that the content of these relationships might differ, that is, the relationship of person A ('Alice') to person B ('Bob') is different than Bob's to Alice's, hence there are two kinds of relationships. This is fine if the relationship in question is between an *officinator* and a *dominus*, but it is far more difficult to establish what kind of relationship existed between social equals, given that social conscious Romans were keenly aware of status and the gradations within particular ranks (Saller 1989: 57). It is possible to indicate strength and direction of ties in social networks analysis, and so in theory we could represent these gradations in rank. For ease of data entry and to lower the risk of misrepresenting the relationships the analysis has been kept at as simple a level as possible: the only information recorded regarding relationships is that a relationship *existed*. This reduces the total maximum relationships by half (Alice's relationship to Bob is the same as Bob's relationship to Alice, so only one tie rather than two is necessary to represent the relationship), to 27 261. Only 3% (870) of these possible ties existed, as evidenced on stamps and from other historical sources (the same way we deduced the network around Rutilius Lupus in 6.2.1). The standard deviation in these ties (a measure of the variation from one actor's ties compared to another's) is 13%. This suggests that in this group of individuals there are a number of subgroups. If two individuals can have as much as 13% difference in the number of ties they have, then some individuals will be poorly connected and others will be well connected, leading to the development of subgroups (following Hanneman and Riddle 2005).

The Shape of the Industry

What kind of subgroups? They could be for instance family groupings, *figlinae* groupings, or even groupings related to geography (e.g., producers from Narni). It could also be that the subgroups are chronological in nature. In a study of 19th century women's reform organisations in the United States it was found that sparse ties between groups corresponded with lulls in activity between periods (Rosenthal *et al* 1985; Emirbayer and Goodwin 1994: 1419). We can examine the relationships within the total brick group to see if there are any 'natural breaks' (places in the network where there are no or very few linkages between people). Any breakages found can be compared against the chronological periods used so far after the fashion of Rosenthal *et al.*'s study. We have found chronological resonances in the geochemistry of bricks; are there any in the sociology of bricks?

Figure 6.4 plots all of the relationships over the entire 1st to 3rd century period. It is a very tangled network. The network may be rationalised by plotting all of the members of a 'block' (where all of the members have

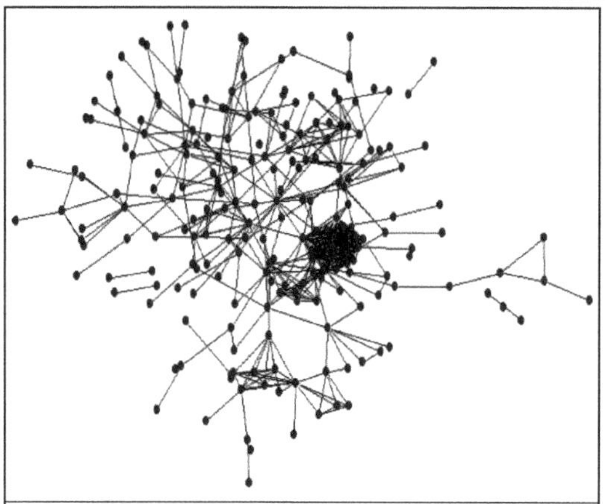

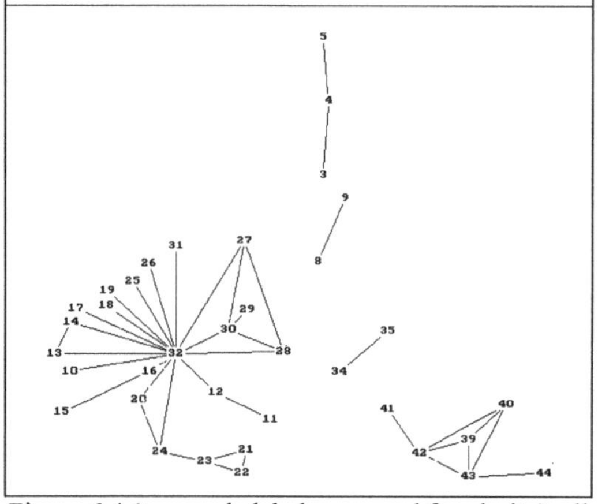

Figure 6.4 (top; node labels removed for clarity), all relationships in brick stamps from the 1st to 3rd centuries

Figure 6.5 (bottom) rationalised depiction of the same. The network is rationalised by plotting as a single block groups of individuals who have the same pattern of ties.

There are 44 such blocks, from 234 individuals. The group centred on block 32 contains individuals from the 2nd century, the group centered on block 39 contains individuals from the Severan period, while the Julio-Claudian period is centred on block 4, and the Flavian period is represented in blocks 8,9,34 and 35.

the same pattern of ties to each other, Hanneman and Riddle 2005) as a single point, as in Figure 6.5. In the rationalised network, the isolated fragments differentiate on the basis of chronology. The 'bridges' which would connect these fragments are ignored in the rationalisation. Figure 6.6 plots the networks by the periods suggested by the rationalisation in Figure 6.5. There is very little spanning between the major clusters suggested in Figure 6.5, and when this occurs, it is within families that come to prominence in the next cluster, especially the *Domitii* and the Flavii. The 2nd century, which up to this point I have divided into two, appears as one mass in this diagram. Socially then the brick industry networks divide as might have been expected, by ruling dynasty. What is interesting is the

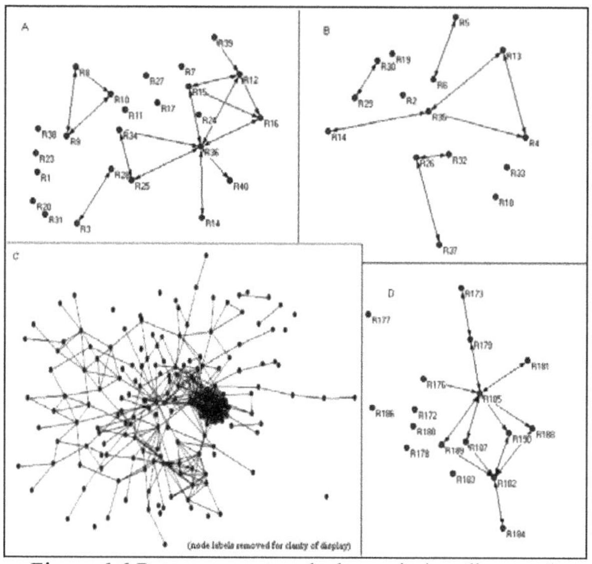

Figure 6.6 Patronage networks by period as discerned through the rationalisation of Figure 6.4. A) Julio-Claudian B) Flavian C) Nerva – Commodus D) Severans

very little continuity between periods, which could be connected to the proscriptions and general chaos amongst the élite during times of dynastic changeover.

The discussion of Rutilius Lupus and the spread of his idea was a discussion of the global dynamics of the brick industry conceived as a network. Alternatively, we can point to an individual, and ask how, precisely, did this individual's links with others affect his or her ability to act? The perspective in this formulation is local.

6.2.4 Structure, Agency, and Small-worlds

At the local level, knowing the network structure in which an individual is embedded is not sufficient for understanding the choices he or she made; structure has implications for agency, but it is not fully determinative (Granovetter, 2002). The mere pattern of linkages is not sufficient for explanation for the particularities of history. For instance, if we plotted the networks of kith and kin in a small town in 1950's Quebec, we would find very few, if any, marriage links between Catholics and Protestant. There is nothing inherent in the network that prevents such a linkage being made (for Catholics and Protestants would be coming face to face with each other frequently) so this *structural hole* can only be explained with reference to the culture of the time (cf. Knoke 1991: 175-6). That is, the cultural level, which emerges from the level of individual interactions, feeds back into that individual level, and affects the patterning of relationships.

Structural holes, the patterning of non-linkages, are important for the dynamics of networks. Someone who fills a hole is called a 'bridge'. Filling a structural hole

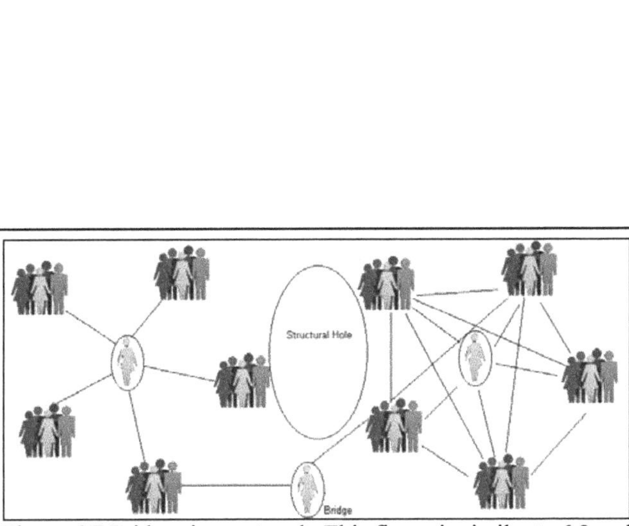

Figure 6.7 Bridges in a network. This figure is similar to 6.2 and 6.3, and indicates how one person can tie together otherwise disparate groups of individuals. Without this individual, the network collapses into isolated clusters.

puts one into a position of power. A bridge connects clumps of actors who would otherwise not be connected at all, and so such a person has the choice over which information goes where. In the discussion concerning the transmission of the idea of consular dating, it was the connection with Plotina which formed the bridge between Rutilius Lupus and Annius Verus, enabling the transmission of the idea, cf. Figure 6.1. The person who forms a bridge to various groups is in a position to turn whatever information comes his way to his or her advantage (Granovetter, 2002). In antiquity, there was this recognition. Dio of Prusa actively suggested that because he was a friend of the emperor and other powerful Romans his fellow citizens should have 'regard' for him (and hence, approach him for his contacts) (Millar 1992: 114-6). The fact that Plotina was a bridge in one network we defined probably indicates that she was a bridge in other networks. She may have had this important role in the networks because she was the Emperor's wife; alternatively, she may have become the Emperor's wife partly due to her already established connections. At any rate, people who function as bridges have pivotal roles in maintaining network structure; such people we are able to identify in antiquity would be worth closer scrutiny.

In a poorly connected network with very few bridges, there might be more conflict when interests clash, but it is unlikely that larger scale phenomena can arise. At the opposite extreme, in the highly connected network there might be much more co-operation. Yet, since everyone occupies much the same position relative to every one else, larger scale phenomena are actually *less* complex than those that arise in the middling situation (Granovetter, 2002). In the middle situation, the bridging of structural holes creates short-cuts across the network (Figure 6.7). The person who acts as the bridge (like Plotina) is in a position to amass power or influence (Granovetter, 2002). It is the situation between the extremes which holds the most interest for us- the 'small-world'.

The discussion in 6.2.1 about the adoption of innovation in the brick industry was in reality a discussion of what happens in a small-world network. Of course, what I am arguing regarding Rutilius Lupus is for a likely sequence of events, based on a theoretical understanding of how disease spreads in a certain network configuration. These theories are predicated on a great deal of empirical modelling, and it is thought that they do capture the essential features of how different network shapes can aid the spread of a new disease – or an idea. If Rutilius Lupus had not been connected to Plotina, and thence to such major hubs as Annius Verus and Domitia Lucilla, the idea of using consular dates in the stamps themselves might never have spread beyond his own *figlinae*. Without Plotina, in fact, the little network centred on Rutilius Lupus would not have been a small-world before 123, which points to the importance of Plotina's role in Roman society not just as wife of the Emperor but also as a landowner. The shapes of the networks are historically contingent: hubs do not suddenly appear, they grow (cf. 6.3.2). The interconnections in the brick industry (whether discerned from stamps or from an archaeometric study of the bricks themselves, cf. 6.3.3) drawn out in 6.2.3 are at present the static snapshots of an evolving industry. The small-world perspective allows us to explore how (and why) the industry assumed these different shapes at different times; it allows us to move from the static to the dynamic.

6.3 The Development of the Brick Industry, As Evidenced from its Patronage and Manufacturing Networks

6.3.1 The Small-World and Complex Systems

Through a series of mathematical explorations, Watts and Strogatz (1998) were the first to identify the characteristics of a small-world, although subsequently other types have been discovered, cf.6.3.2. Essentially, a small-world is a type of network where although the majority of individuals are linked to only a few close neighbours, a few long-distance links to otherwise distant individuals serve to make the entire network close, i.e., whereas formerly it would take tens of links to reach the other side of the network, the long-distance links allow that number to drop to a mere handful (Barabási 2002: 51-53). In the brick industry, most people worked within the confines of one *figlinae*, or owned just the one *figlinae*. Yet the presence of a few *officinatores* who transferred between or worked concurrently in more than one *figlinae*, and *domini* who owned more than one provide the long-distance links, the bridges, which tied the industry together, making it a small-world. Using Watts' formulae (1999: 111-2), one can plot all of the nodes and connections in a network under consideration, and determine whether it conforms to the Watts' version of a small-world (the 'egalitarian' type, cf. 6.3.2). In graph theory (a field of mathematics) the network shapes described above in 6.2.4 can be classified as 'random graphs' and 'ordered

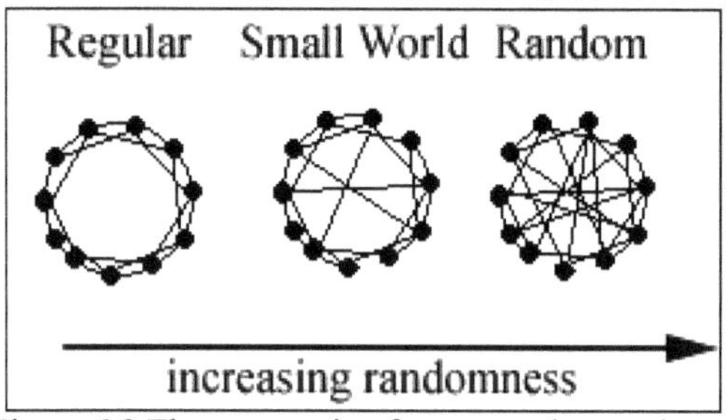

Figure 6.8 The progression from a regular graph to a random graph. After Watts (1999)

graphs'. In a random graph, nodes (actors) are connected at random to each other, while in 'ordered graphs' each node is connected to every other node. Somewhere on the continuum between a completely random graph and a completely ordered graph lies the 'small-world' (Figure 6.8). Specifically, according to Watts, a small-world exists where the average 'degrees of separation' (number of links) of the graph in question is the same as that of a random graph with the same number of nodes and number of links (relationships between nodes), but the density of ties (connectivity) between the nodes is greater by an order of magnitude than that of the random graph; the links 'bunch up', they cluster (114). That is to say, it looks like a random graph but behaves like an ordered graph.

That is the graph-theoretic definition, and it relates specifically to the pattern of relationships in a network. In real life, it means that while the patterns of relationships between individuals appears to be the result of utter chance, certain individuals have more than their fair share, like Annius Verus or Domitia Lucilla. Although this seems intuitive, not every network is a small-world and Watt's definition enables us to measure these properties directly. When a system has a small-world network structure (of either type), it is capable of generating out of its own dynamics higher level phenomena, because the shortcuts act like non-linear feedback loops. Whatever kind of small-world we are dealing with, all of them will have a power-law distribution regarding the number of nodes and the number of connections those nodes have. The power-law distribution is the most significant indicator of the presence of small-worlds (Barabási, pers. comm.). Our networks in the brick industry, while not perfect, are all quite close to being power-laws (and a larger dataset I would predict would be true power-laws) (Figure 6.9).

When a network becomes a small-world, dramatic changes in *global* behaviour result (Watts 1999: 6). Self-organised phenomena can be generated on a small-world network; order crystallises from randomness (Lewin 1993:147). The uniformity of stamp shapes and stamp information in the 2nd century may be an instance of self-organisation. This uniformity emerged out of many individuals' own independent stamping practices as the transmission of ideas about stamping flowed over those bridges between *figlinae*, the *officinatores* who worked in multiple establishments. The corollary is that if the brick industry had not been a small-world, there would have been no such uniformity.

There are other implications to the small-world. The small-world is a similar concept to the colourfully named 'edge of chaos' in complex systems studies. The edge of chaos is that point where a system is in the critical state (self-organised), where a small disturbance can push the system over one way into utterly random or the other way into rigidly fixed behaviour. It was first discussed by the physicist Bak, who used the (now-canonical) analogy of the sand pile (Bak 1991: 28). Imagine a bucket of sand being slowly poured out onto a level surface. The pile gets higher and higher when suddenly a bit more sand causes a small avalanche. That point just before the avalanche, where no more sand can be added to the pile, represents the edge of chaos. Another grain of sand causes an avalanche of unpredictable size, but the range of sizes forms a power law distribution. We have already encountered power laws in the distribution of brick stamp types, and in the distribution of city ranks (cf. 2.2.1, and 2.4.1). A power law is shorthand for stating, the larger the response, the less frequent it is seen (Lucas 2001: 3.6).

If we are not physicists, it can be difficult to see how a sand pile relates to a social system, but the parallel is in the 'communication' between grains of sand (which in the actual physical world would be mediated through internal friction and the angle of repose). The degree to which these avalanches cascade through a system is related to the way individuals are linked to one another. If connectivity is low, the avalanche has little effect

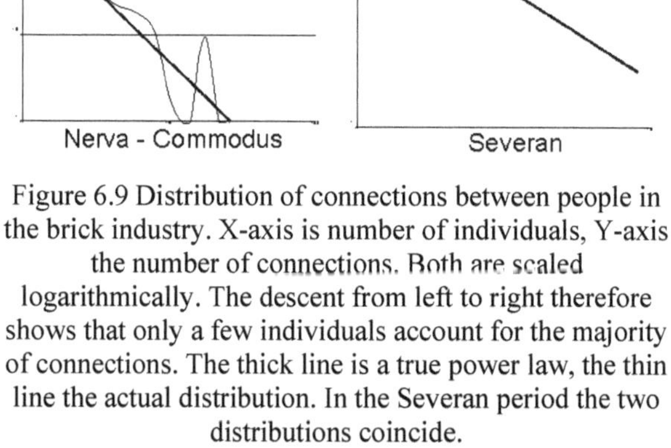

Figure 6.9 Distribution of connections between people in the brick industry. X-axis is number of individuals, Y-axis the number of connections. Both are scaled logarithmically. The descent from left to right therefore shows that only a few individuals account for the majority of connections. The thick line is a true power law, the thin line the actual distribution. In the Severan period the two distributions coincide.

(dropping one grain of sand on the top of the pile does not force many of the other grains to shift position; the angle of repose is quickly reached). If connectivity is high, then small disturbances create large avalanches as the effect 'propagate[s] hectically through the system' (Lewin, 1993: 62). The edge of chaos (or rather, self-organisation or the small-world) is the intermediate stage between these two, where some changes have small effects, while others have drastic effects, and what will happen cannot be predicted. Connections between individuals can mediate avalanches in a way similar to sand piles. In a complex system like society, if something happens to one person who is poorly connected, the effect (the avalanche) swiftly peters out. But if that person is well-connected, the knock-on effect can quickly cascade. Were these to be plotted, they would form a power law distribution. This is why it is significant to find a power-law distribution in brick stamps: it is an indication that the production of bricks itself is a complex system.

The actions of agents (or actors; the terms are interchangeable), based on *imperfect, local* knowledge, after thousands of repetitions can combine to produce unforeseen, potentially undesirable or even completely contrary results to what a given actor may have expected. Complexity theory suggests that internally, a system will progress to the point where, once shortcuts come into existence, positive and negative feedback loops make the system unstable. At this point the system may disintegrate; the system may remain as it is; or the system may evolve to a higher level of complexity. What determines the route is what is called the 'attractor': in the case of humans, this is what influences the choice of actions by a person based on the imperfect knowledge of a huge number of factors. When the system becomes poised on the edge of chaos, an external shock can have counter-intuitive results. A large catastrophe might not wreck the system, but a single perturbation at the right time can bring it crashing down. It is not what you do but when you do it, perhaps even who you do it to, that counts. What matters is the pattern of relationships in which you are embedded. In the brick industry, this means that there is no such thing as a 'typical' *officinator* or *dominus*; the transferrence to a new *figlinae*, the confiscation of property, or the death (whether proscribed or natural) of an individual could have surprising effects. The key is to know what kind of small-world we are dealing with.

6.3.2 Types of Small-Worlds

There are at least two types of small-worlds, the 'egalitarian' and the 'aristocratic' (Buchanan 2002: 118-20). The Watts-Strogatz version of the small-world (1998; Watts 1999) was inherently egalitarian by virtue of Watts and Strogatz's methodology. They began with an already established regular network, where every node had the same number of links, and studied what happened when extra, random links were added. The probability of obtaining one of these random links was the same for all individual nodes (Barabási 2002:54). If the brick industry was 'egalitarian', every individual, whether *dominus* or *officinator* would have roughly the same amount of connections as every other, in rather circumscribed clusters, each *figlina* an island unto itself, as it were, with the occasional long-distance bridge between them.

The 'aristocratic' small-world (also called a 'scale-free network') was discovered when Barabási and his colleagues tried to replicate the patterns of connections they found while studying an existing network (the Internet) (Barabási 2002:55-64; Albert, Jeong, and Barabási 1999: 130-1). In an 'aristocratic' small-world, the network is dominated by the existence of a few, super-connected 'hubs'. If the brick industry is 'aristocratic', then there would be people who have many clients (or indeed patrons), tenants, friends and contacts, over and above what we would normally expect. They are the people with 'their fingers in every pie'. Barabási and his colleagues discovered that there were a few simple mechanisms which accounted for the development of these hubs which give the power-law signature of small-worlds and complex systems (Barabási 2002: 86). The first of these was simply growth. A network grows one node at a time (the 'egalitarian' model did not take this into account). Secondly, they assumed that there was *preferential attachment*: the chance that a new node would connect with another node was a function of how many connections the older node already had. They found that these two assumptions were all that were necessary to grow a network on their computer which demonstrated a power-law in terms of the number of connections per node, and which developed hubs. Barabási calls this the 'theory of evolving networks', and argues that growth and preferential attachment explain the basic features of real networks (Barabási 2002: 90-91; ;but cf Pujol *et al* 2005 for a more sociological take on the question).

In terms of the brick industry, preferential attachment is akin to how patronage worked in the ancient world. One would not become the client of a patron who was not already well-connected (Wallace-Hadrill 1989: 83), thus increasing the number of connections which that patron already. It is a small world, because of the power-law distribution of connections. It could be expected that the brick industry modelled as a network will be an 'aristocratic' one. To investigate which type we are dealing with, I measure the degree to which the real network, the brick industry, matches up against Watts-Strogatz's 'egalitarian' network, and explore the differences (if it is not 'egalitarian' then it is 'aristocratic').

What kind of Small-World is the Brick Industry?

Watts (1999:144) provides some equations for calculating whether a network meets the criteria of an egalitarian small-world. It is more comprehensible to

	Julio-Claudian	Flavian	Nerva-Antonines	Severans
Clustering coeeficient	0.04	0.06	0.02	0.11
Equivalent random graph clustering coeeficient	0.008	0.005	0.02	0.08
Characteristic path length	1.16	1	3.88	2
Equivalent length in a random graph	6.76	4	43	5

Table 6.1 Small Egalitarian Patronage Worlds? An 'egalitarian' small-world exists when the characteristic path lengths of the two networks are similar but the clusering coeefficient of the network under consideration is an order of magnitude greater than the equivalent random network (Watts 1999: 114). Shaded cells indicate 'egalitarian' small-world conditions are fulfilled

explain this equation as a process rather than in maths. For each actor, the number of steps (ties or links between actors) it takes to get to every other actor is counted. Then the average number of steps for each actor is calculated. The median of the averages is then obtained. This gives us the 'characteristic path length' for the network under consideration. The clustering coeeffient of the graph is obtained by dividing the number of ties by the number of possible ties for each actor in its immediate neighborhood. These values (clustering coefficient and characteristic path length) are compared against a random network, which can easily be generated on a computer, especially using a programme such as UCINET (Borgatti 1999). The random network must have the same number of actors and average number of ties between actors as the social network under consideration.

Table 6.1 gives the relevant details for the four patronage networks. The shaded cells come closest to fulfilling the egalitarian small-world conditions. Characteristic path length can be thought of as the distance it takes for information to get from one side of a network to another. Two observations are immediately apparent. In the Julio-Claudian period, when the exploitation of fired brick begins on an industrial scale, the patronage networks are self-organising. That is, there is no outside control over this exploitation. The shortness of the characteristic path length during all periods except the 2nd century suggests that face-to-face communications between actors of different social status were the norm. In the 2nd century, the lengthening of the communication chain perhaps indicates the existence of 'brokers' who mediate the exchanges of information across the network. Importantly, the fact that the 2nd century networks are not 'egalitarian' yet are still small-worlds (on the evidence of their power-laws) suggests that they may in fact be 'aristocratic' small-worlds. How can an 'egalitarian' network be transformed into an 'aristocratic' one?

Egalitarian versus Aristocratic

The different networks correspond to different dynasties over the 1st to 3rd centuries, allowing a series of snapshots of an evolving network. All of these networks give power law responses, which according to Barabási is the most significant indicator of a small-world (pers. comm.; cf. also Barabási 1999). Watts' equations tell us that the first two periods are *egalitarian* networks, while the power laws tell us that they are all small-worlds; hence the 2nd century is characterised by an *aristocratic* network, the existence of hubs (super-connected individuals, who owned more than one *figlinae* or were active producers in multiple establishments) and connectors (the bridges between the hubs, like Plotina). In other real world examples of small-world networks, it has been found that generally 'aristocratic' networks come into being first; later when limitations to growth appear, the network evolves into an 'egalitarian' type.

The airport hub-and-spoke system may be taken as an example (Buchanan 2002: 123-4). In the early days of aviation, the hubs like Heathrow and Chicago's O'Hare Airport offered airlines a place to centralise their operations and also offered a range of destinations within easy reach. Because these airports were so much better connected than the later established airports, airlines, passengers, and flight planners tended to use the older airports more often. This self-reinforcing process turned these airports into hubs (in fact the terminology of 'aristocratic' networks, in using terms such as 'hubs' and 'connectors' harkens back to the airport system for metaphors). However, as the volume of passengers grew, and the links with these airports expanded, in recent years they have become victims of their own success and have not been able to continue their growth. Costs and limitations have appeared in that for example they are hampered by airspace congestion and planning restrictions on expanding their runways. Consequently, delays at these hubs have allowed other lesser airports such as Stansted to expand their own connections, becoming very hub-like themselves. Amaral and his colleagues (2000: 1-2) have found that the airport hub-and-spoke system is not, in actual fact, an 'aristocratic' network anymore: it is egalitarian. To account for this finding they proposed a modification to Barabási's model of network growth: costs or limitations can lessen or negate the effects of preferential attachment. An 'aristocratic' network can grow to become an 'egalitarian' network the kind which Watts and Strogatz first discovered (Buchanan 2002: 125).

In brick, there seems to be the opposite pattern, of an 'egalitarian' network becoming 'aristocratic'. This does not however indicate that the model is wrong; the model as it is formulated is cyclical, with each type giving way to the other in turn. Given that what we are

	Julio-Claudian	Flavian	Nerva-Antonines	Severans
Clustering coeeficient	0.16	0.16	0.06	0.16
Equivalent random graph clustering coeeficient	0.15	0.092	0.03	0.097
Characteristic path length	2.4	2.2	1.5	1.3
Equivalent length in a random graph	8.8	3.8	6.2	3.3

Table 6.2 Small Egalitarian Manufacturing Worlds? An 'egalitarian' small-world exists when the characteristic path lengths of the two networks are similar but the clusering coeefficient of the network under consideration is an order of magnitude greater than the equivalent random network (Watts 1999: 114). Shaded cells indicate 'egalitarian' small-world conditions are fulfilled

observing is a small-world, we may simply be joining the action *in media res*: the structure of the brick industry prior to the mid 1st century ought to have been of the 'aristocratic' kind in order to have evolved to the 'egalitarian' kind measured here. It may have been that in the early days of the brick industry only a few of the total clay resources in the Tiber Valley were known. This would partly have been technological, as the use of brick-faced concrete only began in earnest during the reign of Nero, probably spurred on by the need to rebuild following the Great Fire of 64 (Blake 1959: 10, 55). This ignorance of the true extent of clay sources, coupled with a limited demand prior to 64, likely created a pattern where soon every known source was exploited, erecting a limitation to further growth. The 'aristocratic' network would therefore have been converted into an 'egalitarian' network early on, before the periods considered in this study. However, in the 2nd century, for an 'aristocratic' network to emerge again, this limitation must have been removed. There may have been a sudden surge in the supply of available land, or the discovery that clay was available in more locations than previously known, or a new and sustained demand for the product which allowed new opportunities for growth and preferential attachment.

The mathematics of small-worlds suggests that costs and limitations can create 'egalitarian' networks from 'aristocratic' ones. It follows that the removal of these constraints would turn 'egalitarian' networks into 'aristocratic' ones, which would account for the pattern noted in the brick industry networks. There might be alternatives to this scenario (a general influx of cash into the economy which allowed investment in new enterprises, including clay pits, for instance) but the general picture holds true: a constraint to growth was removed. The picture suggested in 5.3.4 that the year 123 AD represents a land-rush, was based on a completely different line of argument, yet it dove-tails neatly with the evolution of the brick industry as discerned here: a constraint was removed, leading to new growth.

This chapter has, up to this point, been concerned with patronage networks in the brick industry, networks discerned in the connections evident in brick stamps. However, there is another level of networks in the brick industry, that which existed at the level of manufacturing itself, where the connections are between brick makers who shared the same clay sources and so on. What kinds of dynamics are evident at this level, and how do the two levels interact?

6.3.3 The Shape of Manufacturing Networks

In 4.3.1, based on the archaeometry in Chapter 3, I developed a framework for considering all of the relationships in evidence in assemblages of brick, whether those relationships were based on sharing a common clay source, using the same stamp (or having common names in those stamps), or being used at the same site. These relationships in their entirety constitute a network directly connected to the manufacture and consumption of brick. This network is rather more informative concerning the situation 'on the ground' than the patronage networks discussed above because it incorporates social and geographical aspects of both production and consumption.

For network analysis, the patronage network was described in terms of whether or not a relationship existed between two individuals. In the manufacturing networks, on the other hand, I can also indicate the *kind* of relationship. That is, a weighting can be assigned to the most important relationships (however defined). In this case, we might expect that the individual who could exploit more than one source area simultaneously, ship or sell materials to more than one site as being more successful, more powerful, than the individual who could exploit only a single source, or could ship his or her product to a single site. This allows an examination of agency, or the power to act, through the material culture of the brick industry and how this changes over time. The relationships in the network were therefore coded so that production mode 2 relationships ranked higher than mode 3 relationships, both of which ranked higher than mode 1 relationships (cf. 4.3.1 on modes of production).

Figures 6.10 - 6.13 show the manufacturing networks according to the periods discerned in the patronage networks. The different samples have been plotted according to their geographic co-ordinates, so these figures also indicate how the manufacturing networks play across space. The thicker the line, the 'deeper' or more important the relationship. What is clearly evident is that the Tiber is no barrier. The most important relationships in all periods span the Valley. Insofar as the production and consumption of brick is concerned,

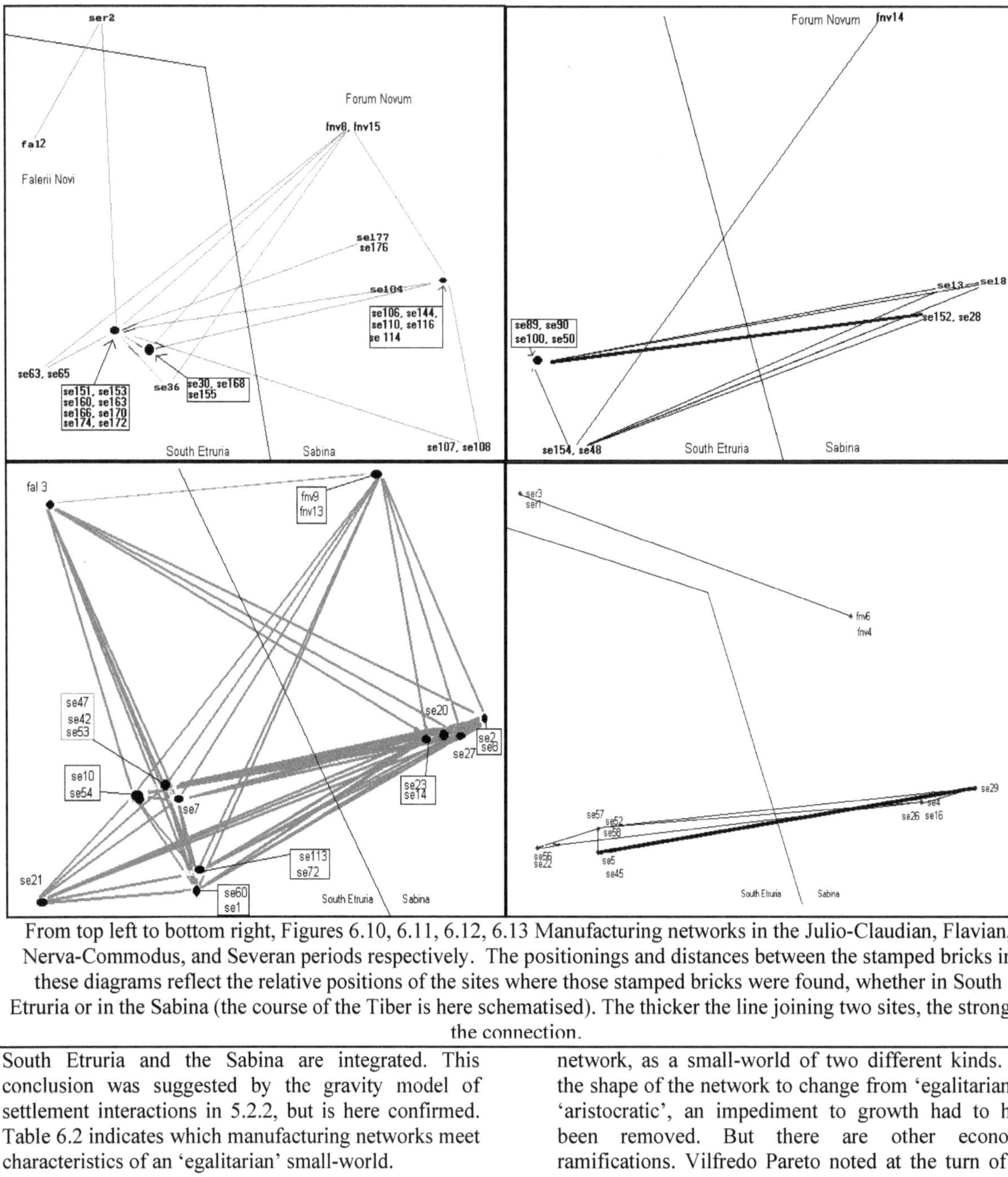

From top left to bottom right, Figures 6.10, 6.11, 6.12, 6.13 Manufacturing networks in the Julio-Claudian, Flavian, Nerva-Commodus, and Severan periods respectively. The positionings and distances between the stamped bricks in these diagrams reflect the relative positions of the sites where those stamped bricks were found, whether in South Etruria or in the Sabina (the course of the Tiber is here schematised). The thicker the line joining two sites, the stronger the connection.

South Etruria and the Sabina are integrated. This conclusion was suggested by the gravity model of settlement interactions in 5.2.2, but is here confirmed. Table 6.2 indicates which manufacturing networks meet characteristics of an 'egalitarian' small-world.

Here there is a slightly different pattern than the one discovered for the patronage networks in 6.3.2, but one which conforms nicely to the idea of growth creating its own limitations. It is a pattern of 'aristocratic' followed by 'egalitarian' repeated once. The limitations to growth in the manufacture of brick might have been in terms of the size of the clay source, or bottlenecks in the distribution network. In 6.4.2 I compare the two levels.

6.3.4 Small-Worlds: The Condensation of Wealth

I have already hinted at some of the economic ramifications of considering the brick industry as a network, as a small-world of two different kinds. For the shape of the network to change from 'egalitarian' to 'aristocratic', an impediment to growth had to have been removed. But there are other economic ramifications. Vilfredo Pareto noted at the turn of the last century that 80% of Italy's wealth was in the hands of 20% of its population (Barabási 2002:66). He found this same distribution in other fields as well, and Pareto's observation has since been elevated to the status of a 'law'. In many systems, but especially the distribution of wealth, the majority will be concentrated in the hands of only about a fifth the population. Pareto's Law is actually a *power law*.

Bouchaud and Mézard (2000: 536) found that on a network, two basic mechanisms could reproduce Pareto's Law of wealth distribution: trade, and investments. (By investments, I recognise that there was no ancient concept which readily equates with our term 'investment'; in using the term I mean the spending of money to improve a piece of land, to buy a ship or part of a cargo for resale, and so on). In their simulation they found that trade redistributed wealth around the network like electricity through the grid, tending to reduce inequalities. The return on investments on the other hand had a more random effect. No two individuals will have the same success rate on their investments; although the majority more or less will break even, some individuals will have a higher degree of luck than the others (and some will lose everything). This it was found drives a 'rich-get-richer' phenomenon. Those with more money, have more money to invest thus increasing their chances of earning a positive return. In this case their wealth grows not through addition, but rather through multiplication. Bouchaud and Mézard found that, even in a world where everyone has an equal amount of money to begin with, random returns on investment were enough to eventually create the enormous disparities of wealth captured by Pareto's Law.

However, like diseases and ideas being transmitted through a network (6.2.1) there exists a 'tipping point' where the irregularities in the returns from investments overwhelm the redistributive effects of trade and taxes (taxes represent a sort of long-distance link between the wealthy and the poor, an enforced trade, Buchanan 2002: 196). With the effects of trade overwhelmed by investment returns, wealth would then 'condense' into the hands of not just 20% of the population, but into less than 1% of the population (195). The social and political ramifications of such a tip would be profound. The condensation of wealth it must be noted could happen in any society, ancient or modern:

> Even though the network model is abstract, this property is also its advantage, for it proves on fundamental mathematical grounds, with few disputable assumptions, that a tipping point of this sort must exist in any economic society. (195)

Some economies in which wealth has condensed into the hands of the few may already exist: in Mexico, forty individuals account for almost 30% of the money supply. Another factor which may contribute to condensation is political instability. After the collapse of the USSR, the taxation system largely failed, nor were companies effectively regulated, allowing huge inequalities to emerge (196).

The networks in the brick industry might be indicating that a similar process was at play. The existence of the 'egalitarian' small-world in the 1st century AD suggests that the balance between the redistributive effects of trade and the condensing effects of investment was roughly equal at that time. Prior to this, the civil wars and interminable proscriptions and purges of the last years of the Republic may have had the effect of levelling the distribution of wealth. Of course, the wealthy were often subject to having their property and estates confiscated, but what we are suggesting here is that such actions, rather than simply replacing one set of the elite with another, had a profound effect on the shape of the Roman economy. There might be a degree of economic literalness in Augustus' self-reference as '*primus inter pares*', first amongst equals.

The sudden transformation from an 'egalitarian' to an 'aristocratic' small-world in the early 2nd century might have allowed investment inequalities to develop as the former impediments to growth were removed. The economy may have been progressing towards a condensed phase as preferential attachment allowed some individuals to increase their knowledge of investment opportunities. The single biggest opportunity in the brick industry may have been, if these hypotheses are correct, the land-rush which we have argued lies behind the brick stamps of 123. With more land available for exploitation, the number and kind of opportunities for investment (whether in production, shipping, building, warehousing, and so on) would expand as well. Not all of these opportunities would necessarily have been successful. Yet for those that were the *dominus* and *officinator* and whoever else was involved would have been enriched, enabling them to take more opportunities as they appeared and propelling the industry towards a condensed phase.

The number of *officinatores* known to have worked with the *Domitii* family (cf. Steinby 1974: 47-58) may be both cause and effect of this family's continuing success in the brick industry. It is easy to imagine how preferential attachment and condensation would have worked for them. This family constitutes a single unbroken chain from the 40s AD to the accession of Marcus Aurelius, a scion of the family, in 161, ending with the death of his son Commodus in 193. As one of the oldest, unbroken families in the industry, preferential attachment suggests that *officinatores* and others would have wanted to have been in their orbit, linked to them. This is because the *Domitii* would have had the long standing social and financial capital necessary to take advantage of opportunities as they presented themselves. This situation would have been particularly true for my speculative land rush of 123. With the continuing process of new individuals seeking to attach themselves to the family, there would have been new ideas and new knowledge of new investment opportunities, allowing the potential for their wealth and power to increase. As wealth was connected with the ownership of land in antiquity, the pattern of *figlinae* becoming concentrated in fewer and fewer hands over the 2nd century (cf. Anderson 1991: 1)

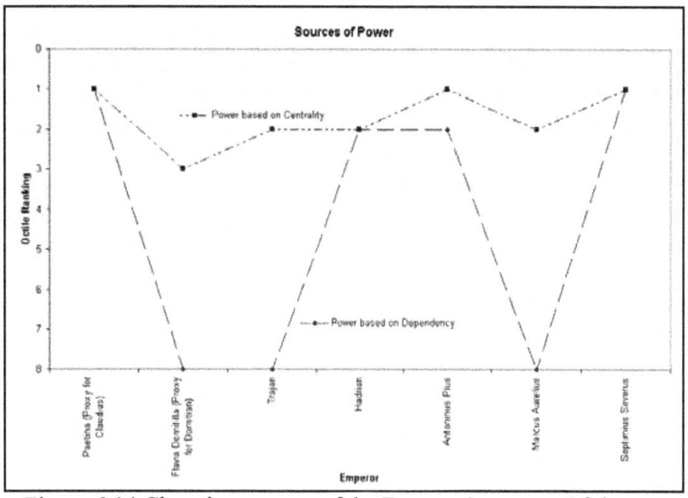

Figure 6.14 Changing sources of the Emperor's power. Of the two types of power that can be measured in networks, whichever one in which the Emperor ranks highest is likely to be the main source for that Emperor at that time.

probably represents the 'hard evidence' for this process of condensation.

Bouchaud and Mézard's model (2000) does not deal with the network effects of removing people (nodes) from the network. However, the changing picture of the brick industry over time suggests that the deaths of individuals could have profound effects on the shape of the economy, and on society. What happened with the death of Commodus, and the end of that line of the *Domitii* and all its established wealth and contacts? In 6.5 I consider that very question, but first I explore the roles different individuals had in the different networks – their social power.

6.4 Social Power

6.4.1 Measuring Power

The flip from an 'egalitarian' small-world to an 'aristocratic' one did not simply 'happen'; fundamentally, the processes which have been identified (preferential attachment, limitations, condensation, trade) all take place at the individual level. This is what complexity theory teaches us: although we can identify higher-level phenomena, these phenomena (preferential attachment, etc.) emerge from the interactions of individuals. Some individuals through their actions will influence (whether consciously or unconsciously) the emergence of the higher-level processes; they will be individuals whose power to do this is a result of their network positioning.

To measure individual positioning in the networks, I turn to the tools of Social Network Analysis. One way to measure the individual power of actors in a network is to count the number of ties to other actors. Somebody with more connections will be more powerful, have greater access to flows of information, than somebody with fewer connections. This is called the Freeman centrality measure of power (Hanneman and Riddle 2005). But of course, the kind of people one is connected to matters, whether they are well connected as well. Alice who is well connected to well connected others is in practice less powerful than Bob, who is connected to many poorly connected others (cf Figure 6.3). This is because, despite the centrality of Alice, there are multiple pathways around her. Bob, who has many connections but to poorly connected others, is in fact more powerful, because the poorly connected others are dependent on him; they have no choice but to go through this single broker. The Bonacich power index is a statistic measuring both centrality as conceived of by Freeman and power in the sense described above (Hanneman and Riddle 2005).

Given the nature of patronage in ancient society, we should be looking at both centrality power and dependency power to measure the sources of power and control of information in these patronage networks. The Bonacich Power Index here was calculated for each actor, giving a measure of centrality and the degree to which each actor's power results from others' dependency (listed in Appendix D). The individual scores were ranked by octile.

One would expect that the Emperor should by rights be the most important individual in all of these networks, and so where he draws his power from may indicate the pattern for élites in general. Figure 6.14 expresses the changing power levels of the Emperor by ranking. A ranking in the first octile indicates that the Emperor is amongst the most powerful individuals. In the Julio-Claudian and Severan periods, on either measure of power (centrality or dependants) the Emperor is most powerful. In other periods it is not so clear cut. Trajan and Hadrian derive more power from being well-connected (though ranking in the second octile) than from having many dependants. Antoninus Pius had as much power from dependants as Trajan and Hadrian had from being well connected. Antoninus Pius' good connections to well connected others (power from centrality) put him in the first octile. Marcus Aurelius on the other hand was more like Trajan and Hadrian. We can draw two conclusions. First, patronage as a system was characterised by both the vertical ties of dependency and the horizontal ones of centrality under the Julio-Claudians and the Severans. Under the so-called 'adoptive Emperors' vertical dependency did not matter as much as the interlocking circles of who-knows-whom which are so clearly indicated in Figure 6.6. Hereditary succession versus adoption of the ablest man can easily be connected with these observations

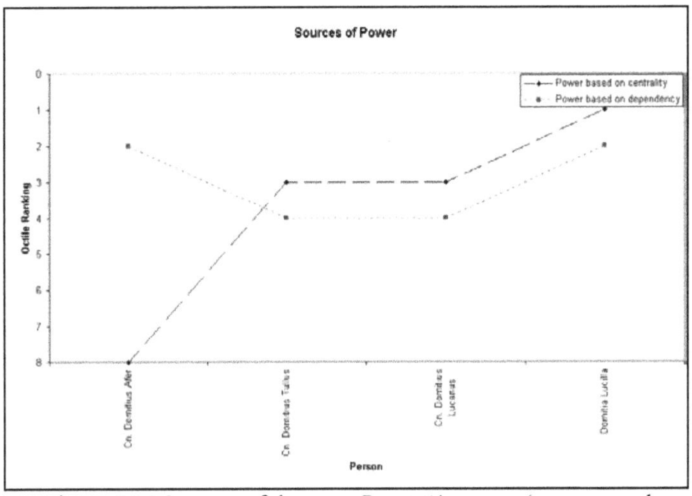

Figure 6.15 Sources of the *gens Domitii's* power (as measured through network positioning)

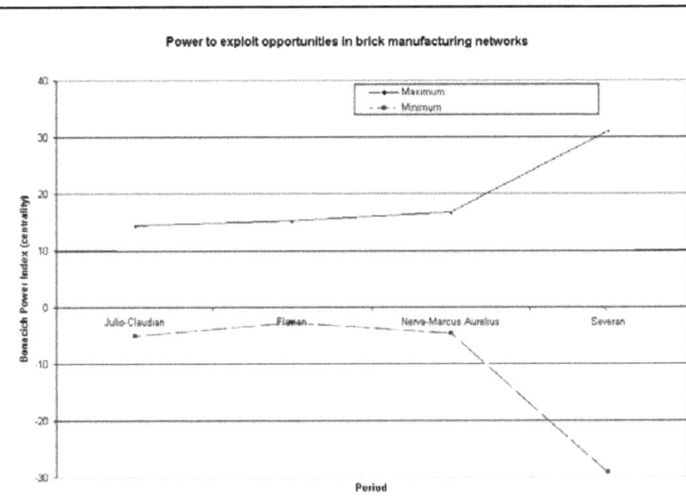

Figure 6.16 Overall power in manufacturing networks

concerning sources of power. If one is accustomed to drawing one's power from below, from within the family and its clients, it would naturally follow that one's successor would come from below. If however one has obtained power by knowing the right people, it would not seem incorrect to adopt an heir from another circle. Given the way wealth and the ownership of land may have been condensing into the hands of a few (cf. 6.3.4) the development of power through knowing the right people may have become a practical necessity.

For the producer of bricks, in earlier days it was enough to establish good connections with one's patron in order to gain access to power. In the 2nd century the intervention of a 'broker' would become necessary to obtain admittance to the various circles, especially if land was becoming concentrated in a limited number of hands. If the Emperor is the most powerful man in the empire, then in these power graphs deduced from land exploitation the people who rank higher than the Emperor probably represent 'power brokers'. That is, they are so well-connected that they can intervene anywhere in the network at will. Domitia Lucilla may be one such person. She ranks in the first octile in this period (see Appendix D for individual rankings), which may both cause and explain the heavy numbers of *officinatores* working for the *gens Domitii* at this period (cf. Wallace-Hadrill 1989: 83 on how social standing in 17th century France depends on having and being seen to have many connections). The evolution of the *Domitii* family's power as a whole can also be viewed through this lens of centrality versus dependency (Figure 6.15). In Domitius Afer's day, power was built by being central and also having dependants, whereas by Domitia Lucilla's it was built through connections. In an 'egalitarian' small-world, power seems to have been built through having dependants. By the 2nd century and the 'aristocratic' small-world, control and access to power was regulated through knowing the right people. Access to land and productive capital was thus more difficult to obtain, but even more problematic would be getting the product to market. Patronage, as Wallace-Hadrill (1989: 72-3) shows, is about manipulating access ro resources; there seem to have been socio-structural reasons for the differential 'marketing access' discussed in 5.2.1.

Given the conception above on which relationships are most important, the measure of power that best fits the situation is one which measures the centrality of connections. Figure 6.16 plots the overall measure of power in manufacturing networks during these periods. In the patronage networks, I used the metaphor of 'knowing the right people' to explain this measure; in manufacturing networks, an appropriate metaphor for having these connections might be agency or 'the power to exploit opportunities'.

6.4.2 Patronage Networks and Manufacturing Networks Compared

One should not expect that the patronage networks and the manufacturing networks would produce exactly the same statistics. The structure of patronage to a certain degree determines what the structure of manufacturing will be, but equally, manufacturing has a role in determining the structure of patronage. This is because of the voluntary role the 'little person' has in deciding to become a tenant (cf. Drummond 1989: 101), and also the decision by the landlord how best to exploit his land. Both structures emerge out of the individual interactions within Roman society. How then do we reconcile what these two tables, 6.1 and 6.2, are telling us? They both indicate 'egalitarian' and 'aristocratic' small-worlds, but at different times at the different levels. Importantly, they agree during the high period of the brick industry, from the Flavians to the end of Commodus.

Why the disagreement during the Julio-Claudian period? I argued in 3.3 that my data-set was representative of the brick industry as a whole, so it may simply be a problem of chronology, where the tested stamped bricks actually date slightly earlier than the names which we incorporated into the patronage network for this period. In that case, the manufacturing network would seem to indicate the 'aristocratic' networks I argued necessarily had to precede the 'egalitarian' one measured. The 'aristocratic' network suggests what would be expected for an industry in its infancy: many different sources in competition of which the eldest and best known have attracted the greatest number of brick makers. The 'egalitarian' network of patronage during the period might not necessarily even be at odds with this picture (and so chronology might not be a problem). If the scale of production was sufficiently small, the countering effects of trade versus investment would still have been in balance, leading to an 'egalitarian' network at the level of patronage. Whatever constraint kept the production of brick on a small scale was the limiting factor preventing the development of an 'aristocratic' network. Part of that constraint, as suggested earlier, is that the demand for brick might simply have been too low. However, the nature of the sources of power in the Julio-Claudian period (power based on dependants) might have contributed to limiting growth as well. It is in a sense a case of Alice and Bob again (6.4.1). Bob is central, but all of his friends know only what Bob knows. As a patron, his knowledge of opportunities is limited.

In the Flavian period the patronage network is still an 'egalitarian' one, but now the manufacturing network is also 'egalitarian'. While demand for brick was low, all of the known sources of clay were sufficient to handle the growth in manufacturing, leading to an 'aristocratic' network. After the Great Fire and the end of the Julio-Claudians, the use of and hence the demand for brick increased; but if I am correct in my interpretation, a bottleneck emerged. It may have been that there were simply not enough clay sources known or exploited to keep up with the demand. Coupled with this is the continuing nature of the patronage system under the Flavians, of power based on having many dependants.

In the 2nd century, the limitations to growth have been surpassed and the nature of patronage power has changed. We might hypothesise that the discovery of suitable parcels of land for brick production would have broken the 'egalitarian' networks. Landowners with newly discovered parcels of land suitable for clay production might have been outside the personal patronage networks of the existing brick *domini*. The new owners, not having the knowledge or the contacts with *officinatores* who could make brick, may have sought out the earlier *domini* to obtain access to brick making expertise. This might have led to *lateral* links between the previously monolithic power structures; of course no single individual's power base would have been monolithic, but these lateral links could account for the interlocking circles of power discussed in 6.4.1. Since the *Domitii* family were one of the long established brick making families, this process is actually one of preferential attachment, and it would certainly have contributed to their later success (ascending the throne via adoption).

Power results from knowing the right people, of favours given and received. In such an atmosphere in the middle of the century we find the construction of the Forum Baths at Ostia by Gavius Maximus, Praetorian Prefect of Antoninus Pius. It has been demonstrated that over 90% of the stamped bricks in this complex were supplied through the contacts Gavius had with the Imperial family, and with Asinia Quadratilla and Flavius Aper (DeLaine, 2002). The structure of the manufacturing network at the time is similar to that under the Julio-Claudians, and for probably similar reasons. Given the numerous large-scale public and private building projects in the 2nd century, being able to supply or withhold building materials, or the knowledge of how to produce bricks, of how to exploit the clay resource effectively, as part of the reciprocal relationships of patronage through interlocking rings of well-connected people necessitates the careful control of one's own supply. In a sense, the products of one's lands have ceased being something one sells, rather they have become 'bargaining chips' in the great game.

Finally, under the Severans, the structure of the networks is the opposite of that under the Julio-Claudians, although the sources of power are the same. The manufacturing network has assumed an 'egalitarian' shape, while the patronage network remains 'aristocratic'. A limit has been reached in the manufacturing network, which might either be in terms of available clay sources to exploit, or in a general lessening in demand for building materials. The manufacturing network, over the years from the Julio-Claudians to the Severans describes a very normal pattern of growth and recession, growth and recession as new constraints are encountered and overcome. At the level of patronage the network is greatly reduced, in numbers of individuals and connections. Significantly, a number of the 'hubs' of the earlier period (especially members of the *Domitii* family) have been excised from the network, allowing others to step into the breaches, thus becoming the loci for preferential attachment. Yet, in the violent confusion of the last years of Commodus' reign and the beginning of Septimius Severus', knowing the right people could suddenly become knowing the *wrong* people, with disastrous results (cf. below 6.5.3).

Catalogue	*CIL* XV.1	Stamp Information	Power index	Percentile	Octile
SE 176	864	CASPR	-11.5	.0%	
SE 172	864	CASPR	-10.97	3.0%	
SE 177	864	CASPR	-9.82	6.0%	
SE 63		ARISTANIUS	-7.85	9.0%	
SE 65		ARISTANIUS	-7.85	9.0%	
SER 2	777	Adivtor	-7.85	9.0%	1ST OCTILE
FNV 8		unstamped	-5.95	18.1%	
SE 106	N.1462 /3a	Q. Sulpicius Sabinus	-4.27	21.2%	
SE 151	1393	Ostorius Scapula	-2.95	24.2%	
SE 163	1393	Ostorius Scapula	-2.95	24.2%	
SE 170	1393	Ostorius Scapula	-2.95	24.2%	
SE 174	1393	Ostorius Scapula	-2.95	24.2%	2ND OCTILE
SE 110	N.1462 /3	Q. Sulpicius Sabinus	-2.39	36.3%	
SE 114	N.1462 /3b	Q. Sulpicius Sabinus	-2.39	36.3%	
SE 144	N.1462 /3a	Q. Sulpicius Sabinus	-2.39	36.3%	3RD OCTILE
SE 108	N.2179 /80	C. Laelius	-2.29	45.4%	
FNV 5		unstamped	-2.03	48.4%	
fnv 9		unstamped	-2.03	48.4%	
fnv 13		unstamped	-2.03	48.4%	
fnv 14		unstamped	-2.03	48.4%	
fnv 15		unstamped	-2.03	48.4%	4TH OCTILE
se 153	1393	Ostorius Scapula	-0.85	63.6%	
se 160	1393	Ostorius Scapula	-0.85	63.6%	
se 166	1393	Ostorius Scapula	-0.85	63.6%	
se 107	2316	Ti. Cl. Censorinus	-0.15	72.7%	6TH OCTILE
se 156	2263	PL (in ligature)	0.47	75.7%	
se 104	1460a	Q. Sulpicius Sabinus	1.04	78.7%	
se 116	N.1462 /3b	Q. Sulpicius Sabinus	1.04	78.7%	
se 36	2263	PL (in ligature)	1.24	84.8%	7TH OCTILE
fal 3		unstamped	3.15	87.8%	
se 155	950	C. Cornelius Natalis	3.26	90.9%	
fal 1		unstamped	4.3	93.9%	
se 30	N.1349 /50.1	P. Ostorious Eros	7.26	96.9%	
se 168	976	Dionysius	7.26	96.9%	8TH OCTILE

Table 6.3 Relative ability to exploit different sources or opportunities in the Julio-Claudian period, as represented by centrality scores (negatively attenuated Bonacich Power Index of each brick's positioning in the manufacturing network of the period; negative numbers indicate greater

Catalogue	*CIL* XV.1	Stamp Information	Power index	Percentile	Octile
SE 48	659 c	Tonneiana de Viccians	-13.2	0%	
SE 18	659 c	Tonneiana de Viccians	-11.6	7.6%	1ST OCTILE
SE 13	659 c	Tonneiana de Viccians	-7.6	15.30%	
SE 50	659 a	Tonneiana de Viccians	-6	23.00%	2ND OCTILE
SE 154	1268	Dama Marci C. servus	-5.6	30.70%	
FNV 6		Unstamped	-5.6	30.70%	3RD OCTILE
SE 152	N.898/ 8	CALP	-4.4	46.10%	4TH OCTILE
SE 28	664 c	Zozimus, Viccians	-1.2	53.80%	
SE 23	913 a	Q. Canusius Praenestinus	0	61.50%	
SE 85	1979	Niceporus Antiochus	0	61.50%	
FAL 2		unstamped	0	61.50%	5TH OCTILE
SE 89	S.431	C. Iulius Neicephorus	2.8	84.60%	
SE 90	S.431	C. Iulius Neicephorus	2.8	84.60%	
SE 100	S.431	C. Iulius Neicephorus	2.8	84.60%	6TH OCTILE

Table 6.4 Relative ability to exploit different sources or opportunities in the Flavian period, as represented by centrality scores (negatively attenuated Bonacich Power Index of each brick's positioning in the manufacturing network of the period; negative numbers indicate greater centrality)

In such a situation, a safer source of power is perhaps having many dependants rather than many good connections (the Army being the ultimate source of dependants?). I will consider in greater depth the transformation in the brick industry, and by extension, Roman society, with the advent of the Severans, in 6.5.

6.4.3 Powerful Manufacturers

In the previous section, I discussed the global properties of the manufacturing networks and how those networks interacted with the higher level patronage networks. In this section I focus on the local properties of each manufacturing network, each manufacturer's network positioning and power. For each period, the centrality scores (the Bonacich Power Index), which indicate in this context the relative ability to exploit different sources or opportunities, were ranked according to octile. The results are presented in Tables 6.3 - 6.6. The sometimes widely divergent ranking of individual bricks using the same stamps perhaps reflects the underlying structure of brick exploitation. When the same bricks with the same stamp types all rank within the same octile, one might interpret that as an indication that the person named as *officinator* (or the person appearing in a single named stamp) as being the person who literally did manufacture that brick. In cases where a named individual appears on bricks which rank to different octiles, it might signify another level of individuals below the named individual, who actually produce the brick.

'Ostorius Scapula' in the one-name stamp *CIL* XV.1 1393 probably does represent a landlord, rather than an *officinator*. The different rankings for his bricks suggest the existence of at least two *officinatores*. By this

Catalogue	CIL XV.1	Stamp Information	Power Index	Percent	Octile
SE 171	App 124d	Sabinus Brutiddius Volusianus	-14	.00%	
SE 113	486a	Cornelia Maiola, Salarese	-12	4.3%	
SE 8	61	M. Statilius Lucifer	-11	8.6%	
SE 72	s. 222	Zozimus	-11	8.6%	1ST OCTILE
SE 2	61	M. Statilius Lucifer	-7	17.3%	
SE 19	unidentified		-7	17.3%	2ND OCTILE
SE 27	528	Salarese (Primigenius)	-5	26.0%	
SE 111	unidentified		-2.75	30.4%	
SE 136	unidentified		-1.5	34.7%	
SE 148	unidentified		-1.5	34.7%	3RD OCTILE
SE 20	731b	Ummidius Quadratus + Annia Faustina	0	43.4%	
SE 42	861	C. Nunnidius Fortunatus	0	43.4%	
SE 47	861	C. Nunnidius Fortunatus	0	43.4%	
SE 53	861	C. Nunnidius Fortunatus	0	43.4%	
FNV 4		unstamped	0	43.4%	4TH OCTILE
SE 14	198	Domitianae Veteres (Succesus)	1.5	65.2%	
SE 7	1106a	Aprilis (gens Domitii)	1.75	69.5%	
SE 10	811f	Anteros	2	73.9%	
SE 29	163	Domitianae Maiores	2	73.9%	
SE 54	811f	Anteros	2	73.9%	
SE 141	unidentified		2	73.9%	6TH OCTILE
SE 21	368	Oceanae Maiores	4	91.3%	
SE 60	1075a	Ti. Cl. Quinquatralis	4	91.3%	
SE 1	368	Oceanae Maiores	6	100.0%	8TH OCTILE

Table 6.5 Relative ability to exploit different sources or opportunities in the Nerva - Commodus period, as represented by centrality scores (negatively attenuated Bonacich Power Index of each brick's positioning in the manufacturing network of the period; negative numbers indicate greater centrality)

reckoning Q. Sulpicius Sabinus ought to be a landlord as well, with at least two *officinatores*. 'Aristanius' bricks on the other hand were produced by one individual, as were the 'CASPR' bricks. Aristanius, CASPR, and 'Adiutor', (significantly, a freedman of the Emperor), all rank in the first octile. The production of brick figured highly in their strategies (including Ostorius and Sulpicius) for the effective exploitation of their land. Most of the Forum Novum unstamped material ranks in the 4th octile, making it the produce of a reasonably middling manufacturer. The lands of the *Tonneianae de Viccians* were most effectively exploited (1st octile) in the Flavian period, with at least four

Catalogue	CIL XV.1	Stamp Information	Power index	Percent	Octile
SE 5	408	Portus Licini	-2.26	.00%	
SE 57	625	L. Aelius Phidelis, Terentianae	-1.65	9.00%	
SE 58	625	L. Aelius Phidelis, Terentianae	-1.65	9.00%	1ST OCTILE
SE 52	625	L. Aelius Phidelis, Terentianae	-1.57	27.20%	
SE 45	408	Portus Licini	-1.4	36.30%	3RD OCTILE
SE 22	1210, 974, or 98		0	45.40%	4TH OCTILE
SER 1	176	Domitianae Minores	2.43	54.50%	
SER 3	758	Ponticulanae	2.43	54.50%	5TH OCTILE
SE 16	759	Dominus Noster (signum: Mercurius)	15.6	72.70%	6TH OCTILE
SE 56	762b	Augustus Noster (signum: Aper)	17.2	81.80%	7TH OCTILE
SE 4	189	Domitianae Veteres (Festus)	26	90.90%	
SE 26	189	Domitianae Veteres (Festus)	32.4	100.00%	8TH OCTILE

Table 6.6 Relative ability to exploit different sources or opportunities in the Severan period, as represented by centrality scores (negatively attenuated Bonacich Power Index of each brick's positioning in the manufacturing network of the period; negative numbers indicate greater centrality)

different individuals making brick (two of them used the same stamps as each other, a third who used a different *signum*, and a fourth who named himself). The same phenomenon is observed in the 2nd century (e.g., M. Statilius Lucifer likely had others working beneath him because bricks carrying stamps which name him alone have different rankings while C. Nunnidius Fortunatus, whose stamps all rank to the same octile probably represents the entire outfit) and also under the Severans (*Portus Licini* and *Terentianae* both have bricks ranking in different octiles, implying that there are different *officinatores* hiding behind the same stamps).

In the Julio-Claudian and Flavian periods, individuals who were connected to the Emperor also ranked in the first octile (one *domina* of the *Tonneianae de Viccians* under the Julio-Claudians was Paetina, wife of Claudius, which suggests that the Flavians would have inherited at least part of this estate). In the 2nd century however, in the manufacturing network many of the bricks connected to the major *figlinae* or the *Domitii* tend to rank in the bottom four octiles. Why is that? In the 2nd century the nature of patrons' power changed from being based on dependants to knowing the right people (who in turn also know the right people, and so on...). Also, both the patronage network and the manufacturing network of the period were 'aristocratic' small-worlds. With the expanding scope of the markets and the available supply of brick, these manufacturers were less free to do as they wanted because they had to accommodate their patrons' demand for brick (remembering the argument in 6.4.2 concerning using building materials and knowledge about resources and so on as 'bargaining chips' in the games of patronage). While producing over and above most others, and being able to ship it to the city, exactly how much they produce, what they produce, and where they send it to has been decided elsewhere.

What this ranking is suggesting is that the smaller, less significant producers had more freedom to act in this period than the tenants in the major estates or of the *Domitii*. Asinia Quadratilla and Flavius Aper for instance ranked in terms of their social power in the 8th octile while their *officinator* C. Nunnidius Fortunatus ranked in the 4th octile in the manufacturing network. The obvious corollary is that the less powerful one's patron, the more 'free' the dependent. In the Severan period, the pattern marks a return to a pattern similar to 1st century manufacturing, though the overall mean centrality score (and hence freedom to act) has plummeted. All the statistics which I have generated in this chapter point to a drastic change at the end of the 2nd century, and the rise of the Severans. The final part of this chapter deals with the social and economic ramifications of this changeover.

6. 5 When Somebody Dies: Network Robustness, Collapse, and Transformations

The patronage networks discernable in the brick industry are related to the exploitation of the land. Land was fundamental to political and social status. These patronage networks then also reflect the wider political development in Rome. With that being the case there is one further implication of the presence and absence of the different small-world types. The implication has to do with the ability of a small-world to withstand stress. What happens when a small-world network loses individuals? What can be inferred from the cessation of brick stamping after the reign of Caracalla? It is dangerous to argue from silence, but I will examine the trajectory leading to that point, and imagine what might have happened next.

6.5.1 Achilles' Heel: Strengths and Weaknesses in a Small-World

Researchers have found that randomly connected networks or even ordered, lattice-work networks are easily susceptible to 'attack'. That is, after a certain number of nodes are removed at random the network begins to fragment, disintegrating quickly (Barabási 2002: 115). Small-worlds on the other hand are special when it comes to robustness. In small-worlds, random attacks have a far lesser effect. Because most nodes or individuals in a small-world have only a few connections, the chances of a random attack removing a critical individual are rather remote (114).

In the brick industry, the loss of a node would only occur when, for whatever reason, the individual removed him or herself entirely; no more manufacturing; no more discussing likely parcels of land at which bricks might be made. A more likely, and effective removal would occur when an *officinator* or a *dominus* died. In the brick industry, the more or less random deaths of various individuals would be like a random attack on the network. Yet even deaths would not disrupt the network too much because of the Roman concern for taking care of inheritance and the continuance of family lines.

However, the very property of a small-world which gives it strength is also its greatest weakness. The pattern of connections in a small-world makes it robust to the various depredations which may randomly affect it, because on average, none of these will affect the 'hubs'. However, if a concerted attack is made against a few of the hubs, then the entire network can fragment. This is the Achilles' Heel of small-worlds (117). If the right hub is selected, or the right number of hubs, a chain reaction can be set off, called in networks a 'cascading failure'. One of the main properties of small-worlds is that there are multiple pathways to and from any given set of nodes. When a number of these pathways go through a particular hub, the removal of that hub means that alternate paths have to be found. The 'load' for that hub is defused into the network, but this means that other hubs, other pathways have to take up the load. If the new load outstrips the ability of a node to carry it, that node can fail, diffusing the load again. How far the cascade, the avalanche, reaches depends on the positioning and capacity of the first nodes to fail (119-120; cf. also Albert, Jeong and Barabási 2000).

In human terms, the obligations and responsibilities of patronage, if say a few key *domini* were murdered, would be taken up by other members of those families. But if the new individuals were not as well connected as their predecessors, their ability to manage their families' fortunes, of generating new links, of taking advantage of new opportunities, would correspondingly be lessened. However the demands made of them would not and they would fail, starting the sequence over again. This property of complex systems, cascading failure, might be behind some of the more spectacular 'failures' in the wider network of Roman society in general: the Emperors Caligula, Nero, and Commodus, men who were not able to handle the excess load created by the deaths of their predecessors. Let us look at Commodus in particular.

6.5.2 The Death of Commodus

After the death of Commodus, civil war engulfed the empire. Septimius Severus won these internecine wars by defeating the Governor of Britain, Clodius Albinus. Herodian writes that Severus:

> ...turned his full anger on Albinus' friends in Rome. The head of Albinus was sent to Rome with orders that it should be publicly displayed on a pole... He then bitterly attacked the friends of Albinus, by producing some of their secret letters, others he charged with sending gifts to Albinus. Each one was indicted for a crime; all who were prominent at that time in the Senate, or who were richer and more noble in the provinces were destroyed ruthlessly. He pretended that he was furious with his enemies, but in fact his dominant motive was to gain their wealth. There never was an emperor so obsessed with money.
> BkIII.8

When Septimius Severus came to power he eliminated many amongst the ruling élite, which was not an unknown way for a new Emperor to behave. For Herodian, his motivation was greed, but as an alternative I might argue that his motivation resulted from structural issues. In the plot of relationships in the preceding Antonine period (cf. Figure 6.6), the major hubs are in a single tightly bound core of people. Septimius Severus has only one tenuous link with this ruling core, a distant relation in the person of Flavius Aper (Setälä 1977: 144). As an outsider, elected by his troops, the establishment was already arrayed against him; he did not have access to the power brokers. This may also have been a motivation for Commodus' proscriptions as well. As the biological heir of Marcus Aurelius, rather than becoming heir through adoption into the ruling circle, he too was a sort of outsider. In the face of nearly one hundred years of tradition he became Emperor through blood ties. He may have perceived his power base as being weak, built on his dependants rather than the rarefied circles that Marcus Aurelius had moved in. Commodus' reaction to the 'excess load', the burden of being Emperor, and his own perceived weakness, set off a cascading failure. The easiest way to gain access to power (first for the short-lived Commodus and then the more successful Severus) was to eliminate the power brokers and put his own people in place. This created structural holes in the core of the network. People participate in other networks besides this patronage network focused on land, so for every person eliminated, holes appeared in multiple networks. As each individual was eliminated, their heirs would take on the social and economic obligations of the recently deceased. Not only would they be in danger of not being able to cope with the strain, but the shifting of the load to these new individuals would make them correspondingly more visible in the social fabric, especially to those doing the proscribing. By force of arms Commodus began and Severus finished the re-arrangement of the power structure of Rome.

What Severus did was not so much different from earlier dynastic changeovers. Only one connection spans the Julio-Claudian to Flavian succession (Domitius Afer - Domitius Lucanus and Tullus), and likewise only one spans the Antonine to Severan change (Flavius Aper - Septimius Severus). The difference between the earlier and later periods is that in the 'aristocratic' small-world of the 2nd century, the targeted removal of hubs (the power-brokers) could initiate cascading failures of greater magnitude than in the earlier 'egalitarian' networks. The earlier period was characterised by a patronage network that structurally was an 'egalitarian' small-world, and so it was better able to absorb the shocks of proscription. In the Julio-Claudian period, 12% of actors can be removed from the network before more or less complete (95%) disintegration occurs. In the Flavian period, 38% of actors could be removed before the same degree of disintegration occurs. The network could sustain damage because the wiring was more random. In later periods, a small number of hubs emerged, centred on only a few families. Consequently the network was no longer able to absorb the shocks in the same way. In the Nerva to Commodus period, the removal of only 8% of actors results in 95% disintegration. In such a situation, the cascade set off by traditional behaviours (such as eliminating one's enemies upon accession to the purple) could have disastrous effects. The structure of the network, and therefore the system, had changed. (In the Severan period, 24% of actors could be removed before 95% disintegration of the network occurred).

Given this scenario, I would like to advance a tentative hypothesis concerning the end of the industry in the form which relied on epigraphic stamps. The cessation of brick stamping, using epigraphic stamps at the turn of the 3rd century may be related to this collapse of the centralised control. It is important to remember that however advantageous one's position in the network, one still has to act. After Caracalla, while the network was still 'aristocratic' there evidently was not anyone willing or strong enough to leverage the network or fill the holes. When the strength to maintain this swiss-cheese order was removed, failures quickly cascaded, collapsing it into randomness. That is of course an argument from silence. It may be instead that the vertical power relations evidenced under the Severans continued, but instead of dependent tenants (who had at least some degree of legal freedom from their patron) there was a much less free agricultural workforce.

Stamping would no longer be necessary if the agricultural workforce had little choice in what or when or how to produce or ship. It is no coincidence perhaps that anepigraphic stamps continued to be used in the 3rd century well after the discontinuation of epigraphic stamps. Anepigraphic stamps, it will be remembered, are likely tied to internal divisions in the kiln and therefore have no message external to the productive unit.

Construction and the manufacture of bricks certainly did not *stop* during the 3rd century. For instance, Alexander Severus seems to have had a building programme in place (Ramsay 1935). Aurelian built the Thermae Aurelianae and other structures, probably in conjunction with his triumph over Palmyra (Palombi 1999: 48-9). The short-lived Emperor Trajan Decius built the largest structure on the Aventine, a bathing complex, between 249 and 251 (La Folette 1999: 51-53). Finally of course there are the Aurelianic walls, built towards the end of the century, but of their total length of nearly 19 km, only a tenth of that incorporated existing buildings or employed re-used building material, which must have left a significant part of their mass of brick faced concrete (on average, roughly 8 m high and 10 m wide) built from new bricks (Sartorio 1996: 290-299) Whatever may have actually happened regarding the brick industry, a clue exists in the structural differences of power relationships between the 2nd and early 3rd century.

6.6 Chapter Summary

Complexity Theory studies the way large scale phenomena (from traffic-jams to culture) emerge from the interactions of independent actors. Emergence happens when the pattern of interactions and feedback has the shape of a small-world network. We can find traces of these individual interactions, whether recorded epigraphically in a brick stamp, or geochemically in the fabric of the brick itself, and develop a picture of them. The findings in Chapter 3 led to the suggestion in Chapter 4 that the brick industry could be understood in terms of tenancy; tenancy operated firmly in the context of patronage. For this reason, and also because I am concerned with the ownership of land by the élite of Rome, the brick industry can serve therefore as an indicator for wider patterns in high (and low) Roman society.

In this chapter I considered patronage as a system. The connections found epigraphically between different individuals formed the skeleton, the network, for this system, mediating individuals' ability to act. The same could be done at the level of manufacturing itself, drawing the geographical and geochemical connections between the tested bricks. Plotting these networks over space and time, exploring their own internal dynamics, the dynamics between preceding and following periods, and the dynamic interplay between these two levels, allowed an exploration of the historical trajectories and dynamics of the exploitation of the hinterland of Rome. While the archaeometry of the tested bricks indicated a certain continuity of exploitation of the same sources in the industry in all periods, in the social networks of patronage and manufacturing I found that there were discontinuities, breakages, and abrupt transformations.

I first explored the network shape during the ten or so years of Rutilius Lupus' career, to explain how his idea of using consular dates in stamps could spread into the brick industry at large. I drew on the findings of epidemiology, substituting 'idea' for 'disease'. For a disease, anything which increases the chances of it infecting more than one other person pushes the progress of the disease as a whole closer to 'the tipping point'. In certain network shapes, especially ones characterised by the existence of hubs and connectors, epidemiological models suggest that the disease will always be tipped; the density of connections always leads to more than one other infection. In the case of the idea of consular dating, the presence of hubs and connectors in Rutilius Lupus' immediate network guaranteed that the idea would spread (the case of Rutilius Lupus is discussed in Graham, forthcoming).

The way diseases and ideas spread in a network, the tipping point, is a property of the network itself, not the individuals who make up the network. When all of the individuals in all periods for whom connections can be plotted were drawn into the network, it was found that the network thinned out considerably during times of dynastic changeovers. Redrawing the one network into four, it was then possible to consider each picture as a 'snapshot' of a dynamically evolving network, the brick industry. The resulting snapshots, the Julio-Claudian patronage network, the Flavian, the Nerva to Commodus network, and the Severan were interrogated from two distinct view points. Using the tools of social networks analysis, I considered the positioning of individuals in these networks; this was in a sense the static approach. The dynamic approach considered the emergent properties of each network as a whole.

The pattern of connections for each network demonstrated power-law distributions, indicating that their structure followed a 'small-world' pattern. A small-world is one in which the majority of individuals have only local connections, but the existence of a few long-distance connections enables communication across the entire network in only a few short steps, typically less than six. The networks were found to be of two small-world types, the 'egalitarian' and the 'aristocratic'. These two types are related to each other on evolutionary grounds. 'Aristocratic' networks tend to emerge first, as growth and preferential attachment give a sort of 'first-mover' advantage to the earliest established individuals, eventually turning them into the hubs to which everybody else is connected. Limitations to growth prevent the early hubs from enjoying this advantage in perpetuity, allowing later individuals to become rich in connections as well, creating the 'egalitarian' type of small world. In the brick industry, preferential attachment may have manifested itself in *officinatores* seeking to lease the lands of the *domini* who early on recognised the advantages of their estates for brick production. Limitations to growth might have been a lack of capital or a lack of knowledge concerning the extant of the clay resources of the Tiber Valley. Low demand for brick in the early years would also have limited growth. However, after the Great Fire and the need to rebuild the city, demand for brick would have increased significantly, yet the 'egalitarian' shape of the manufacturing network in the Flavian period suggests that all available capacity was used up.

By the beginning of the 2nd century, both the patronage and manufacturing networks display an 'aristocratic' pattern, which suggests that earlier limitations had been surmounted. The shift in the source of social power in this period away from having many dependants to being connected to many well-connected others could have resulted or even have caused the removal of the limitations to growth. For landowners who previously had not known that their land contained suitable clays for bricks, the discovery of such clays and the desire to profit from the discovery might have led to new lateral ties with families who already knew how to exploit clay effectively. The less-well established families, needing 'bargaining chips' to play the games of patronage, or to maintain their own standing, could not afford to leave their lands under-productive. The lateral ties they established shifted the sources of power, revealed the existence of new clay deposits, and provided fresh capital to exploit the opportunity. As the century progressed, wealth may have tended to condense in the hands of the oldest-established families (principally the *Domitii*) because their knowledge of the industry and their ties with just about every other producer led to increased investment opportunities e.g. new lands to develop by establishing kilns or clay pits, knowledge of plans for buildings to be built, and so on

Drinkwater argues that the reason the Emperors were blind to the advantages of trade was because they did not tax it creatively (Drinkwater 2001: 306). In economic simulations played on networks, trade and investment tend to work against each other in the distribution of wealth. Trade redistributes it (and taxation can be considered a form of indirect trade) while inequalities in investment incomes tend to concentrate it. If the balance between the two 'tips' towards returns generated by investments, all of an economy's wealth can condense in the hands of a very small fraction of the population. In the 2nd century, wealth may have been beginning to condense in the hands of the *Domitii*, and perhaps this had a significant role in the adoption of Marcus Aurelius by Antoninus Pius as heir (not to mention family connections and natural capacity). Marcus Aurelius, brick baron extraordinaire and 'investment banker'? Taxation is not the key to creating wealth, it is investment. The structure of the network in the 2nd century, a structure which favoured increasing returns on investments for the hubs suggests that the élite were well aware of the advantages to be had in controlling this particular industry.

Networks are very robust regarding random failures. However if and when a failure should happen to a node, the redistribution of the load can initiate a series of cascading failures. In human terms, this means that a normal death would not overly affect the continuation of the network, but a series of targeted removals, perhaps proscriptions, could cause the entire network to fall apart. A cascading failure might explain the dynastic changeover from the Antonines to the Severans. It may be that Marcus Aurelius, in selecting his own biological son to be heir rather than selecting from the circles of power put Commodus in an unenviable position, setting him up for disaster. Unable to cope with the obligations and responsibilities left to him by his father, his 'failure' set off a cascade, leading to civil war and the rise of the Severans. In the proscriptions of the period, hubs were targeted, setting off new cascades, utterly changing the brick industry in the process.

For those working the land under the Severans, while having a nominal freedom to act, the actual freedom as measured in the manufacturing network of the time was rather minimal. Brick stamping, in the form known for two centuries (which can be connected to various choices made by the manufacturer and the landowner) ceased early in the 3rd century. The trajectory leading up to that point suggests that perhaps, with such minimal freedom to act, it may simply have been a case that epigraphic stamps no longer served any real purpose. Hence they were discontinued. The ownership of the estates which produced brick had condensed into the hands of the Emperor alone.

Brick and tile represent power ossified. Careful archaeometric study of the fabric of the bricks themselves, and the elucidation of the structures surrounding their production and consumption, enables us to study how this power played out across the Tiber valley. The final section of this study presents my conclusions.

Chapter 7: Conclusions

7.1 Introduction

This study has been an investigation into the organisation of the brick industry, aimed at answering three principal questions: what is the relationship like between Rome and its hinterland in the Tiber valley, as evidenced by the brick industry?; Why is it like this?; What does this mean for our understanding of Roman society and history?

To answer these questions the study examined the way the Tiber valley, as the immediate hinterland of Rome, functioned in terms of its economic and social geography. It used the brick industry as an indicator of the social and physical networks which stretched out from the city, intersected with and overlaid the Valley. Through an archaeometric approach to the bricks themselves, coupled with a social networks analysis approach to the patterning of social and physical connections represented by the bricks and their associated stamps, the study arrived at an understanding of the social and economic relationships which characterised the city-hinterland relationship. The patterns of land exploitation were studied by pinning down likely areas of production for bricks carrying the stamps of various *figlinae* and *praedia* (and indeed for unstamped bricks as well). The different patterns of production discerned archaeometrically suggested particular methods of land-tenure, which in turn allowed the exploration of the sources of social power. How these sources of power changed over time seems to agree with larger historical patterns; the discovery that the second century networks were transformed into the 'aristocratic' small-world type points to the conscious manipulation of social and physical networks in the manner of Drinkwater's 'Edward III Woolmonger Extraordinary' (2001: 306, cf 1.2): 'Marcus Aurelius, Network Supremo'. The obvious correlate of this is that at one level (the most visible) the Tiber Valley was exploited. But at a lower level, the Valley was inherently productive. The dynamics between these two levels, that creative tension, is what truly characterises the relationship between the city of Rome and its hinterland.

7.2 The Main Findings and Arguments

In 1.3, I asked a series of questions, the answers to which if they could be found, would meet the objectives. I summarise the answers as follows.

7.2.1 Questions of Production

There were certain fundamental questions which had to be answered from the outset, in order to provide the foundations for this study. I began by asking where the exploited sources of clay were, and what these patterns of exploitation suggested for land-tenure in the area. A consideration of what other archaeometric studies of ceramics in the Tiber valley have discovered suggested the likely patterns of clay exploitation that would be found in the brick industry. By including modern, sourced samples in our archaeometric study of the SES collection, clusters of ancient brick were able to be sourced to particular areas of the Tiber valley. It was not necessary, given my methodology, to be able to point to *the* exact spot of production, but simply to be able to indentify clusters of reasonably similar composed bricks, and the general area from which they came. The cluster mapping visualization of the interrelationships represented the samples in a schematic, two-dimensional plotting based on differences in the trace element chemistry and the mineralogy in each individual sample. The inclusion of the modern samples formed the 'pegs' which pinned this cluster map to real-world geography. Some samples were probably from the area just north of Rome, towards Veii; others were most similar to the Fiano Romano/Lucus Feroniae modern sample; and others corresponded best with the sample from Orte. In the Sabina, the majority were most similar to the Monterotondo scalo sample, although a few likely came from further up the valley, along the Aia tributary. A number were also most similar to the example from Narni scalo, in Umbria. (3.2, 3.6, 4.2.1).

The word *figlina* has been understood in the past to mean 'clay-district' or 'brickyard' but as archaeometry indicates, several geographically dispersed sources could be united under the banner of a single *figlina* (cf. 4.2.2). Alternatively, one source could be exploited in the name of several *figlinae*. In traditional Latin usage, *figlina* simply means 'potter's workshop' and *officinator*, 'one who keeps a workshop'. The *figlinae Domitianae* then are merely the workshops on the Domitianae estate, and there is nothing inherent in this formulation which would indicate that *all* of the workshops had to be in the exact same place. The pattern of land tenure has implications for how that land is exploited. There are a few hints in brick stamps. The word *conductor* used in place of *officinator* (which is found in some stamps) gives the lie: the person has 'taken away' the workshop, which was 'placed' for use by the *dominus*, or *locator* (cf. 4.2.2). This is the exact opposite understanding of brick stamps to Steinby's interpretation (cf. 2.2.2), but it better explains both the patterning of land holdings (cf 4.2), the appearance of consular dates in stamps (which are tied to the need to pay the *merces* or land rent, cf. 5.3.4 and 4.2.2), and the usage of *signa*. (cf. 5.3.1, 5.3.2)

By establishing the relationships in brick assemblages at and between sites in terms of whether the bricks shared the same clays, or had the same stamps, I was able to discern three modes of production. In the first mode, the bricks at a site have a common origin and travelled through the same distribution channels; in the second, the brick maker had multiple kiln-sites throughout the region; in the third brick makers shared clay resources, or at least exploited the same general clay body. Finally, the methodology allowed the observation that in certain cases the consumer could choose from a wide variety of suppliers. Over time, mode 1 was never very important, and modes 2 and 3 seemed to mirror each other over the first three centuries AD. Mode 3 implied the existence of rural *societates*, while Mode 2 suggested relatively wealthy individuals; the trends in the different modes over time suggest that a watershed was reached in the mid 2nd century after which the ownership of land began to consolidate. While that observation was always suspected based on information in the stamps themselves, here we observed it in the archaeometric patterning of brick assemblages. (3.2.5, 4.3)

One cannot put a price tag on a batch of bricks, but there are elements in the archaeometry of brick and the modes of production which suggest that brick had real value, and that attempts were made to lower the costs of production thus increasing the relative value. There is the probable parallel siting of a brick kiln with a lead-working establishment (as suggested by the lead content in the bricks of Ostorius Scapula), which would have lowered production costs for both products by being able to share labour (the two are seasonal and complementary) and distribution costs. Mode 2 may have lowered costs by allowing the producer to achieve economies of scale and to shift production between the various kilns. Mode 3 may have lowered costs by allowing many brick makers to band together and defray the costs of fuel for the kiln, transportation, and perhaps also to fill large orders more successfully, generating further commissions, and so on. So long as *officinatores* could get access to the Tiber, the cost of shipping the bricks was more or less uniform no matter where in the Tiber valley the *officinatores* were based; however, if they could not get that access, transportation costs could be considerable. How far the bricks could be transported overland would depend therefore on keeping other costs low, and the price consumers would have been willing to pay. The mineralogy of Tiber valley bricks suggested to a modern brick maker interviewed for this study that the volcanic content might have prevented the bricks from collapsing under their own weight as they were stacked for drying, thus allowing larger tiles to be made than otherwise might have been the case. This property of Tiber valley brick might have made them inherently valuable. Finally, the evidence of shipwrecks in the Mediterranean suggests that bricks were often a proper paying cargo; there was only one wreck which had bricks that were probably taken on board as ballast while there were many more where the displacement of the wreck suggested that brick had been the major cargo for that voyage. (4.3, 4.5)

7.2.2 Questions of Meaning

With the physical nature of bricks, and the patterns of land exploitation established, the perennial question of what a brick stamp represents could be addressed. There are two kinds of brick stamp, epigraphic and anepigraphic. Anepigraphic stamps (i.e., having no text in them) are probably related to the problems of arranging a kiln effectively, and keeping track of production within the productive unit. Slave-name stamps of *Salarese* and *Quintanensiae* are probably the 'missing-link' between epigraphic and anepigraphic stamps. Epigraphic stamps, which in their most developed form contain the name of the *officinator*, the *dominus*, the consular date, and *signum*, are in this interpretation initially related to the problems of distribution, and secondly the paying of the *merces*, the land-rent, (because the *officinator* is a tenant on the land of the *dominus*). The ethnographic parallel with the 19th century Ottawa valley timber industry indicated ways in which stamping could spontaneously develop in response to the difficulties of using a river as infrastructure. It also indicated that it only in later phases do stamps become quasi-accounting devices. (2.3, 4.4)

In their most developed form, in this interpretation an epigraphic stamp serves as confirmation that the land-rent has been paid for a given year. This meaning of stamps was probably a secondary development, while the basic purpose of *epigraphic* stamps is likely connected to distribution. I argued that brick had inherent value, and so it would have been worth the while to invest in, to make, and to sell brick. Transportation costs from the different clay bodies were largely uniform so long as access to the river could be guaranteed; profit margins were therefore more dependent on the mode of production than on how the bricks were distributed. The pattern of interconnections between sites which had access to stamped brick suggests that there was a dedicated system for distribution (5.2.2); access to this system varied. Some were more successful than others in gaining access and were able to market their products to the metropolis. At the building sites in the countryside, some builders were able to build with the same materials as were used in Rome. In statistical tests of associations between the elements in a stamp and where that stamp was found (whether it tended to be distributed to 'rural' sites or 'urban' sites, as determined by the comparative index in 5.2.1), I found that bricks destined for 'urban' sites tended to carry stamps that were orbicular in shape, having either a *signum* or a consular date, while 'rural' bricks tended to carry stamps that were rectangular in shape, and had neither a *signum* nor a consular date. This led me to conclude that, once an *officinator*'s land rent had been paid (these bricks presumably going to Rome or the *dominus*' other properties in the Valley), an *officinator* could sell the rest of his product wholesale in Rome (these stamps would not have consular dates, but would be orbicular with a *signum* if going to the specialised docks) or on the local market (these would be the rectangular ones). Orbicular stamps without a *signa* may have been shipped along the river to another part of the *dominus*' estate. (5.3)

If this interpretation regarding the meaning and purpose of stamps is correct, then the relationship between stamped and unstamped brick ought be one where the stamped brick is the most visible in any batch or stack of bricks; for every stamped brick there would be a straightforward ratio to unstamped ones. To determine what that ratio actually is would have required far more effort than it would have been worth in the context of this study; a study which dealt exclusively with the problems of unstamped brick might find it a profitable question to explore (cf 7.3.3). However, in this study I was able to relate unstamped and stamped bricks together on the basis of their chemistry and mineralogy. By incorporating unstamped material from Falerii Novi and Forum Novum into the cluster map, insight into the patterns of supply for these two centres was possible. As far as it goes, Falerii Novi seems to have been supplied in a similar fashion to the major villa site at the Mola di Monte Gelato, from the area between Veii and Rome. Forum Novum on the other hand seemed to have been supplied principally from the other side of the Valley, near Fiano Romano. Stamped brick associated with this source carried the name of the *Tonneianae*; there might be the faintest suggestion that the Tonneii, after whom the *figlinae* are presumably named, may have had something to do with the construction of the Forum Novum villa. I base this argument on apparent co-incidences in the building history of the villa and the ownership history of the *figlinae*. Finally, it seems to be possible to date unstamped brick and tile on the basis of their chemistry and mineralogy. In the dated material, a slow progression to other clay bodies or faces within a general area can be identified, and it is probably this fact which allows unstamped bricks to be dated: in each period, slightly different clays were exploited, and the cluster mapping coupled with a discriminant analysis can pick up these slight differences. (5.3, 5.4)

7.2.3 Questions about the Place of this Industry in Society

Having developed an interpretation of what brick stamps represent, and how they were used, I was able to start exploring how the distribution of bricks was effected. Rihll and Wilson's (1991) gravity settlement model was adapted for this study to study the interconnections between sites which used stamped brick in the Tiber valley. The model suggested an interconnected pattern of hubs and spokes, where it took only four steps to get from one end of this network to another. The inter-connections were based on geographical proximity and the likelihood of interaction between any two sites. Given past successes of the model to suggest historically plausible sequences in the development of villages into the later Greek Classical city-states (for which the model was originally developed), in the context of this study the model is probably a better depiction of site interactions in the Tiber valley than the usually deployed Thiessen Polygons. I imagined that the four links in the network might correspond with a process involving movment from site to warehouse to transhipment centre to warehouse to site.

In the consideration of how the Tiber worked as infrastructure (using the ethnographic parallel of the Ottawa valley timber industry in the 19th century), I suggested that *signa* in stamps might be related to making sure each batch of bricks ended up on the proper dock. *Signa* are usually seen as a sort of heraldic device related to the *officinator*. Yet I found that *signa* were associated with 'urban' bricks, and that often a particular *signum* might be used by otherwise unrelated individuals. Instead of being heraldic devices, *signa* indicate the destination of that batch of bricks. I suggested that there may have been docks in the river ports and at Rome which specialised in the handling of brick cargoes (much as in a later period did the docks in the Port of London specialise in particular cargoes). In the argument, these specialized docks, dealing in the same commodity over a period of years, might have been indentified by the use of a particular *signum*. Anepigraphic stamps are not connected with the problems of distribution, but rather relate to internal divisions within the kiln. However, anepigraphic stamps are by and large on *bessales*, the small 8 inch square variety of brick. Arranged 3 x 3, nine *bessales* cover the same area as a *bipedalis*, so perhaps they could have been distributed in a 'sandwich' arrangement, layered between *bipedales*.

I also inquired into the role of the *dominus* in the industry. In this interpretation, the *dominus* is concerned to exploit his land effectively. By letting out parcels of land for brick production, the *dominus* gains at several levels. In using tenancy, the *dominus* obtains a regular return on the land, with very little input on his part. Once the *merces* was paid, which in my opinion was more often in kind than in money, the landlord would have had a number of bricks which he could then sell on, essentially at 100% profit. I suggested that the control of river infrastructure was probably in the hands of the various *domini* on whose lands these ports and so on were located. This would probably have been another way for the landlord to profit. Because bricks had inherent and added-value (as I have argued) the *dominus* probably did take an active role in promoting whatever aspect of the trade to which he was connected (whether as owner of the clay deposits, or the ports, or the warehouses, and so on). (4.5, 5.2)

Because the *dominus* does seem to have had a role in the industry, it was natural to wonder if any one *dominus* had an undue influence: was the industry under the overall control of any one individual or family, or indeed the government? The evidence of the year 123 has led in the past to arguments regarding government intervention in the brick industry, which presupposes that the industry was subject to outside control of some sort. In this study, I have shown how the internal dynamics of the brick industry, the connections *domini* and *officinatores* had to each other, could be enough to self-generate the patterns we observe. In other words, there was no outside control of the industry; control was internally generated. The rate of stamping, of stamp types to number of examples of each type, indicates that the year 123 was no different, in terms of stamping practice, than any other year. I suggested instead that the evidence of 123 pointed to a land rush. In roughly five-year periods, there were 'echoes' of this boom in the rate of stamping, suggesting that tenancy agreements (normally in antiquity of five years) entered into in 123 had expired and were renewed, but each time with fewer and fewer people involved. It is a pattern similar to the Ottawa valley timber industry: after an initial rush, the industry experience periodic shake-outs as the less able manufacturers withdrew, or the poorer resources were no longer economical to exploit. There are of course instances where certain *officinatores* seem to have stamped the consular date in consecutive years. Such stamped bricks were not available for testing in this study, but it may be that if such bricks were tested, I would predict that the *officinator* was exploiting multiple sources, in which case the interpretation would be that he exploited each source under a different contract signed in a different year. (2.3.2, 5.3.4, 6.4)

7.2.4 The Wider Implications

All of these findings, when considered in total, have implications for our understanding of the human landscape in the Tiber valley, and by extension, the relationship between the city and its hinterland. I argued in chapter 2 that understanding the hinterland and its relationship to Rome had to be predicated on an understanding of individual people, and of the networks they built around themselves and which mediated their actions.

I argue that the economy of the Tiber Valley was a productive economy, because the economic exchanges which characterise the brick industry are *reciprocal* in Erdkamp's sense (2001; cf 2.4.1): production above that level necessary to pay the land rent in order to earn money. This is however not exactly what Erdkamp had in mind. Because ultimately the economy depended on the production of food surpluses, which largely were redistributed through élite households and the bonds of patronage, the economy was a consumer economy. Erdkamp writes that even if the aggregate demand of peasants and other rural folk was sufficiently large, it was dispersed over a wide area, thus diluting overall demand in the countryside (2001: 350). Non-food productive enterprises therefore remained focussed primarily on the city. Food surpluses enabled non-food production to take place in the countryside, but this production was in practice little different from production actually sited within the boundaries of a city or town. According to Erdkamp, we would have to demonstrate that non-food production 'profited significantly from dealings with the rural population' (2001: 344) (that is, there was a reciprocal relationship) in order for a productive economy to have existed.

However, one crucial aspect that Erdkamp neglected is the differential effect of scale. The size of the market in a village as compared to Rome means that the impact of a certain amount of trade in the village could have a far greater social and economic effect than that same amount in the city. In the British countryside today, the people in many villages fear that the closing of their post office (and the trade which goes on there) might mean the death of their community. The closing of a single post office in London would on the other hand scarcely be noticed. It is not absolute size which matters, rather it is *magnitude*.

With regard to brick and tile production in the Tiber valley, while the local rural producer may only have been a 'drop in the bucket' at Rome, his position in the local market could have been profound. In the Tiber Valley, the use of tenancy as the form for the exploitation of the land for brick would have stimulated over-production, in my interpretation. The tenant not only had to produce enough brick to meet the land rent, he also had to produce enough to sell to meet his own basic needs. The paucity of sites in the Tiber Valley where stamped brick was found (cf 5.3.3), compared to the overwhelming majority of sites with unstamped brick and tile, combined with my arguments concerning the nature and purpose of stamps (cf 4.4), suggests that much of their produce was locally consumed while their land rent went to build Rome. That many of these *officinatores* are known to have produced brick for several years suggests that their strategies were by and large successful. They were able to pay their land rent and yet still generate enough income to meet their own needs. One could argue therefore that they did indeed profit significantly in their dealings with their fellow rural dwellers, whether paid in money or in kind: the Tiber valley economy was a *productive* economy.

This is not at odds with asserting that the city was primarily a consumer city, or that the Roman economy as a whole was a consumer economy. This is because we are dealing with different levels of complexity. I demonstrated in Chapter 6 how this could work, where different interrelated levels could be characterised independently, yet together create another dynamic entirely (cf. 6.2). In fact, it may well be that there are, in the multitude of 'micro-regions' (to borrow a concept from Horden and Purcell, 2000) which compose the Empire and beyond, many productive economies (cf. the papers in Mattingly and Salmon 2001); however, what happens at one level of complexity is no indicator of what may happen at another.

Patronage-as-a-system is a *complex* system, open to other systems (other urban hinterlands), nested within larger systems (the socio-economy of Rome), encompassing smaller systems (families, farms, *vici*, *municipia*). Settlements of all types emerge from the overlapping and interconnected networks of relationships, friendships, families, economics, and politics (cf. 2.3.2). However, the detail of any other settlement or system is the result of contingent experiences, the specific historical trajectories which lead to the moment being observed. Individual experiences will differ. Individual instances of patronage will happen across discrete points in space, creating different types of settlement, ie. urbanisation. Some settlements will mostly produce things, others will mostly consumer them, some will be the arena for aristocratic display, others will merely ship things onwards, and every one will have characteristics of most every other.

That an overall consumer economy should emerge from myriad productive economies is truly remarkable. Why this should be the case can be related to the patronage system, to Romanisation as understood by Millett (1990: 7-8), or urbanisation as conceived by Laurence (2001b: 91), but should form a major direction of future research. Relationships between individuals are the key to understanding the different levels of social complexity in the ancient (and modern) world.

Finally, these results should demonstrate a way forward in the study of the ancient economy. The recognition that the economy can be both 'consumer' and 'productive', but at different levels of complexity, should allow us to move beyond the old arguments over the nature of the ancient economy. An approach which considers the particularity of the region/city/industry under consideration and explores the interactions of the different levels will provide a holistic picture. Complexity theory and evolving networks provide mathematically grounded tools and models to explore these interactions. Understanding the relationships in our materials is what archaeologists are very good at; situating these relationships within the framework of complexity is then a rather small leap. We provide the relationships; the network dynamics people tell us what can happen when interacting components interact in that particular pattern.

By studying the networks in evidence in the brick industry, whether of patronage as discerned from the stamps, or of manufacturing as discerned archaeometrically, I was able to examine the social structures in which individuals lived. I considered these networks to be *evolving*, which allowed me to use the frameworks currently being developed by physicists, economists, epidemiologists, and sociologists. Certain configurations of the networks produce higher level phenomena, which we as archaeologists and ancient historians can translate into the terms of Roman social and economic history. This was the programme for Chapter 6. The evolving shape of the networks I uncovered suggested for instance that proscriptions could radically transform the shape of society; they suggested why such proscriptions might take place. The fact that land was being concentrated in the hands of the Imperial House has long been known, but in this study I found evidence for how and why that could happen: the 'aristocratic' shape of 2nd century relations, and the inequalities of investment returns probably 'tipped' the distribution of wealth inordinately into the hands of an exceedingly tiny minority of people. Perhaps Herodian was right after all to ascribe greed as the motivation behind Septimius Severus' proscriptions.

I found that at the level of manufacturing, the networks continually flipped between the two small-world types, while at the level of patronage it flipped only once. Explaining the dynamics between the two levels, and how at each level the evolution progressed allowed us to write social and economic history from the archaeology. The theory of evolving networks is a quickly developing field, and no doubt in coming months new processes in the evolution of networks will be uncovered (see for instance Pujol *et al* 2005). These, when translated into the brick industry or other aspects of ancient history, will allow new avenues to be explored, new dynamics to be uncovered, offering new insights into ancient culture.

7.3 Some Further Directions

7.3.1 The Environment

In this day and age of environmental sensitivity, the discovery of evidence of lead pollution (cf. 4.5.3) preserved within the fabric of the bricks themselves suggests a very different avenue for further research. Brick could be tested with an eye towards developing a proxy map of heavy-industrial siting and environmental degradation in central Italy. Compared with other environmental indicators such as ice-cores, and pollen and phosphate analysis, it might be possible to study the general long term environmental effects of Roman rule on a very specific region.

The exploitation of brick itself must have had a profound impact on the ecology and environment in the

Tiber valley. At Minety, in north Wiltshire, evidence for filled-in clay pits has been discovered in the form of parch marks, ranging in diameter from 20 to 60 metres across (Darvill and McWhirr 1982: 142). It is remarkable that, despite the enormous number of bricks and tiles used in the Tiber valley (including Rome), the major clay pits and extraction sites have never been identified or explored. Extraction can have long term consequences, as evidenced at the Minety site: the farmer has never been able to grow a crop on top of the filled-in pits. Similarly infertile parcels of land in the Tiber valley, especially in the neighbourhoods identified in this study as the locations of the likely clay sources, might bear further investigation as a clue towards pin-pointing the *exact* extraction sites (cf. 3.6).

7.3.2 Production Sites and Technical Issues

Being able to locate production sites is a holy grail amongst those who study brick and tile in the Tiber Valley. The cluster-mapping methodology developed in this study is one possible avenue (cf 3.6). But one of the main reasons behind wanting to know the actual location of production is to understand the relationships between different producers, of knowing whether they shared the same clay, or whether a certain producer worked the same source as a named *figlina*. In these respects, cluster-mapping or multi-discriminant analysis are perfectly suited to the task, and so as long as the relative positionings are clear, it is not necessarily required to know the exact geographic location of the clay source. Of course, the addition of modern sourced samples to the analysis does permit a general geographic localisation; with the amount of archaeometry done by other researchers already, and with more samples of known provenance included (which in this case will principally be modern), there is no reason why we should not be able to (quickly) resolve this question once and for all.

The methodology of cluster-mapping could itself be further refined for other purposes. A developed form of the methodology may be able to answer questions of a more technical nature. These might be aimed at understanding the physical properties of bricks (e.g. compression strength or heat retention) and comparing them with bricks produced elsewhere in the Empire. Such a study, combined with geographic distribution studies (on land and at sea, in the form of shipwrecks), could transform our understanding of brick as a trade item.

7.3.3 Unstamped Brick

This study has primarily been concerned to sort out the various problems of stamped brick, but it did consider unstamped material to a certain degree (cf. 5.4). However it is now unstamped material which urgently needs a proper, thesis length evaluation. Unstamped brick from a variety of sites and site types in the valley could for instance be quantified and tested to work out the relative contributions of the various sources. Then the relationships between different sites could be studied after the fashion employed in chapter five to work out the inter-site dynamics. When this information is coupled with the information from stamps, a far more complex picture of the hinterland of Rome could be developed, indicating other places where 'intensification' in the various networks creates 'urban' type places. Then the interesting empirical questions would be to observe the flows of people, material, and capital through these networks, and the concomitant social and cultural changes over time and space (cf. 2.3.2).

7.3.4 Places and Spaces

The maps created in 5.2.1, considering the differential market access for different producers and highlighting the most intense connections with Rome, are maps of distinct *places*, where local life intersects with the metropolis, rather than being simply rural isolated localities (cf. Favro 1996: 13; 2.3.2). This approach to the hinterland allows the distribution of stamped brick and the information in those stamps to be used as an indicator of the influence of different families, of different individuals, at different places in the Tiber Valley. It is an approach which attempts to confront the landscape from the point of view of a person living at that time. Because these highlighted places are associated with the individuals who have the best access to Rome, they represent the intensifications in the networks which I argued lead to urban development (cf 2.3.2). Urban geographers-cum-anthropologists consider that the creation of a place, as distinct from just a space, is part of a relationship between travel and memory (Curtis 2001: 55). That is to say, 'people make places and places make people' (Borden *et al.* 2001: 5). For a rural person living in the nearby environs, these places would represent points where their local productive economy directly interacted with the global consumer economy, where Rome directly had an impact on his or her life. Further research could be aimed at discovering the nature of these urban places, if they are similar to other élite interventions in the landscape (e.g. municipal euergetism), and their interactions with the surrounding productive economy. How do these places change the way people live?

Travelling up or down the valley, or across it, a person's experience of the economic nature of the world would alternate between the different levels of complexity, as they entered and exited these different zones of intensification. This may have had ramifications for how the Romans perceived geography. Situating Forum Novum in this framework may lead to an understanding of why it failed to develop as an urban centre. It may have been too remote, not just in literal terms of distance, but also in conceptual terms as well. The more 'backward' the perception of a place, the more difficult for an urban dweller to imagine going there (an experience anybody who has ever lived in the 'bush' can relate of their fellow urban citizens). The pattern of interactions between different sites connected with the brick industry (cf. 5.2.2) suggests that the importance of road versus river travel changed over time, and that places were sometimes avoided. Overall, in the Tiber valley the main geographic division was not as is usually expected one of east and west, but rather one of north and south (cf. 5.2.2). Indeed, east and west had strong links in the productive economy (cf. 6.4.2), but within these two northern (upper valley) and southern (lower valley) groupings. Further research therefore may reconsider the results of field surveys in the Tiber Valley on the basis of 'upper' and 'lower' rather than the traditional 'South Etruria' (eastern) and 'the Sabina' (western).

7.4 Conclusion: History and Archaeology

Most agricultural products are perishable; the indestructibility of brick on the other hand has petrified patterns of agricultural exploitation and land holding over time. The importance of landed wealth for political and social power in Rome is a commonplace; the relationships which can be discerned archaeometrically and epigraphically in brick therefore mirror the political and social life of not only the élite, but also of their clients and tenants as well. Much of the patterning of élite relationships could be drawn out from the numerous prosopographical studies in existence. However this sort of work has not been attempted, certainly not from a social networks analysis point of view. This line of research, which combines complex systems with an evaluation of the social network has the potential to not only revolutionise our understanding of élite and common life, but also to enable us to connect what we find on the ground, archaeologically, into wider social and historical patterns. Ancient historians should never again be able to write phrases like:

> It [archaeology] has only occasional contributions, on quite specific topics, to make to the larger project...it has not yet been able to provide results across a broad enough spectrum from which useful generalisations could be inferred and analogies drawn.

(Horden and Purcell 2000: 48)

A Complex Systems perspective attunes us to the way large scale phenomena emerge from the interactions of independent actors when these interactions and feedback have a 'small world' pattern. Archaeology finds the traces of these interactions; it is up to archaeologists to knit these traces together. In doing so it becomes possible to study the potentials and constraints in the various networks which were created by, but also mediated, individuals' ability to act. Individuals do matter. When we plot these networks over time and space paying special attention to how the networks evolve, we can write history from archaeology.

Appendix A: Catalogue of the SES Collection of Stamped Bricks

Throughout this study I have referred to the stamped bricks in the SES collection using an abbreviated form of their catalogue numbers. The full catalogue number refers to the crate in which the bricks are kept at the BSR, hence SE 1 = se225-1 (South Etruria Crate 225, Brick Stamp #1). Identifications were made using Steinby 1987. LSO = Lateres Signati Ostienses. (This catalogue is not the definitive catalogue of the SES collection; that version will be published at a later date.)

Each entry in the catalogue is arranged in the following manner:
Catalogue Number. (as recorded on brick). TVP site reference #. **Shape** of the stamp. If it is orbicular, semicircular, or circular: **Dimension** in cm across the orbicular stamp (+- indicates the radius has been doubled to obtain the dimension); **Dimension** in cm across the orbiculus itself (+- indicates the radius has been doubled to obtain the dimension).

If the stamp shape is rectangular or *litt. cavis*: **length** in cm of rectangular stamp (+ indicates an incomplete stamp, marg. vid. indicates that the measurement has been taken from the writing itself because there is no stamp margin); **height** in cm of stamp (+ indicates an incomplete stamp, marg. vid. indicates that the measurement has been taken from the writing itself).

height in cm of the letters, starting from the outside on orbicular stamps, from the top on rectangular stamps; **number** of auxiliary lines between lines of text; **brick least thickness** (cm); **brick greatest thickness** (cm); **brick average thickness** (cm); ***CIL* XV.1; LSO; volume of PBSR** if already published in the South Etruria Survey reports; **visible text on stamp; expanded transcription; date** (reference to Steinby 1974 if the stamp is discussed there).

Bricks from which samples were taken for archaeometric study:

SE 1 SE 2 SE 4 SE 5 SE 7 SE 8 SE 10 SE 13 SE 14 SE 16 SE 18 SE 19 SE 20 SE 21 SE 22 SE 23 SE 26 SE 27 SE 28 SE 29 SE 30 SE 36 SE 37 SE 42 SE 45 SE 47 SE 48 SE 50 SE 51 SE 52 SE 53 SE 54 SE 56 SE 57 SE 58 SE 60 SE 63 SE 65 SE 72 SE 85 SE 89 SE 90 SE 100 SE 104 SE 106 SE 107 SE 108 SE 110 SE 111 SE 113 SE 114 SE 116 SE 136 SE 141 SE 144 SE 148 SE 151 SE 152 SE 153 SE 154 SE 155 SE 156 SE 160 SE 163 SE 166 SE 168 SE 170 SE 171 SE 172 SE 174 SE 176 SE 177 SER 1 SER 2 SER 3 FAL 1 FAL 2 FAL 3 FNV 4 FNV 5 FNV 6 FNV 8 FNV 9 FNV 13 FNV 14 FNV 15 MOD 1 MOD 2 MOD 3 MOD 4 MOD 5 MOD 6 MOD 7 MOD 8 MOD 9

SE 1. (SE 225-1) TVP 2672 **orb**(9+-; 3.5) **let** (1.0, 0.8) **lin** (1,1,2) **brick** (2.4, 2.7, 2.6).
***CIL* XV.1** 368; LSO 347
]PV[.]E/[...]ESN OP[.]D
[ex fic(linis) ocea(nis) mai(oribus)] caes(are) n(ostri) op[us] d[o(liare)] [Q. Perusius] Pu[d]e(ntis)
Antoninus Pius (Steinby 1974: 70)

SE 2. (SE225-2) TVP2414 **orb** (?;?) **let** (1.2) **lin** (1,2) **brick** (2.6, 3.0, 2.8)
***CIL* XV.1**61; LSO 92 *PBSR* 33 (1965)
]CIFER
[sta(tius) marcius lu]cifer
late Trajanic

SE 3. (SE225-3) TVP3171 **orb** (?,?) **let** (2.0) **lin** (1,1) **brick** (2.5, 2.5, 2.5).
***CIL* XV.1**1201 *PBSR* 40 (1972)
*I*V[
*i*u[*l*i*]
Flavian?

SE 4. (SE225-4) TVP2337 **orb** (?,?) **let** (1.2,1.0) **lin** (2,1,1) **brick** (3.0, 3.7, 3.4).
***CIL* XV.1**189 *PBSR* 33 (1965)
...]R[...]/[...]D D N[..
opus doliare ex prae]d d n[/ ex fig vete]r[es]
193-198 (Steinby 1974: 39)

SE. 5 (SE225-5) TVP 452 **orb** (?,2.1) **let** (1.2,1.0) **lin** (?,1) **brick** (2.6, 2.6, 2.6)
***CIL* XV.1**408a-d LSO 385.386.387 *PBSR* 36 (1968)
OP[...]ANTO/NI[...]IC
op[(us) dol(iare) ex pr(aedis) m(arci) aureli anto/ nini aug(usti) n(ostri) port(u) l]ic(ini)
212-217 (Steinby 1974: 77)

SE 6 (SE225-6) TVP3173 **orb** (10.5,4.5) **let** (1.3,1.0,0.7) **lin** (1,2,2) **brick** (3.2,3.8,3.5).
***CIL* XV.1**861 LSO 711 *PBSR* 40 (1972)
...]ASINIAE QV[...]O D C NV/[...]VCIO/[...]
ex fig(linae) asinae quadratilla o(pus) d(oliare) c(ai) nunidi fortunati lucio quadrato co(n)s(sul)
142 (Steinby 1974: 66)

SE 7 (SE225-7) TVP 452 **orb** (9.0, 4.1) **let** (1.3,1.1,1.1) **brick** (5.0, 5.0, 5.0)
***CIL* XV.1**1106a LSO 873 *PBSR* 36 (1968)
A[...]MITI/[...]BVLI/DOL
aprilis cn domiti / agathobuli / dol(iare)
Trajan/Hadrian (Steinby 1974: 55)

SE 8 (SE225-8) TVP2414 **orb** (?, ?) **let** (1.2) **lin** (1,2) **brick** (3.0, 3.0, 3.0).
***CIL* XV.1**61 LSO 92 *PBSR* 33 (1965) [...]VCIFER[...] [sta(tius) marcius l]ucife[r] **late Trajanic** (Steinby 1974: 32)

SE 9. (SE225-9) TVP2317 **orb** (11.3, 4.3) **let** (1.2,1.2,1.2) **lin** (1,2,1) **brick** (4.5, 4.8, 4.6)
***CIL* XV.1**359 LSO 340
DOL EX PRAED CAES N C AQUILI A[...]S / PATINO ET APRONIANO / COS
doli(are) ex praed(is) caes(ar) n(ostri) c(aius) aquili a[prili]s / paetino et aproniano / cos
123 (Steinby 1974: 70)

SE 10. (SE225-10) TVP1547 **orb** (7.8, 4.5) **let** (1.1) **lin** (1,1) **brick** (2.4, 2.4, 2.4)
***CIL* XV.1**811f LSO 684 *PBSR* 36 (1968)
DOL ANTEROT SEVER CAES
dol(iare) anterot(is) SEveri(ari) caes(aris)
ca 123

SE 11. (SE225-11) TVP1596 **orb** (?, ?) **let** (1.3) **lin** (1,?) **brick** (3.6, 4.0, 3.8)
CIL **XV.1**S.269, 992 LSO 769 *PBSR* 36 (1968)
...]STIDV[...
[calli]sti du[orum domitior]
Domitian (Steinby 1974: 50)

SE 12. (SE225-12) TVP1423 **orb** (9.3, 4.7) **let** (1.3,1.0) **lin** (1,2,2) **brick** (3.5, 4.5, 4.0)
CIL **XV.1**1029a LSO 795 *PBSR* 36 (1968)
OP D D[...]O[...]LVCIL/P[...]R COS
op(us) d(oliare) dionys domit(ia) P. f. Lucil(la) / p[(aetino) (et) (ap])r(ilis) co(n)s(ul)
123 (Steinby 1974: 52)

SE 13. (SE225-13) TVP3592 **semi-circ** (?) **let** (1.5,0.8) **let** (1,1) **brick** (2.7, 3.0, 2.8)
CIL **XV.1**659c LSO 565 *PBSR* 33 (1965)
TO[...]/[...]CCIA[..]
to[nnei de figl(i)n(is)] / [vi]ccian[ns]
mid 1st century (Steinby 1974: 96)

SE 14. (SE225-14) TVP2322 **orb** (?,?) **let** (1.3,1.3) **lin** (1,1,1) **brick** (3.8, 3.8, 3.8)
CIL **XV.1**198 LSO 219 *PBSR* 33 (1965;there unidentified)
[...] GIL / DOL S[...]
[ex figil veteris opus] / dol s[ucces]
Commodus (Steinby 1974: 39)

SE 15. (SE225-15) TVP3186 **orb** (9+-, 4.5) **let** (1.3,1.1) **lin** (?) **brick** (4.6, 4.6, 4.6)
CIL **XV.1**714 LSO 614 *PBSR* 40 (1972)
...]VALLIVS PRCLVS F / EX PRAEDIS
[faustinae l] vallius pr(o)clus f(ecit)
140-160

SE 16. (SE225-16) TVP2337 **orb** (9.5, ?) **let** (1.3,1.0) **lin** (2,2,1) **brick** (2.5, 2.5, 2.5)
CIL **XV.1**759 *PBSR* 33 (1965)
OPVS DOLIARE[...]DIS / DOMIN[...]TRI
opus doliare[m ex prae]dis / domin[i nos]tri
212-217 (Steinby 1974: 44)

SE 17. (SE225-17) TVP3582 **semi-circ** (11.7) **let** (1.5,1.0) **lin** (2,2) **brick** (2.5, 3.0, 2.8)
CIL **XV.1**659c LSO 565 *PBSR* 33 (1965)
TONNEI DE FIG[...] / VICCIANS
tonnei de figl(i)n(is) viccians
mid 1st century (Steinby 1974: 96)

SE 18. (SE225-18) TVP2357 **semi-circ** (?) **let** (1.5) **lin** (1,?) **brick** (2.6, 3.0, 2.8)
CIL **XV.1**659c LSO 565
...]ONNEI DE[...
[t]onnei de [figl(i)n(is) / viccians]
mid 1st century (Steinby 1974: 96)

SE 19. (SE225-19) TVP2414 **orb** (8.4+-, ?) **let** (1.0, 0.8) **lin** (1,1,1) **brick** (2.8, 2.8, 2.8)
Unidentified
O[PV]S DOLIAR[...] / [...]
Severan?

SE 20. (SE225-20) TVP2357 **orb** (?, ?) **let** (1.1,1.1) **lin** (1,2,?) **brick** (2.8, 3.0, 2.9)
CIL **XV.1**731b LSO 627 *PBSR* 33 (1965; there an incorrect transcription)
EX PR VM [... / ...] XAP[...]
op dol] ex pr vum[i qvad et an / faus ex fi sex ap silv
140-160

SE 21. (SE225-21) TVP1732 **orb** (8.6, 3.8) **let** (0.8,0.8) **lin** (1,2,2) **brick** (4.0, 4.5, 4.2)
CIL **XV.1**368 LSO 347 *PBSR* 45 (1977)
[...]C OCEA MAI CAES N OP DO / [...]Q[...]DE[...]
[ex fi]c(linis) ocea(nis) mai(oribus) caes(aris) n(ostri) op(us) do(liare) / Q. Perusi Pude(ntis) .
Antoninus Pius (Steinby 1974: 70)

SE 22. (SE225-22) TVP1547 **semi-circ** (?) **let** (1.0) **lin** (2,2) **brick** (2.3, 2.3, 2.3)
CIL **XV.1**947 (?) *PBSR* 36 (1968)
[...] C C O [...
[...] c c o [...
Flavian-Trajan?

SE 23. (SE225-23) TVP2322 **orb** (8, ?) **let** (1.5) **lin** (1,?) **brick** (1.8, 2.5, 2.2)
CIL **XV.1**913a *PBSR* 33 (1965)
Q CANV[...]
q. canu[si praenestini]
mid 2nd century

SE 24. (SE225-24) TVP2324 **orb** (9.7, 1.5) **let** (1.2,1.0) **lin** (?) **brick** (2.0, 2.7, 2.4)
CIL **XV.1**625 LSO 530 *PBSR* 33 (1965)
OP DOL [...]AVG N FIG TERE / [...]
op(us) dol(iare) [ex pr(aedis)] aug(usti) n(ostri) fig(linis) tere/ [nt(iae) l aelio phidele
212-217 (Steinby 1974: 94)

SE 25. (SE225-25) TVP1837 **orb** (?, ?) **let** (1.3) **lin** (1,?) **brick** (4.4, 4.4, 4.4,)
unidentified
[...]LI[...]/[...] [...]li[...]/[...]
2nd century?

SE 26. (SE225-26) TVP2337 **orb** (11+-) **let** (1.2,1.1) **lin** (1,2,1) **brick** (2.3, 2.6, 2.4)
CIL **XV.1**189 *PBSR* 33 (1965)
[...]DOLIA[... / ...] FIG VET[...]
[opus] dolia[r ex praed d n / ex] fig vete[eres]
193-198 (Steinby 1974: 39)

SE 27. (SE225-27) TVP2391 **orb** (?, ?) **let** (0.9) **lin** (1) **brick** (3.0, 3.0, 3.0)
CIL **XV.1**528 *PBSR* 33 (1965)
[...]SAL PR[...] sal pr[imig]
Trajan/Hadrian (Steinby 1974: 84)

SE 28. (SE225-28) TVP2304 **orb** (9.5, 6.6) **let** (1.1,0.8) **lin** (1,1,1) **brick** (3.0, 3.5 3.2)
CIL **XV.1**664 c LSO 570 *PBSR* 33 (1965)
ZOSIMI L IVLI RVFI / VICCIANA
Zosimi l iuli rufi / vicciana
Vespasian (Steinby 1974: 97)

SE 29. (SE225-29) TVP2413 **orb** (11.5, 2.0) **let** (1.4,1.0) **lin** (1,2,1) **brick** (4.0, 4.7 4.4)
CIL **XV.1**163 LSO 190 *PBSR* 33 (1965)
O[...]AV[...]FL DOMITI / A[...]
o[p(us) d(oliare) ex pr(aedis)] au[g(usti) n(ostri)] fl domiti / a[nas maiores]
Septimus Severus (Steinby 1974: 38)
SE 30. (SE225-30) TVP420 **orb** (8.2, 4.0) **let** (1.5) **lin** (1,1) **brick** (2.5, 3.3, 2.9)
N.1349/50.1 *PBSR* 36 (1968)
P OSTORI EROT
p ostori erot
Flavian?

SE 31. (SE225-31) TVP2822 **rect** (5.4+, 4.7) **let** (4.7) **brick** (2.5, 2.5, 2.5)
CIL **XV.1**2120 (?)
[...]E[..]

[r]e[m...]
(letter E is a series of impressed dots)
1st century BC?

SE 33. (SE225-33) TVP3262 **rect** (8.4+, 2.8+) **let** (1.0+, 1.1) **brick**(2.8, 3.0, 2.9)
CIL **XV.**12194
CRV[...] / FECIT C IVLIVS
c ru[tili felicis] / fecit c iulius [dama]
1st century

SE 34. (SE225-34) TVP2391 **rect** (8.3+, 6.3+) **brick** (1.7, 2.1, 1.9)
Not a stamp; impressed lattice-work grid

SE 35. (SE225-35) TVP932 **orb** (?,?) **let** (1.4, 1.0) **lin** (1,?,?) **brick** (3.0, 3.5, 3.2)
CIL **XV.1**745 = S.587 corr 636
[...] CAES [...] /[...]N
[ex fig] caes [n / ab coccei aug lib primige]n[i]
mid second century?

SE 36. (SE225-36) TVP2650 **rect** (2.6, 5.0) **let** (3.2) **brick** (2.0, 2.5, 2.2)
CIL **XV.1**2263 *PBSR* 36 (1968)
PL
p l
1st century

SE 37. (SE226-37) TVP1495 **circ**(6.3) **let** (1.0) **lin** (1,1) **brick** (2.6, 2.6, 2.6)
CIL **XV.1**1574 b *PBSR* 36 (1968)
OF DOM
[of(ficina) s(ummae) p(rivatae)] of(ficina) dom(itiana)
Diocletianic

SE 38. (SE226-38) South Etruria site ID: 837532 **semi-circ** (?) **let** (1.3) **lin** (?) **brick** (3.2, 3.5, 3.4)
unidentified
[...]T. QV[...]
1st century?

SE 39. (SE226-39) TVP2371 **circ** (?) **let** (1.2) **lin** (?) **brick** (3.2, 3.3, 3.2)
CIL **XV.1**1244 b LSO 950 *PBSR* 33 (1965)
[...]LICI[...]
[c]lici[ni donacis]
1st century

SE 40. (SE226-40) South Etruria Site ID:867557 **orb** (?, 4.4+-) **let** (1.3) **lin** (?) **brick** (2.6, 2.8, 2.7)
CIL **XV.1**585 a *PBSR* 36 (1968)
[...]VLP[...]
[c cul dia s]ulp[...]
110-120? (Steinby 1974: 90)

SE 41. (SE226-41) TVP1290 **circ** (7.3) **let** (1.2) **lin** (1) **brick** (2.6, 3.1, 2.8)
CIL **XV.1**1581 a pbsr 36 (1968)
OF S DOM[...]
of(ficina) s(ummae vel summarum) dom[i(tiana) s(aturnini)]
Diocletianic

SE 42. (SE226-42) TVP417 **orb**(10.8+-, 5+-) **let** (1.2, 0.9, 0.6) **lin** (1,1,1) **brick** (3.5, 4.2, 3.8)
CIL **XV.1** 861 LSO 711 pbsr 36 (1968)
[...]DCNV[.] / [...]O / [...] COS
[ex fig(linae) asiniae qvadratillae o d c nun/nidi fortunat lucio/ quadrato cos
142 (Steinby 1974: 66)

SE 43. (SE226-43) TVP2324 **orb** (?, ?) **let** (1.3) **lin** (?) **brick** (3.0, 3.3, 3.2)
CIL **XV.1** 1380 *PBSR* 33 (1965)
[...]ONTI C[...]
[c p]onti c[rescentis]

SE 44. (SE226-44) South Etruria Site ID: 638522 **semi-circ** (7.7) **let** (1.2, 0.7) **brick** (3.2, 4.1, 3.6)
Unidentifed
[...] SEPTIMI I / PRIMIGEN
[...] Septimi i / primigen
Flavian?

SE 45. (SE226-45) TVP452 **orb** (11+-, ?) **let** (1.2,?) **lin** (?) **brick** (2.6 3 2.8)
CIL **XV.1** 408 a-e LSO 385.386.387 *PBSR* 36 (1968)
[...]L[...]MAVREL[...] / [...]VG N PORT[...]
[op(us) do]l[(iare) ex pr(aedis)] m aurel[i anto / nini a]ug(usti) n(ostri) port(u) [lic(ini)]
212-217 (Steinby 1974: 77)

SE 46. (SE226-46) TVP3320 **orb** (10.5, 3.3) **let** (1.0, 1.0) **lin** (1,?,?) **brick** (3.3, 3.3, 3.3)
CIL XV.1 2173
EX P T[...]L[...]LI[...] / DOMI[...]INI
ex p(raedis) t[i. Iu]l[i iu]li[ani opus] / domi[ti ruf]ini
Trajan/Hadrian?

SE 47. (SE226-47) TVP417 **orb** (10.4, 4.4) **let** (1.4,0.9 ,0.7) **lin** (1,2,1) **brick** (3.2 4.1 3.6)
CIL **XV.1** 861 LSO 711 *PBSR* 36 (1968)
EX FIC[...]IAE O D CNVN / NIDI FOR[...]LVCIO / QVADR[...]COS
ex fic(linis) [asiniae quadratill]ae o d c nun/nidi for[tunat(i)] lucio / quadrato co(n)s(ul)
142 (Steinby 1974: 66)
143

SE 48. (SE226-48) TVP2678 **semi-circ** (?) **let** (1.6,?) **brick** (2.9, 3.2, 3.0)
CIL **XV.1** 659c LSO 565
TONNE[...]
tonne[i de figl(i)n(i)s / viccians
mid 1st century (Steinby 1974: 96)

SE 49. (SE226-49) TVP1493 **cir** (?) **let** (1.2,?) **lin** (1,?) **brick** (2.1, 2.2, 2.2)
CIL **XV.1**1581a *PBSR* 36 (1968)
[...]DOM[...]
[of(ficina) s(ummae vel summarum)] dom[i(tiana) saturnini]
Diocletianic

SE 50. (SE226-50) TVP452 **semi-circ** (10.2) **let** (1.5, 0.9) **brick** (2.7, 3.1, 2.9)
CIL **XV.1** 659a LSO 563 *PBSR* 36 (1968)
TONNEI DE FIGLIN VICCIANS
tonnei de figlini(is) / viccians
mid 1st century (Steinby 1974: 96)

SE 51. (SE226-51) TVP1495 **circ** (6.2) **let** (1.2) **lin** (1,1) **brick** (3.6, 4.1, 3.8)
CIL **XV.1** 1552a *PBSR* 36 (1968)
OFSPOFBO
of(ficina) s(ummae) p(rivatae) of(ficina) bo(coniana)
Diocletianic

SE 52. (SE226-52) TVP454 **orb** (10.2, 1.6) **let**(1.2,0.8) **lin** (1,2,2) **brick** (3.2, 3.7, 3.4)
CIL **XV.1** 625 LSO 530 *PBSR* 36 (1968)
OP[...]PR AVG N FIG TERE / NT L AELIO PHIDELE
op[(us) dol(iare) ex] pr(aedis) aug(usti) n(ostri) figl(inis) tere/nt(iae) l aelio phidele
212-217 (Steinby 1974: 94)

SE 53. (SE226-53) TVP417 **orb** (10.0+-, ?) **let** (1.3, 0.9, 0.7) **lin**(1,2,2) **brick** (3.6, 3.9, 3.8)
CIL **XV.1** 861 LSO 711 pbsr 36 (1968)
EX FI[...] / NID[...] / QV[...] ex fi[g(linae)
asinae qvadratillae o(pus) d(oliare) c nun] / nid[i fortunat(i) lucio] /

qu[adrato co(n)s(ul)
142 (Steinby 1974: 66)

SE 54. (SE226-54) TVP1547 **orb** (?, ?) **let** (1.0) **brick** (4.1, 4.3, 4.2)
***CIL* XV.1** 811f LSO 684 *PBSR* 36 (1968)
DOL AN[...]
dol(iare) an[terot(is) Sever(iari) caes(aris)
ca 123

SE 55. (SE226-55) TVP399 **orb** (8.4, 3.4) **let** (1.1,0.9) **lin** (1,?,?) **brick** (3.4, 3.6, 3.5)
N. 1415/6 *PBSR 36* (1968)
APR[...]/EX P SCAPVL[...]
apr[o(niano) et pae(tino) co(n)s(sule)] ex p(raedis) scapul[rubr p d epi]
123

SE 56. (SE226-56) TVP1547 **orb** (10.0+-, ?) **let** (1.1,1.0) **lin** (1,2,1) **brick** (2.2, 3.0, 2.6)
***CIL* XV.1** 762 b LSO 647 *PBSR* 36 (1968)
[...]ARE EX[...] / [...] I N AV[...
[opus doli]are ex [pr(a)e(dis) / domin]i n(ostri) aug(usti)
212-217 (Steinby 1974: 78)

SE 57. (SE226-57) TVP454 **orb** (10.2, ?) **let** (1.2,0.8) **lin** (?,2,?) **brick** (3.4, 4.0, 3.7)
***CIL* XV.1** 625 LSO 530 *PBSR* 36 (1968)
OP DOL EX PR AVG N FIG TERE / NT L AELIO PHIDELE
op(us) dol(iare) ex pr(aedis) aug(usti) n(ostri) fig(linae) tere/nt(iae) l aelio phidele
212-217 (Steinby 1974: 94)

SE 58. (SE226-58) TVP454 **orb** (10.2, 1.6) **let** (1.2,0.8) **lin** (?,2,2) **brick** (3.5, 4.5, 4.0)
CIL XV.1625 LSO 530 *PBSR* 36 (1968)
OP DO[...]VG N FIG TERE/NT L AELIO PHIDELE
op(us) do[l(iare) ex pr(aedis) a[ug(usti) n(ostri) fig(linae) tere/nt(iae) l aelio phidele
212-217 (Steinby 1974: 94)

SE 59. (SE226-59) TVP3111 **rect** (4.7+, 2.6) **let** (1.8) **brick** (3.5, 3.5, 3.5)
unidentified
P SVLP
p sulp
Julio-Claudian?

SE 60. (SE226-60) TVP2672 **orb** (9+-, 4) **let** (1.0,1.0) **lin** (2,2,2) **brick** (2.4, 2.6, 2.5)
***CIL* XV.1**1075 a LSO 834
OPV DOL E[...]VDI / QVIN[...] R
opu(s) dol(iare) e[pagathus claudi] / quin[qua ser(vus)]
Trajan (Steinby 1974: 53)

SE 61. (SE226-61) TVP423 **orb** (8.8+-, ?) **let** (1.2, 1.1,1.0) **lin** (1,2,2) **brick** (2.7, 3.0, 2.8)
***CIL* XV.1** 263 LSO 273 *PBSR* 36 (1968)
[...]PHIM AGA[...] / [...] EL[.]C [...] / DL
[tro]phim aga[thobuli domit]/[ia]e l[u]c[illae do(ilare)]/ d l
123 (Steinby 1974: 52)

SE 62. (SE226-62) TVP1463 **orb** (8.8+-, ?) **let** (1.0, 0.9) **lin** (1,1,2) **brick** (2.3, 3.1, 2.7)
***CIL* XV.1** 1002 LSO *PBSR* 36 (1968)
[...]HOBVLI / [...] ITI TVLL
[agat]hobuli/ [dom]iti tull
93/94 (Steinby 1974: 51)

SE 63. (SE226-63) TVP1758 **rect** (7.7+, 3.3) **let** (2.3) **brick** (2.2, 2.5, 2.4)
unidentified *PBSR* 45 (1977)
[...] TANI
aris[tani]
1st century

SE 64. (SE226-64) TVP1675 **rect** (5.4+, 3.7+) **let** (1.5,?) **brick** (2.2, 2.4, 2.3)
***CIL* XV.1** S.324 LSO 940 *PBSR* 45 (1977)
C IVNIVS / DIOGENES
c iunius diogenes
1st century

SE 65. (SE226-65) TVP1758 **rect** (5.2+, 2.4+) **let** (1.9+) **brick** (2.9, 3.3, 3.1)
unidentified *PBSR* 45 (1977)
[...]RIS[...]
a[ris]tani
1st century

SE 66. (SE100-66) TVP1155 **orb** (?, ?) **let** (1.3,?) **lin** (?) **brick** (2.9, 2.9, 2.9)
unidentified *PBSR* 45 (1977)
EX [...] / A[...]
2nd century

SE 67. (SE176-67) TVP1862 **orb**(?,?) **let** (1.2) **lin** (?,1) **brick** (2.5, 2.5, 2.5)
***CIL* XV.1** 960
[...]A C A T C [...]
[lannio largo c prast p]acat c[os ex of(ficina) c calpeta(ni)/ pannychi op(us) f(iglinum) ex pr(aedis) cosin(iae) gra(tillae?)]
147

SE 68. (SE88-68) TVP1713 **rect** (4.3+, 1.7+) **let** (1.2+) **lin** (2.4, 2.4, 2.4)
***CIL* XV.1** S.324 LSO 940
CIV[...]/[...]
c iu[nius] / [diogenes]
1st century?

SE 69. (SE225-69) TVP463 **orb** (9.0+-, 1.5) **let** (0.9,?) **lin** (1,?) **brick** (3.0, 3.0, 3.0)
***CIL* XV.1** 203 *PBSR* 36 (1968)
OP[...]NOV/ [...]
op [dol(iare) ex pr(aedis) aug(usti) n(ostri) figl(inae)] nov(ae) / [sabinia ingenua]
212-217 (Steinby 1974: 40)

SE 70. (SE227-70) findspot unknown **orb** (9.6+-, 3.9) **let** (1.2,1.2) **lin** (1,?,?) **brick** (3.6, 3.9, 3.8)
***CIL* XV.1** 1440 a or b
IMP[...]/ DPQSPDO[...]
imp [antonino ii et balbin (or brttio) cos] / d p q s d o [arabi]
139 (Steinby 1974: 68)

SE 71. (SE227-71) findspot unknown **orb** (8.6,?) **let** (1.5) **lin** (1,?) **brick** (2.5, 2.9, 2.7)
***CIL* XV.1** 862 LSO 712 *PBSR* 23 (1955)
[...]NNFOR[...]
c nunn(idi) fort prim/ p ( ) ( )
ca 142 (Steinby 1974: 66)

SE 72. (SE227-72) TVP2673 **orb** (8.2+-, 4.2+-) **let** (0.9, 0.7) **lin** (1,2,2) **brick** (4.7, 4.8, 4.8)
***CIL* XV.1** S.222 LSO 675
[..]IAN COS / [...] S MAUS
[paetin(o) et apron]ian co(n)s(ul) / [zozimu]s m a(nni) v(eri) s(ervus)
123 (Steinby 1974: 78)

SE 73. (SE227-73) findspot unknown **circ** (6.4+-) **let** (1.4) **lin** (?) **brick** (2.3, 2.7, 2.5)
unidentified
[...]TITHALAN[..]
1st century?

SE 74. (SE227-74) findspot unknown **orb** (?, ?) **let** (1.2,1.0) **lin** (?) **brick** (3.0, 3.3, 3.2)
***CIL* XV.1** 521 *PBSR* 23 (1955)

[...]PRAEDISLIVL[...] / [...] S [...]
[de] praedis l iuli [ursi valeri flac] / [salare]s[e a taurione]
post 123 (Steinby 1974: 85)

SE 75. (SE227-75) findspot unknown **orb** (10.3, 3.8) **let** (1.4, 0.9, 0.9) **lin** (?,1,1,1) **brick** (3.5, 3.7, 3.6)
***CIL* XV.1** 375 LSO 355
L B RVT[...]ALIS FEC / OPV[...]AESN / [...] BI
brut[ttidi augst]alis fec(it) / opu[s dol(iare ex fic(linis) c]aes(aris) n(ostri) / [prop et am]bi (consul)
126 (Steinby 1974: 71)

SE 76. (SE227-76) findspot unknown **semi-circ** (8.4+-) **let** (1.4) **brick** (2.7, 3.1, 2.9)
***CIL* XV.1** 666 LSO 573 *PBSR* 23 (1955)
[...] S SPVRILI[...] F R[...]
[de figvlini]s spuriliae f(lo)r(i)
Nero (Steinby 1974: 96)

SE 77. (SE227-77) findspot unknown **circ** (?) **let** (1.3,1.0) **lin** (1,2,1) **brick** (2.3, 2.7, 2.5)
N.S. 318/1205
[...]OFIC[...] / [...]YM[...]
[ex] offic [c iuli] / [h]ym[enis]

SE 78. (SE227-78) findspot unknown **semi-circ** (5.3) **let** (1.2,1.5) **brick** (2.7, 3.1, 2.9)
***CIL* XV.1** 2408 LSO 1236
PHILET [...] / C
c philet(i)
1st century?

SE 79. (SE227-79) findspot unknown **semi-circ** (8.4+-) **let** (1.3,0.9) **brick** (2.6, 2.8, 2.7)
unidentified
L SEP [...] / PRIM[...]
l sep[...] / prim[...]
2nd half of 1st century

SE 80. (SE227-80) findspot unknown **orb** (10.2+-, ?) **let** (1.2,1.1) **lin** (1,2,2) **brick** (4.5, 4.7, 4.6)
***CIL* XV.1** 408a-d LSO 385.386.387
[...]RMAVR[...] / [...]VGNPOR[.]L[...]
[op(us) dol(iare) ex p]r(aedis) m aureli anto / [nini au]g(usti) n(ostri) pror(t)(u) lic(ini)
212-217 (Steinby 1974: 75)

SE 81. (SE227-81) findspot unknown **rect** (8.0+-, 4.8+-) **let** (1.2) **brick** (5.7, 6.0, 5.8)
app. 160 *PBSR* 23 (1955)
[...] CIV[..] FELI[...]
c iul(ius) feli(cis) **Flavian?**

SE 82. (SE227-82) findspot unknown **orb** (10.2+-, 4.6+-) **let** (1.0, 1.0, 1.2) **lin** (1,2,2) **brick** (3.7, 3.9, 3.8)
***CIL* XV.1** 1228a LSO 943 [...]ETRTRAN / [...] M BIBO / COS
[pila her iun(ius) sulp(icianus) c(aius) p]etr(onius) tran( ) / [vero iii et a]mbibo / co(n)s(ul)
126

SE 83. (SE227-83) findspot unknown **orb** (9.9, 2.2) **let** (0.9) **lin** (1,2) **brick** (2.8, 3.1, 3.0)
***CIL* XV.1** 624
OP DOL EX PR AVG[....]NTIA/ [AELI FELICIS]
op(us) dol(iare) ex pr(aedis) aug(usti) [n(ostri) fig(linis) terent]ia / [aeli felicis]
212-217 (Steinby 1974: 84)

SE 84. (SE246.2-84) South Etruria Site ID:830519 **rect** (8.5, 3.4) **let** (1.5,1.3)**brick** (2.2, 3.4, 2.8)
CIL XV.1 1183 a
[...] C HERI / SECUNDIONI
c heri secundioni
1st century?

SE 85. (SE246.2-85) TVP 420 **rect** (4.5+, 2.2) **let** (0.9, 0.9) **brick** (3.2, 3.6, 3.4)
App. 79 = N 2191/2 = *CIL* XV.1 1979 *PBSR* 36 (1968)
FICIL[...]/M PV[...]
ficil[ilina nicepori] m pu[pi antioci]
Vespasian (Steinby 1974: 96)

SE 86. (SE246.2- 86) South Etruria Site ID:8765643 **rect** (6.2+, 2.4) **let** (2.0) **brick** (2.3, 2.7, 2.5)
N 1450/1
P.S[...]
p s[osius]

SE 87. (SE246.2-87) TVP545 **rect** (12+, 3.1) **let** (1.1,1.1) **brick** (2.9, 3.2, 3.1)
***CIL* XV.1** 310 LSO 304
[...] FIGLINIS MARCIANI[...]/[...]MARCI RABBAEI
[de] figlinis marciani[s] / [sta] marci rabbaei
Vespasian (Steinby 1974: 64)

SE 88. (SE246.2-88) TVP564 **orb** (?,?) **let** (1.2,?) **brick** (2.6, 2.6, 2.6)
unidentified
[...]ATE[...]/[...]MITIAE[...]

SE 89. (SE246.2-89) TVP452 **rect** (9.6+, 4.0) **let** (1.5,1.4) **brick** (2.4, 3.0, 2.7)
***CIL* XV.1** S.431 LSO 1175 PBSR 36 (1968)
[...]LINEIS/[...]EICEPH
[fig]lineis / [c iuli n]eiceph
1st century?

SE 90. (SE246.2-90) TVP 452 **rect** (11.2+, 4.0) **let** (1.5,1.5) **brick** (2.8, 2.9, 2.9)
***CIL* XV.1** S.431 LSO 1175
FIGILINEIS/CIVLINEICEPH
figlineis / c iuli neiceph
1st century?

SE 91. (SE246.2-91) South Etruria Site ID:715468 **circ** (2.3) **brick** (2.5, 3.2, 2.8)
unidentified
round impress

SE 92. (SE246.2-92) TVP872 **orb** (10.3, 4.1) **let** (2.3) **brick** (3.7, 4.0 3.9)
***CIL* XV.1** 1118b
NIEPOS[...]ITIT[...]IM[...]
niepos [cn dom]iti t[roph]im[i]
ca. 123 (Steinby 1974: 57

SE 93. (SE246.2-93) TVP558 **orb** (?, ?) **let** (1.4, 1.4) **brick** (2.4, 2.4, 2.4)
unidentified [...]NAR[...]/[...]DV[...]
SE 94. (SE246.2-94) South Etruria Site ID:897356 **rect** (4.8+, 1.5) **let** (0.9)
unidentified P.EMF[...]
SE 95. (SE246.2-95) TVP1124 **rect** (4.2+, 2.4+) **let** (1.1,0.6+) **brick** (2.5, 2.6, 2.6)
unidentified *PBSR* 45 (1977)
[...].ACTI/[...]VSF

SE 96. (SE246.2-96) TVP2623 **orb** (9.8+-, ?) **let** (1.4) **lin** (?,2) **brick** (4.6, 4.6, 4.6)
***CIL* XV.1** 275 or 1108 LSO 282,876
AGATHOB/[...]
ca. 123 (Steinby 1974: 55)

SE 97. (SE246.2-97) TVP1732 **rect** (8.3+, 3.2) **let** (1.2, 1.2) **brick** (2.1, 2.2, 2.2)
unidentified
PBSR 45 (1977)
L CIVLI[.]CTI/ENIVS F

l c iuli [a]cti / enius f(ecit)
1st century?

SE 98. (SE246.2-98) TVP2623 **orb** (9.4+-, ?) **let** (1.4) **lin** (?) **brick** (2.8, 3.0, 2.9)
***CIL* XV.1** 841
[...]VLIAEP[...]
[q artic]ul[e]i aepagati
2nd century

SE 99. (SE246.2-99) TVP565 **semi-circ** (?) **let** (1.5, 1.3) **lin** (2,1,2) **brick** (2.5, 2.5, 2.5)
***CIL* XV.1** 118 a LSO 158
IANVARI.E.F[...]/VQF
[t grei] ianuari ex f [c d d] / v q f
Domitian (Steinby 1974: 34)

SE 100. (SE246.2-100) TVP452 **rect** (7.9+, 3.9) **let** (1.4, 1.4) **brick** (3.1, 3.4, 3.2)
***CIL* XV.1** S.431 LSO 1175
FI[..]/C IVLI [...]
fi[glineis]/ c iuli [neiceph]
1st century?

SE 101. (SE246.2-101) findspot unknown **rect** (3.0+, 2.2+) **let** (1.4,?) **brick** (2.7, 2.9, 2.8)
***CIL* XV.1** 2445 = S. 302 (?) d
[...]GI[...]
[dia]gi[zia m fulvi s(ervus) f(ecit)
Tiberius-Claudius (Steinby 1974: 34)

SE 102. (SE246.2-102) findspot unknown **rect** (3.2+,5+; 2nd impress 1.2+, 3.6) **brick** (3.6, 3.7, 3.7)
Unidentified
double impress, very deep, no letters preserved

SE 103. (SE244-103) TVP3183 **rect** (8.1+, 4.6) **let** (1.5,1.5) **brick** (3.0, 3.2, 3.1)
***CIL* XV.1** 1000c,e, or f LSO 774
PRIMIG[..] / DOMITI
late Domitian (Steinby 1974: 50)

SE 104. (SE244-104) TVP3592 **rect** (6.4+, 3.2) **let** (2.3) **brick** (2.7, 3.0, 2.8)
***CIL* XV.1** 1460a
Q SVLP[...] q sulp[(i)ci(us) sabin(us)]
1st century

SE 105. (SE244-105) TVP3177 **rect** (10.9+, 4.2) **let** (2.6) **brick** (3.4, 3.6, 3.5)
***CIL* XV.1** 1315 var or N1331/2 0.989
CNAEV[...]
c naevi
Augustan (or a little later) (Steinby 1974: 67)

SE 106. (SE244-106) TVP2414 **rect** (7.3+, 3.3) **let** (2.2) **brick** (3.7, 3.7, 3.7)
***CIL* XV.1** 1460a
Q SVLP[...]
q sulp[(i)ci(us) sabin(us)]
1st century

SE 107. (SE244-107) TVP3284 **rect** (12+, 3.0) **let** (1.7) **brick** (3.2, 3.6, 3.4)
***CIL* XV.1** 2316
[...]CLCENSOR[...]
[ti] cl censori[ni]
1st century

SE 108. (SE244-108) TVP3284 **rect** (9.5, 3.1) **let** (2.3) **brick** (2.9, 3.5, 3.2)
N. 2179/80 1176
CLAE[...]
c laeli(us)

SE 109. (SE244-109) South Etruria Site ID:015628 **rect** (4.7, 1.6) **let** (1.2) **brick** (2.1, 2.4, 2.2)
unidentified
[...]VI

SE 110. (SE244-110) TVP2414 **rect**(9.7+, 3.3) **let** (2.1) **brick** (3.3, 3.3, 3.3)
N1462/3 a or b
Q SVLPIC.S[...]
q sulpic[i(us)] s[abin(us)]
1st century

SE 111. (SE244-111) TVP2322 **rect** (3.9+, 2.1) **let** (1.4) **brick** (2.7, 2.7, 2.7)
unidentified
AR[...]

SE 112. (SE244-112) TVP2382 **rect** (3.0+, 3.3) **let** (2.2) **brick** (2.8, 3.4, 3.1)
***CIL* XV.1** 1460a
Q S [...]
q sulpic[i(us)] s[abin(us)]
1st century

SE 113. (SE244-113) TVP2673 **litt. cavis** (8.9+mag.vid., 2.6) **let** (1.2) **brick** (3.5, 3.5, 3.5)
***CIL* XV.1** 486 a LSO 445
APR ET PAET CO[...]/CORM[...]
123 (Steinby 1974: 83)

SE 114. (SE244-114) South Etruria Site ID:091663 **rect** (3.9+, 2.5) **let** (1.6) **brick** (3.1, 3.4, 3.2)
***CIL* XV.1** 1460b
[...]ABIN
[q sulp(i)ci(us)] sabin(us)
1st century

SE 115. (SE244-115) South Etruria Site ID:878840 **rect** (4.3+, 2.4) **let** (2.1) **brick** (2.7, 2.9, 2.8)
unidentified
PS[...]

SE 116. (SE244-116) TVP2414 **rect** (4.3+, 4.6) **let** (2.3) **brick** (3.4, 3.7, 3.6)
N1462/3b
[...]ABN
[q sulp(i)ci(us)] sab(i)n(us)
1st century

SE 117. (SE244-117) TVP3175 **litt. cavis** (10.4+marg.vid., 2.7 marg. vid.) **let** (2.7) **brick** (3.9, 4.3, 4.1)
unidentified
COIV

SE 118. (SE247-118) TVP 2314 **rect** (5.1, 2.4) **let** (1.6) **brick** (2.3, 2.5, 2.4)
***CIL* XV.1** 864 I-II = 785; 865 I-II LSO 714,715
CASPR
1st century?

SE 119. (SE247-11) findspot unknown **rect** (10.7+, 3.5) **let** (2.7) **brick** (2.8, 3.1, 3.0)
***CIL* XI** 6689.150b
[...]ALI.GA[...]
[m]ali ga[lli]
1st century?

SE 120. (SE247-12) findspot unknown **rect** (10.6+, 4.0) **let** (1.5, 1.3) **brick** (2.8, 2.8, 2.8)
***CIL* XI** 6689.145
[...]LVXSI/PRIMIGENI
[cn] luxsi / primigeni
1st century?
SE 121. (SE247-121) findspot unknown **rect** (7.4, 1.4) **let** (1.1)

brick (3.7, 3.9, 3.8)
App 244, *CIL* VIII 22636.21= *CIL* XV6081
SALONI

SE 122. (SE247-12) findspot unknown **litt. cavis** (6.3+, 2.1 marg vid) **let** (2.1) **brick** (3.8, 3.8, 3.8)
***CIL* XV.1** 1393 LSO 1030
RDPRIIIOSTSC
r ( ) d(e) pr(aedis) (trium) ost(oriorum) sc(apularum)
mid 1st century

SE 123. (SE247-123) findspot unknown **rect** (6.4+, 2.9+) **let** (2.2) **lin** (1,?) **brick** (2.3, 2.5, 2.4)
Agg. 103 (d) or N1376/7
SEX.F[...]

SE 124. (SE247-124) findspot unknown **rect** (1.1+, 1.7) **let** (?) **brick** (3.4, 4.0, 3.7)
unidentified
only a corner of the stamp, no letters preserved
SE 125. (SE247-12) findspot unknown **litt. cavis** (3.8+ marg vid, 2.5 marg vid) **let** (2.5) **brick** (4.0, 4.1, 4.0)
N676/7.2 = CIL XV.1 2096 =*CIL* XV.1 1980
DNO[...]
d no[m p cor her

SE 126. (SE247-12) findspot unknown **rect** (8.2+, 3.2) **let** (1.9) **brick** (2.6, 3.9, 3.2)
unidentified
MO[...]

SE 127. (SE247-127) findspot unknown **orb?** arc is 13.5+ **brick** (2.8, 3.4, 3.1)
unidentified
unidentifiable

SE 128. (SE246.1-128) TVP624 **orb** (?, ?) **let** (1.1, ?) **lin** (1,2,?) **brick** (2.6, 2.9, 2.8)
***CIL* XV.1** 414 LSO 390
[...]/LICINI[...]
[opus doliari]/licini [venusti licini felicissi]
193-198 (Steinby 1974: 73)

SE 129. (SE246.1-129) TVP1227 line of impressed circles, 8.7+ long, intersected by second line 8.0 long, each circle diametre 1.8 **brick** (3.6, 3.9, 3.8)

SE 130. (SE246.3-130) TVP1227 **rect** (5.5+, 3.3) **let** (2.3) **brick** (3.2, 3.7, 3.4)
unidentified
[...]NI
aristanius
1st century

SE 131. (SE246.3-131) TVP1227 **rect** (10.0+, 3.0) **let** (2.2) **lin** (1,1) **brick** (3.3, 3.8, 3.6)
unidentified
[...]ARIST[...]
aristanius
1st century

SE 132. (SE246.3-132) TVP1227 **rect** (4.5+, 3.3) **let** (2.2) **lin** (1,1) **brick** (3.1, 3.6, 3.4)
unidentified
[...]ANI
aristanius
1st century

SE 133. (SE246.3-133) TVP1227 **rect** (7.7+, 3.2) **let** (2.3) **brick** (2.2, 2.5, 2.4)
unidentified
[...]ANI
aristanius
1st century

SE 134. (SE246.3-134)South Etruria Site ID:066537 **rect** (3.1+, 2.0+) **let** (?) **brick** (3.2, 3.2, 3.2)
unidentified
only one bar of a letter is visible

SE 135. (SE246.3-135) TVP1227 **rect** (7.3+, 3.3) **let** (2.4) **lin** (1,?) **brick** (2.8, 3.5, 3.2)
unidentified
[...]TANI
aristanius
1st century

SE 136. (SE246.3-136) TVP2391 **litt. cavis** (7.8+ marg vid, 2.3) **let** (2.3) **brick** (3.6, 3.9, 3.8)
unidentified
AC.PR

SE 137. (SE246.3-137) South Etruria Site ID:078654 **rect** (9.6+, 2.5) **let** (1.4) **brick** (3.5, 4.4, 4.0)
N 1462/3a
[...]SVLPCISABIN
[q] sulp(i)ci(us) sabin(us)
1st century
SE 138. (SE246.3-138) findspot unknown **rect**(12.4+, 3.5) **let** (2.0) **brick** (3.8, 4.4, 4.1)
unidentified
[...]NISVM

SE 139. (SE246.3-139) TVP1227 **rect** (3.9+, 4.9) **let** (4.9) **brick** (2.3, 2.4, 2.4)
unidentified
D

SE 140. (SE246.3-140) TVP1227 **rect** (10.8+, 3.1) **let** (2.2) **lin** (1,1) **brick** (3.3, 3.9, 3.6)
unidentified
[...]RISTANI
aristanius
1st century

SE 141. (SE246.3-141) TVP2413 **rect** (10.0+, 3.3) **let** (2.1) **brick** (3.3, 3.3, 3.3)
unidentified
N[...]S.L[...]

SE 142. (SE246.3-142) TVP2316 **rect** (3.3+, 2.3) **let** (1.8) **brick** (2.4, 2.4, 2.4)
***CIL* XIV** 5308.18b
DA[...]
da[psiliis]

SE 143. (SE246.3-143) TVP2324 **rect** (5.5+, 1.9) **let** (?) **brick** (3.1, 3.3, 3.2)
unidentified
SV[...]

SE 144. (SE246.3-144) TVP2414 **rect** (11.9, 2.3) **let** (1.5) **brick** (3.1, 3.7, 3.4)
N 1462/3a
QSVLPCISABIN
q sulp(i)ci(us) sabin(us)
1st century

SE 145. (SE246.3-145) TVP2410 **rect** (4.8+, 2.5) **let** (1.7) **brick** (3.0, 3.3, 3.2)
***CIL* XV.1** 1460
QS[...]
1st century

SE 146. (SE246.3-146) TVP2414 **rect** (4.3+, 2.7+) **let** (2.2) **brick** (2.6, 2.8, 2.7)
App 229
[...]I.SER

SE 147. (SE246.3-147) South Etruria Site ID:074658 **rect** (7.2+, 4.1) **let** (1.3,1.4) **lin** (?) **brick** (3.1, 3.3, 3.2)
App 253
[...]RBANI/[...]ODI

SE 148. (SE246.3-148) TVP2391 **rect** (1.9+, 2.3) **let** (1.9) **brick** (2.8, 2.9, 2.8)
Unidentified
C[...]/E[...]
1st century

SE 149. (SE246.3-149) TVP2419 **rect**(13.1+,3.3) **let** (2.2) **brick** (3.6, 3.8, 3.7)
Unidentified
maybe same as SE 141

SE 150. (SE246.3-150) South Etruria Site ID:055647 **rect** (8.3+, 3.0) **let** (2.3) **brick** (2.8, 3.1, 3.0)
N 1462/3b
[...]PIC SABN
[q sul]pic(ius) sab(i)n(us)
1st century

SE 151. (SE245-151) TVP1547 **litt. cavis** (11.5+marg.vid., 2.2 marg. vid) **let** (2.2) **brick** (4.0, 4.0, 4.0)
CIL **XV.1** 1393 LSO 1030
RDPRIIIOSTSC
r ( ) d(e) pr(aedis) (trium) ost(oriorum) sc(apularum)
mid 1st century

SE 152. (SE245-152) TVP2304 **rect** (6.6+, 3.8+) **let** (2.3+) **lin** (1) **brick** (3.0, 3.4 3.2)
N 898/8
[...]ALP
[c]alp

SE 153. (SE245-153) TVP1547 **litt. cavis** (4.6+marg. vid., 2.3 marg. vid) **let** (2.3) **brick** (3.7, 4.0, 3.9)
CIL **XV.1** 1393 LSO 1030
RDPRIIIOSTSC
r ( ) d(e) pr(aedis) (trium) ost(oriorum) sc(apularum)
mid 1st century

SE 154. (SE245-154) TVP 2678 **rect** (8.3+, 2.4) **let** (1.5) **brick** (3.0, 3.5, 3.3)
CIL **XV.1** 1268 = S.333
[...]A.MARCI.CS
[dam]a marci c s
1st century?

SE 155. (SE245-155) TVP 420 **rect** (3.6+, 2.4) **let** (0.6,0.7) **lin** (?,?,1) **brick** (2.1, 2.4, 2.3)
CIL **XV.1** 950
[...]/NATA[...]
[c cornelius] / nata[lis f]
1st century?

SE 156. (SE245-156) TVP 1547 **rect** (2.8, 4.4) **let** (3.2) **brick** (2.9, 3.3, 3.1)
CIL **XV.1** 2263
PL
1st century?

SE 157. (SE245-157) South Etruria Site ID:179596 **rect** (10.8+, 2.4) **let** (1.9) **brick** (4.0, 4.2, 4.1)
CIL **XV.1**1460
Q SVLPICI[...]
q sulpici[(us) sabin(us)]
1st century

SE 158. (SE245-158) TVP 3085 **rect** (3.9+ 3.7 / 2nd stamp 2.9+3.2+) **let** (1.4, 1.4 / 2nd stamp 2.5)
brick (3.4, 3.4, 3.4)

SE 159. (SE245-159) TVP3275 **rect** (12.5+marg. vid, 2.9) **let** (2.3) **brick** (3.6, 4.0, 3.8)
cf. **CIL XV.1** 2180
M LATINI.DIPA
m. latini diop(h)a(nis)

SE 160. (SE245-160) TVP1547 **litt. cavis** (7.5+ marg.vid., 2.2 marg vid) **let** (2.2) **brick** (4.0, 4.0, 4.0)
CIL **XV.1** 1393 LSO 1030
RDPRI[...]
r ( ) d(e) pr(aedis) (trium) ost(oriorum) sc(apularum)
mid 1st century

SE 161. (SE245-161) TVP3177 **litt. cavis** (5.2+ marg vid, 2.7 marg vid) **let** (2.7) **brick** (3.7, 3.7, 3.7)
CIL **XV.1** 578 a
[...]I.CLA[...]
ti cla (bla sul)
Trajan (Steinby 1974: 90)

SE 162. (SE245-162) TVP616 **rect** (3.0+, 1.9) **let** (1.1) **brick** (2.9, 3.4, 3.2)
CIL **XV.1** S.360 or 1383a,b,c
L P C [...]

SE 163. (SE245-163) TVP 1547 **litt. cavis** (10.4+ marg vid, 2.2 marg vid) **let** (2.2) **brick** (3.8, 4.2, 4.0)
CIL **XV.1** 1393 LSO 1030
RDPRIIIO[...]
r ( ) d(e) pr(aedis) (trium) o[st(oriorum) sc(apularum)]
mid 1st century

SE 164. (SE245-164) TVP2654 **rect** (8.7+, 5.0) **let** (1.1, 1.1) **brick** (2.9, 2.9, 2.9)
CIL **XV.1** 999 a=c?
[...] DOMITIOR[...]/[...] ET TVLLI
[fortunati] domitior[um]/[lucani] et tulli
Domitian (Steinby 1974: 50)

SE 165. (SE245-165) TVP3204 **rect** (5.5, 2.7) **let** (1.9) **brick** (2.7, 2.7, 2.7)
CIL **XV.1** 864 LSO 714
CASPR
caspr
1st century

SE 166. (SE245-166) TVP1547 **litt. cavis** (14.5 marg vid, 2.2 marg vid) **let** (2.2) **brick** (4.1, 4.1, 4.1)
CIL **XV.1** 1393 LSO 1030
RDPRIIIOSTSC
r ( ) d(e) pr(aedis) (trium) ost(oriorum) sc(apularum)
mid 1st century

SE 167. (SE245-167) South Etruria Site ID:869597 **rect** (6.7, 2.5) **let** (0.6, 0.7) **lin** (1,1,1) **brick** (2.3, 2.6, 2.5)
CIL **XV.1** 950
CCO[...]/NATALIS[...]
c co[rnelius]/natalis [f]
1st century?

SE 168. (SE245-168) TVP420 **rect** (3.5+ marg vid, 3.6+) **let** (1.3, 1.3) **lin** (?) **brick** (2.6, 2.7, 2.7)
CIL **VIII** 22636.1(d) = *CIL* XV.1 976
DION[...]/FVL[...]
dionysiu[s]/fulvi m s[er]
1st century?

SE 169. (SE245-169) findspot unknown **rect** (1.5+, 1.9+) **let** (0.8+) **lin** (?) **brick** (2.7, 2.7, 2.7)
impossible to identify, only fragment of letter leg preserved

SE 170. (SE245-170) TVP 1547 **litt. cavis** (9.0+ marg vid, 2.2 marg vid) **let** (2.2)**brick** (3.9, 4.3, 4.1)
***CIL* XV.1** 1393 LSO 1030
[...]RIIIOSTS[...]
[r ( ) d(e) p]r(aedis) (trium) ost(oriorum) s[c(apularum)]
mid 1st century

SE 171. (SE245-171) TVP 2673 **rect** (2.1+, 3.9) **let** (1.3, 1.3) **lin** (1) **brick** (3.3, 3.5, 3.4)
App 124(d)
[...]VT/[...]AE
[sabinus br]ut[/tid volusian]ae

SE 172. (SE245-172) TVP1547 **rect** (5.6, 2.4) **let** (1.7) **brick** (2.9, 2.9, 2.9)
***CIL* XV.1** 864 LSO 714
CASPR
1st century?

SE 173. (SE245-173) TVP1172 **rect** (6.7+, 2.4) **let** (1.9) **brick** (2.9, 3.2, 3.1)
***CIL* XI** 6689.150a
[...]GALVS
[p malius p l] gallus

SE 174. (SE245-174) TVP1547 **litt. cavis** (7.3+ marg vid, 2.2 marg vid) **let** (2.2) **brick** (4.3, 4.3, 4.3)
***CIL* XV.1** 1393 LSO 1030
[...]IOSTS[...]
[r ( ) d(e) pr(aedis)] (trium) ost(oriorum) s[c(apularum)]
mid 1st century

SE 175. (SE245-175) findspot unknown **rect** (5.9+, 4.0) **let** (1.3, 1.3) **brick** (3.3, 3.5, 3.4)
CIL XV.11015a LSO 782
PRIMITI[...]/[...]LVCILL
primitivi/domitiae lucill
118-120 (Steinby 1974: 53)

SE 176. (SE244-176TVP) South Etruria Site ID:0403772623 **rect** (5.6, 2.4) **let** (1.7) **brick** (2.9, 3.0, 3.0)
***CIL* XV.1** 864 LSO 714
CASPR
caspr
1st century

SE 177. (SE244-177TVP) South Etruria Site ID:0403772623 **rect** (5.6, 2.4) **let** (1.7) **brick** (2.9, 2.9, 2.9)
***CIL* XV.1** 864 LSO 714
[...]ASPR
caspr
1st century

SE 178. (SE244-178TVP) South Etruria Site ID:0403772623 **semi-circ?** (5.2+, 2.6) **let** (1.6) **brick** (2.6, 3.1, 2.9)
unidentified
[...]MRC[...]

SER 1 (seripola 1) South Etruria Site ID:868059 **circ** (10.0) **let** (0.9, 0.7, 0.7) **lin** (1,2,2,1) **brick** (4.4, 4.2, 4.3)
***CIL* XV.1** 176 LSO 205 *PBSR* 72 (2004)
FIGLIN OPVS DOLI[...]AVG N/DOMITI[...]NOR[...]/NUM[...]TI
opus doli[iare ex praed] praed aug n figlin/domiti[anas mi]nor[es]/num[eri ius[ti]
Commodus (Steinby 1974: 38)

SER 2 (seripola 2) South Etruria Site ID:868059 **rect** (6.4+, 4.4) **let** (1.7, 1.5) **lin** (1,2,1) **brick** (2.9, 2.9, 2.9)
***CIL* XV.1** 777 LSO 655 *PBSR* 72 (2004)
[...]TORIS/[...]GL
[adiu]toris/[au]g l
1st century?

SER 3 (seripola 3) South Etruria ID:868029 **orb** (9.2+-,?) **let** (1.2) **lin** (?) **brick** (4.4., 4.4, 4.4)
***CIL* XV.1** 758 *PBSR* 72 (2004)
[...]SESP[...]
[ex prae aug]ses p[on eli]
169-176 (Steinby 1974: 73)

Appendix B: XRD Peak Heights above Background for Tested SES Bricks

For mineral peaks selected and the methodology to convert peak heights (measured in mm) to semi-quantified relative amounts of minerals, see 3.3

Sample	Quartz	Augite	Haematite	Gehlenite	Calcite	Analcime	Muscovite	Dolomite	Anorthoclase	Sanidine	Albite
se 1	58	70	14			10	34	36	105		
se 2	16	83	5		43	9		23			66
se 4	89	36	17	36	70		7	14		55	61
se 5	42	94	8				10	40		65	55
se 7	20	102	9			15	10	30	85	25	
se 8	50	14	10	20	75		10		56		
se 10	50	60	16	20			42	48		50	45
se 13	87	25	24	30	55	19	14	13	56		
se 14	32	104	9	3		20	10	40	85		
se 14	38	106	12		5	17	15	47		60	65
se 16	86	28	14	50	86	25	26	22	103		
se 18	38	71	17		32		3	28	95		
se 19	38	92	9		50			30			100
se 20	106	41	23	23	32	16		24	96		
se 21	23	100	7				11	34		105	
se 22	36	106	6	40		7	42		72		
se 23	26		3		124		12				12
se 26	68	33	8	18		24		23	70		
se 27	45	93	17		32		3	40			75
se 28	76	44	14	33	66	14	6		65		
se 29	21	95	8	7	95		7	32		50	60
se 30	82	44	12	6			6	17		35	
se 36	52	90	8		43			39	95		
se 37	97	48	24	33	12	22	24	20	95		
se 42	56	105	11			17		40		114	90
se 45	29	92	10				5	31	55		
se 47	47	110	11			20		40		88	96
se 48	41	16	7	7	80		8	12		30	35
se 50	88	22	12	17		14	17			50	50
se 51	40	110	8		10	15	10	42	84		88
se 52	84	50	13	48	60		21	26		60	55
se 53	40	115	11					27		70	

Sample	Quartz	Augite	Haematite	Gehlenite	Calcite	Analcime	Muscovite	Dolomite	Anorthoclase	Sanidine	Albite
se 54	35	65	11				17	20		80	40
se 56	112	53	19	33			24				102
se 57	44	92	10		15		13	32		73	
se 58	38	85	6	11	30		11	40		65	55
se 60	65	88		30	64		15	25	85		
se 63	109	24	20	16							79
se 65	85	40	20	20			16	23			105
se 72	30	107	3		65	29		30	40		
se 85	91	64	14	9			8	16		22	30
se 89	64	41	20	9			10	12			68
se 90	112	27	23	20			23		91		
se 100	22	32	11	4				6			80
se 104	36	58	9	40	54	12	5	18	92		
se 106	49	104	13	33	99	39		38			55
se 107	16	21	7	50	11	6				8	16
se 108	107	64	23	52				24			63
se 110	86	30	17	74	75	16	20		26		33
se 111	17	40	15	10			15	17	42		47
se 113	39	15	6	62	108		6	35	20		
se 114	83	49	15	66	97			17	62		
se 116	47	55	18	53	105	5		17	61		
se 136	89		11	37	104		19		26		
se 141	63	40	19	33	100	15		12	53		
se 144	77	10	8	43	109	11	17	11	30		
se 148	37	90			51	15	9	46			75
se 151	21	20	23		2						87
se 152	53	87	8	11	84		12	22	84		
se 153	75		11		10	12	13			24	
se 154	33	84	10	2			8	30	66		
se 155	83	46	12	15	40	24	10	26			100
se 156	47	54	10				9	21	115		
se 160	45	24	27			11					90
se 163	48	23	11		40	16		15	110		
se 166	90	26	9		45			17		17	
se 168	117	85	18	60	10	25	22	21	100		
se 170	33	23	31		11			12			117
se 171	63	26	13	46	100	13	15		25		70
se 172	107	45	31					16	76		103

Sample	Quartz	Augite	Haematite	Gehlenite	Calcite	Analcime	Muscovite	Dolomite	Anorthoclase	Sanidine	Albite
se 174	30	29	34				11				109
se 176	33	55	7	2	14			10		30	85
se 177	65	36	10	13	74			10		35	70
fal 1	105		10	60	30	26	33		90		
fal 2	28	63		6	120	36		24	66		
fal 3	92	54	20	90	76	20	9		85		
fnv13	43	87	7	30	60	15			42		60
fnv14	92	91	13		110	9		32	70		
fnv15	76	25	20		100	16			24		
fnv4	64	25	6		72		9	3			27
fnv5	52				94		14	4	30		
fnv6	72	22	10		75		16		23		
fnv8	36	108	7		63	50		35			55
fnv9	14	98	6		72	8	7	23			77
mod 1	113	55		54		15	9		98		
mod 2	37	120	7	43		10		27	24		
mod 3	112	22	24	71	96	12	17		62		
mod 4	16			5	127					5	
mod 5	52	96	10	49		9		27	53		
mod 6	57	105	6	60		6		20	36		
mod 7	37			38	124		16				25
mod 8	105			20	100	20	20		32		
mod 9	11		3		54		3	4			3
ser 1	89	38	9	12	79		12	28			70
ser 2	97	18	14	8	60		18	7		36	50
ser 3	90	46	10	6	90		19	19	45	35	

Appendix C: XRF Raw Scores for Tested SES Bricks

These values have not been normalised, and are presented as measured.

Majors (% wt)

SAMPLE	Na_2O	MgO	Al_2O_3	SiO_2	P_2O_5	K_2O	CaO	TiO_2	MnO	Fe_2O_3
SE1	0.94	4.13	15.89	58.16	0.2	3.65	11.28	0.71	0.11	6.63
SE2	0.95	3.95	12.82	49.43	0.21	2.08	17.49	0.7	0.11	6.36
SE4	0.78	3.35	13.16	52.3	0.28	2.44	16.26	0.67	0.11	6.08
SE5	1.17	4.18	15.2	58.98	0.25	3.77	11.45	0.69	0.11	6.41
SE7	1.17	3.97	15.36	57.08	0.18	3.68	12.45	0.73	0.12	6.66
SE8	0.73	3.26	14.44	52.29	0.23	3.33	14.94	0.71	0.1	6.57
SE10	0.92	3.23	16.02	58.41	0.18	3.74	10.55	0.73	0.11	6.62
SE13	0.87	3.59	15.04	54.83	0.21	3.08	13.45	0.71	0.13	6.88
SE14	1.17	4.19	15.21	56.94	0.15	3.24	13.18	0.7	0.12	6.62
SE16	0.76	3.52	14.57	52.77	0.26	3.37	15.12	0.7	0.11	6.46
SE18	0.93	3.9	14.61	55.31	0.15	2.95	14.89	0.71	0.11	6.63
SE19	1.04	3.81	14.51	53.88	0.25	3.34	15.52	0.69	0.1	6.26
SE20	1.02	2.86	14.74	58.68	0.18	2.3	10.39	0.71	0.16	6.48
SE21	1.22	4.4	15.63	56.47	0.22	3.61	12.1	0.71	0.11	6.62
SE22	1.33	4.9	15.24	56.4	0.18	2.83	13.85	0.71	0.11	6.68
SE23	0.56	4.43	12.26	43.88	0.2	2.48	20.62	0.62	0.08	5.52
SE26	0.92	3.42	15.53	57.02	0.25	3.8	10.58	0.71	0.12	6.76
SE27	1.12	4.01	14.75	55.82	0.18	3.35	14.87	0.66	0.1	6.17
SE28	0.86	3.61	14.31	52.06	0.44	3.11	15.52	0.69	0.14	6.62
SE29	1.13	3.4	13.9	51.7	0.17	2.69	18.66	0.67	0.11	6.21
SE30	0.97	3.15	16.2	61.14	0.25	2.83	7.3	0.79	0.16	7.32
SE36	0.91	3.81	15.18	54.59	0.19	3.45	14.77	0.7	0.12	6.62
SE37	0.82	3.34	16.16	58.22	0.26	3.36	10.25	0.76	0.13	7.05
SE42	1.32	4.57	15.08	57.24	0.17	2.75	13.24	0.7	0.1	6.69
SE45	1.21	4.54	15.43	57.88	0.17	3.38	12.47	0.71	0.12	6.63
SE47	1.39	4.52	14.9	57.78	0.16	2.69	13.52	0.7	0.1	6.64
SE48	1.02	3.71	12.78	50.87	0.18	2.91	17.74	0.62	0.09	5.75
SE50	1.09	4.28	15.31	60.37	0.21	3	8.92	0.7	0.09	6.46
SE51	1.15	3.82	15.23	55.97	0.16	2.55	15.1	0.72	0.12	6.77
SE52	1.05	4.19	13.94	52.9	0.17	3.43	14.87	0.67	0.11	6.24
SE53	1.31	4.44	14.75	57.76	0.15	2.74	13.88	0.69	0.11	6.58
SE54	0.92	3.84	16.09	57.52	0.16	3.48	12.2	0.71	0.12	6.59

SAMPLE	Na_2O	MgO	Al_2O_3	SiO_2	P_2O_5	K_2O	CaO	TiO_2	MnO	Fe_2O_3
SE56	0.88	3.52	16.31	59.01	0.17	3.94	9.31	0.74	0.11	6.83
SE57	1.3	4.45	14.98	55.7	0.15	3.03	14.36	0.71	0.11	6.48
SE58	1.08	4.21	14.3	51.94	0.17	3.06	16.36	0.69	0.11	6.31
SE60	1.06	3.94	13.84	53.11	0.25	3.33	16.38	0.65	0.1	6.15
SE63	0.79	3.15	16.72	59.59	0.2	2.77	8.72	0.79	0.11	7
SE65	0.88	3.53	16.36	58.97	0.36	2.77	10.63	0.78	0.11	6.99
SE72	1.98	5.07	13.51	49.77	0.15	1.47	20.12	0.61	0.1	5.72
SE85	0.81	3.11	15.62	61.41	0.19	2.99	6.27	0.73	0.14	6.75
SE89	1.47	3.51	16.32	58.63	0.34	2.66	9.47	0.77	0.14	7.08
SE90	1.33	3.15	16.41	59.72	0.33	2.82	8	0.77	0.16	7.05
SE100	1.42	2.68	15.49	58.44	0.24	2.46	10.76	0.75	0.14	6.88
SE104	1.12	4.47	13	51.28	0.17	1.73	19.15	0.67	0.12	6.14
SE106	1.38	5.1	12.41	49.84	0.19	1.33	20.8	0.63	0.09	5.57
SE107	0.52	2.3	12.33	43.31	0.23	2.08	26.57	0.65	0.07	5.8
SE108	0.84	3.3	15.81	58.13	0.24	2.98	10.42	0.78	0.15	7.29
SE110	0.75	3.86	12.6	47.55	0.66	2.16	20.18	0.66	0.12	5.94
SE111	0.94	3.95	16.52	57.99	0.35	3.56	8.55	0.78	0.11	7.03
SE113	0.88	3.32	11.91	43.88	0.22	2.18	24.77	0.63	0.09	5.85
SE114	0.83	4.01	12.18	48.33	0.49	2.04	20.02	0.66	0.1	5.83
SE116	0.93	4.54	12.98	48.68	0.17	1.88	21	0.66	0.11	5.85
SE136	0.74	2.65	14.6	52.31	0.2	2.44	15.18	0.74	0.15	6.75
SE141	0.84	4.21	12.5	47.49	0.18	1.88	20.42	0.66	0.1	5.76
SE144	0.63	4.39	11.57	47.11	0.27	2.01	20.01	0.63	0.09	5.53
SE148	1.2	4.02	14.74	53.73	0.19	2.93	15.52	0.7	0.1	6.54
SE151	1.19	2.5	17.88	57.71	0.27	2.94	9.29	0.81	0.17	7.41
SE152	1.03	3.94	14.88	53.06	0.18	3.23	15.54	0.72	0.1	6.55
SE153	0.96	2.6	17.12	57.56	0.29	2.96	7.77	0.78	0.17	7.24
SE154	1.14	3.96	15.47	58.32	0.22	3.35	11.87	0.71	0.11	6.55
SE155	1.4	3.68	14.41	57.24	0.25	2.92	13.21	0.67	0.1	6.2
SE156	1.01	4.39	15.77	58.66	0.15	3.09	11.65	0.73	0.11	6.97
SE160	1.04	2.69	18.11	58.85	0.22	2.83	8.76	0.83	0.17	7.8
SE163	1.16	2.53	16.98	55.63	0.26	2.84	11.1	0.78	0.17	7.13
SE166	0.78	2.31	17.22	56.9	0.25	3.2	7.78	0.8	0.16	7.28
SE168	1.00	3.54	15.28	58.13	0.35	3.27	12.1	0.7	0.11	6.54
SE170	1.14	2.56	17.66	57.35	0.21	2.85	10.23	0.79	0.17	7.41
SE171	0.74	3.33	12.74	47.83	0.18	3.19	18.5	0.65	0.1	5.69
SE172	0.75	3.08	16.67	59.06	0.27	3.08	9.07	0.78	0.12	7.02

SAMPLE	Na_2O	MgO	Al_2O_3	SiO_2	P_2O_5	K_2O	CaO	TiO_2	MnO	Fe_2O_3
SE174	1.15	2.53	17.25	57.56	0.26	3.02	9.6	0.78	0.17	7.16
SE176	0.92	3.36	15.91	56.75	0.15	2.99	13.15	0.74	0.11	6.79
SE177	0.85	3.13	15.12	54.95	0.17	2.93	14.12	0.73	0.12	6.91
FAL1	1.55	4.37	14.72	57.61	0.43	3.02	9.42	0.66	0.11	5.55
FAL2	1.77	4.2	12.93	47.75	0.15	1.7	21.62	0.58	0.09	4.94
FAL3	0.82	4.1	14.12	49.25	0.28	2.7	19.81	0.69	0.11	6.24
FNV13	0.93	4.73	15.07	49.36	0.24	1.58	17.08	0.73	0.11	6.49
FNV14	0.53	4.36	14.4	52.18	0.33	1.56	16.68	0.67	0.12	6.07
FNV15	0.43	2.14	14.44	47.19	0.47	1.45	16.6	0.69	0.1	6.17
FNV4	0.55	2.75	12.61	45.52	0.23	2.42	19.81	0.63	0.08	5.56
FNV5	0.4	3.24	13.18	44.04	0.27	2.33	20.05	0.68	0.1	5.97
FNV6	0.56	3.5	13.36	48.45	0.6	2.62	15.85	0.69	0.12	6.16
FNV8	1.68	4.6	13.35	53.27	0.34	1.53	17.14	0.62	0.12	5.67
FNV9	1.52	4.79	15.45	50.55	0.17	2.17	17.45	0.71	0.1	6.11
MOD1	0.99	3.63	14.13	54.67	0.13	2.67	16.49	0.65	0.1	6.21
MOD2	0.99	4.67	12.62	48.66	0.14	2.32	24.02	0.59	0.09	5.57
MOD3	0.91	2.62	13.75	53.49	0.14	2.34	15.7	0.67	0.14	6.08
MOD4	0.2	3.72	9.55	36.6	0.14	1.75	30.91	0.53	0.11	4.89
MOD5	1.22	5.67	14.05	50.35	0.14	2.55	18.83	0.67	0.11	6.04
MOD6	1.07	6.56	13.47	47.69	0.13	2.46	21.31	0.63	0.1	5.87
MOD7	0.66	3.66	11.33	44.48	0.13	2.5	20.88	0.59	0.09	4.98
MOD8	0.61	3.45	13.42	51.44	0.13	2.56	17.96	0.64	0.12	6.08
MOD9	0.09	1.85	9.85	37.69	0.1	2.22	32.75	0.51	0.09	3.36
SER1	0.79	2.92	14.36	53.45	0.16	3.41	14.07	0.67	0.11	6.49
SER2	1.17	3.39	14.92	56.84	0.15	3.17	10.2	0.7	0.1	6.38
SER3	0.87	3.64	15.42	54.56	0.15	3.47	12.7	0.74	0.11	6.84

Trace Elements (ppm)

SAMPLE	V	Cr	Co	Ni	Cu	Zn	Pb	Rb	Sr	Y	Zr
SE1	117	137	17	75	31	108	48	198	588	34	177
SE2	144	102	17	58	36	98	55	216	466	41	243
SE4	98	103	19	49	35	92	50	204	413	43	219
SE5	124	128	16	74	33	102	53	293	603	36	183
SE7	137	136	18	80	32	109	45	345	589	44	192
SE8	115	122	18	73	33	105	41	211	494	32	153
SE10	124	134	19	65	32	92	88	223	644	35	185
SE13	108	150	24	98	41	118	32	194	455	33	155
SE14	103	147	22	83	40	105	45	268	562	32	180
SE16	126	135	13	69	33	103	35	210	537	32	149
SE18	105	154	19	89	40	114	31	173	457	36	158
SE19	112	141	18	79	35	108	37	201	533	56	164
SE20	89	106	19	71	37	95	40	168	445	40	224
SE21	112	151	17	80	36	107	40	229	557	35	189
SE22	112	132	18	78	34	107	36	309	529	39	181
SE23	127	83	15	57	30	89	34	271	447	32	130
SE26	109	109	24	80	33	108	38	229	574	65	179
SE27	104	138	20	73	30	105	39	181	505	37	170
SE28	129	126	17	72	29	107	37	195	537	35	169
SE29	126	138	15	66	30	101	51	267	635	31	153
SE30	136	129	24	79	40	97	57	273	556	41	259
SE36	103	144	20	86	44	108	40	193	564	32	136
SE37	126	124	16	67	27	111	37	222	516	40	206
SE42	127	174	20	82	32	110	32	212	449	33	175
SE45	126	138	17	74	31	103	46	342	589	36	183
SE47	127	170	17	86	37	138	33	200	448	35	181
SE48	100	125	16	66	26	88	30	150	415	39	171
SE50	99	126	14	76	28	106	31	195	359	31	181
SE51	129	149	18	71	30	112	35	274	575	40	168
SE52	119	137	17	70	32	96	35	190	588	34	158
SE53	120	167	15	83	30	110	30	195	443	33	177
SE54	123	145	16	67	33	103	49	216	653	33	189
SE56	119	147	14	72	34	104	44	222	554	39	173

SAMPLE	V	Cr	Co	Ni	Cu	Zn	Pb	Rb	Sr	Y	Zr
SE57	119	143	17	77	35	108	43	224	518	34	171
SE58	122	143	18	73	39	107	30	183	580	33	168
SE60	117	124	19	73	30	93	37	195	562	38	173
SE63	142	138	21	82	38	112	39	258	462	36	184
SE65	122	155	20	87	41	115	34	250	446	45	183
SE72	90	106	15	63	29	99	40	276	506	43	167
SE85	102	107	21	70	33	99	43	233	421	43	241
SE89	103	103	18	72	41	107	56	446	556	44	253
SE90	110	103	18	74	44	107	56	378	570	45	259
SE100	139	120	21	73	44	107	60	301	511	43	265
SE104	102	116	16	58	33	85	44	351	434	41	213
SE106	97	102	15	54	32	77	40	225	464	42	213
SE107	106	82	18	72	45	106	40	220	665	34	154
SE108	113	94	22	84	49	118	41	220	466	43	216
SE110	99	86	12	52	37	94	43	220	421	35	177
SE111	128	142	18	98	40	102	35	321	489	38	173
SE113	86	77	16	55	37	92	36	241	583	35	159
SE114	108	98	18	52	40	85	42	266	448	33	207
SE116	89	105	17	52	33	86	46	198	414	36	194
SE136	116	103	19	73	40	107	46	204	453	38	219
SE141	90	93	17	47	31	83	48	181	443	37	216
SE144	98	82	18	47	32	75	46	246	433	36	204
SE148	115	144	16	78	34	110	33	269	523	35	165
SE151	135	101	22	66	47	117	80	311	719	47	330
SE152	129	147	17	77	34	107	27	182	476	35	156
SE153	130	105	18	75	48	108	75	434	670	49	314
SE154	99	152	16	80	34	110	40	186	510	31	166
SE155	104	136	13	81	35	104	34	208	456	41	181
SE156	111	154	20	93	38	119	35	170	446	35	161
SE160	110	118	17	75	46	116	72	316	668	45	302
SE163	116	104	19	64	46	103	78	318	750	48	327
SE166	161	99	22	72	50	111	72	448	727	52	297
SE168	105	139	16	69	34	116	47	250	534	38	183
SE170	142	108	19	64	48	112	79	295	690	45	334
SE171	102	98	11	61	29	92	36	174	591	29	161
SE172	108	162	22	107	39	110	37	246	407	39	152
SE174	143	102	21	63	44	107	77	366	741	46	337

SAMPLE	V	Cr	Co	Ni	Cu	Zn	Pb	Rb	Sr	Y	Zr
SE176	125	172	22	102	40	120	26	172	400	35	148
SE177	124	171	22	99	41	121	33	170	404	35	130
FAL 1	80	86	11	58	31	91	45	485	497	40	254
FAL 2	84	92	13	54	22	83	42	503	648	40	201
FAL 3	119	113	16	61	28	97	35	177	580	35	165
FNV13	112	145	15	72	37	104	46	134	334	35	168
FNV14	90	121	18	57	30	85	46	91	334	41	197
FNV15	112	124	9	58	27	66	44	107	275	40	186
FNV4	105	94	16	54	24	89	25	130	298	32	155
FNV5	114	82	15	58	29	95	35	125	328	37	157
FNV6	100	102	15	59	31	101	35	136	312	35	168
FNV8	87	107	12	55	36	124	45	156	417	37	218
FVN9	108	118	16	62	41	155	42	130	557	36	230
MOD1	128	142	18	82	35	115	32	142	449	36	149
MOD2	129	125	13	58	24	98	35	135	526	30	124
MOD3	127	105	23	70	35	99	39	155	387	34	211
MOD4	111	88	8	46	45	77	36	81	829	27	108
MOD5	132	130	17	62	32	94	32	158	341	33	163
MOD6	127	120	14	64	28	101	29	147	336	37	141
MOD7	104	92	11	46	28	70	30	119	268	26	151
MOD8	142	125	19	74	35	117	64	135	419	32	167
MOD9	102	103	16	56	34	90	26	112	528	28	82
SER1	108	120	14	67	43	101	45	174	531	34	176
SER2	122	132	15	80	29	100	36	212	424	33	221
SER3	99	150	21	89	38	111	43	183	530	39	174

Appendix D: Individual Power Rankings

- When the Bonacich Power Index is attenuated negatively, higher negative scores correspond to higher centrality-based power.
- When it is attenuated positively, higher positive scores correspond to higher dependents-based power.

Julio-Claudian

Centrality based power (negative attentuation of Bonacich Power Index)

Network #	Person	Power Index	Ranking	Percent	Octile
R12	Rubria	-6	1	.00%	
R15	Tonneius Apolinaris	-6	1	.00%	
R16	Spurilia Flori	-6	3	.00%	
R36	Paetina	-6	3	.00%	1st Octile
R25	Calpurnia Corvini	-2	3	16.60%	
R34	Ti. Iulius Optatus	-2	3	16.60%	
R39	C. Calvisius Primus	-2	3	16.60%	
R40	Q. Aelius Tubero	-2	3	16.60%	2nd Octile
R1	Ti Iulius Iulianus	0	3	33.30%	
R7	A Caesannius Gallus	0	3	33.30%	
R8	Ceionius Pamphil	0	3	33.30%	
R9	C Ceionius	0	3	33.30%	
R10	Hilarus Ceonius	0	3	33.30%	
R11	Grattia	0	3	33.30%	
R17	Sex Quit	0	3	33.30%	
R20	M Antoichus	0	3	33.30%	
R21	Mestria	0	3	33.30%	
R22	Q Vaerlius Cato	0	18	33.30%	
R23	L Arruntius	0	18	33.30%	
R24	C Asinius Pollio	0	18	33.30%	
R27	L Cornelius Priscus	0	18	33.30%	
R31	L Faenius Rufus	0	22	33.30%	
R38	L Sestius P.f Albanius Quirinalis	0	22	33.30%	3rd Octile
R3	M Allius Clemens	2	22	95.80%	
R28	Cn. Domitius Afer	2	22	95.80%	8th Octile

Dependents-based power (positive attenuation of Bonacich Power Index)

Network #	Person	Power Index	Ranking	Percent	Octile
R36	Paetina	94	1	100.00%	
R39	C. Calvius Primus	18	2	95.80%	1st Octile
R8	Ceionius Pamphil	1	3	83.30%	
R9	C Ceionius	1	3	83.30%	
R10	Hilarus Ceonius	1	3	83.30%	
R3	M Allius Clemens	0.67	4	75.00%	
R28	Cn. Domitius Afer	0.67	4	75.00%	2nd Octile
R1	Ti Iulius Iulianus	0	5	25.00%	
R7	A Caesannius Gallus	0	5	25.00%	
R11	Grattia	0	5	25.00%	
R17	Sex Quit	0	5	25.00%	
R20	M. Antiochus	0	5	25.00%	
R21	Mestria	0	5	25.00%	
R22	Q Valerius Cato	0	5	25.00%	
R23	L Arruntius	0	5	25.00%	
R24	C Asinius Pollio	0	5	25.00%	
R27	L Cornelius Priscus	0	5	25.00%	
R31	L Faenius Rufus	0	5	25.00%	
R38	L Sestius P.f Albanius Quirinalis	0	5	25.00%	6th Octile
R15	Tonneius Apolinaris	-18	6	16.60%	
R16	Spurilia Flori	-18	6	16.60%	7th Octile
R25	Calpurnia Corvini	-30	7	8.30%	
R34	Ti Iulius Optatus	-30	7	8.30%	
R12	Rubria	-34	8	4.10%	
R40	Q Aelius Tubero	-46	9	.00%	8th Octile

Flavians

Centrality based power (negative attentuation of Bonacich Power Index)

Network #	Person	Power Index	Rank	Percent	Octile
R2	Avita	0	1	.00%	
R18	M. Varienus	0	2	.00%	
R19	Q Allius Maximus	0	3	.00%	
R33	Ti Iulius Iulianus	0	4	.00%	1st Octile
R4	Canuleia Crispina	2	4	28.50%	
R5	L Cassius	2	6	28.50%	
R6	P Cassius Cae	2	6	28.50%	
R13	Rutilia L.f Ocrati	2	6	28.50%	
R29	Cn. Domitius Lucanus	2	6	28.50%	
R30	Cn. Domitius Tullus	2	6	28.50%	3rd Octile
R32	Flavia Domitilla	4	6	71.40%	
R37	Plotilla	4	12	71.40%	7th Octile
R14	Iulia Dynamis	4.67	12	85.70%	
R26	Cocceius Nerva	6	12	92.80%	
R35	L Iulius Rufus	7.33	12	100.00%	8th Octile

Dependents-based power (positive attenuation of Bonacich Power Index)

Network #	Person	Power Index	Ranking	Percent	Octile
R35	L Iulius Rufus	2.44	1	100.00%	
R26	Cocceius Nerva	2	2	92.80%	1st Octile
R4	Canuleia Crispina	0.67	3	50.00%	
R5	L Cassius	0.67	3	50.00%	
R6	P Cassius Cae	0.67	3	50.00%	
R13	Rutilia L.f Ocrati	0.67	3	50.00%	
R29	Cn. Domitius Lucanus	0.67	3	50.00%	

R30	Cn. Domitius Tullus	0.67	3	50.00%	4th Octile
R2	Avita	0	9	7.10%	
R18	M. Varienus	0	9	7.10%	
R19	Q Allius Maximus	0	9	7.10%	
R32	Flavia Domitilla	0	9	7.10%	
R33	Ti Iulius Iulianus	0	9	7.10%	
R37	Plotilla	0	9	7.10%	
R14	Iulia Dynamis	-0.22	15	.00%	8th Octile

Nerva-Marcus Aurelius

Centrality based power (negative attentuation of Bonacich Power Index)

Network #	Person	Power Index	Rank	Percent	Octile
R164	Servilius Firmus, P	-35	1	.00%	1st Octile
R166	Vibius Pudens	-31	2	.50%	
R139	Caetennius Magnio	-30	3	1.10%	
R84	Coelius Fortunatus	-26	4	1.70%	
R85	Coelius Philetus	-26	5	1.70%	
R113	Q. Aburnius Caedicianus	-23	6	2.80%	
R126	T. Stat. Max	-17	7	3.40%	
R79	Antoninus Pius	-17	8	3.90%	
R127	T. Stat. Max. Sev. Had	-16	9	4.50%	
R129	Ti. Claudius Maximus	-15	10	5.10%	
R82	C. Iulius Apollinaris	-14	11	5.60%	
R167	Vismatius Fortunatus	-13	12	6.20%	
R153	Ocius Antiochus, P	-13	13	6.80%	
R191	Arria Iulia Lupula	-11	14	7.30%	
R81	Arria_Fadilla	-9	14	7.90%	
R168	Ulpius Anicetianus, M	-8	16	8.50%	
R157	Pettius Proculus	-8	17	9.00%	
R73	Plotina	-8	18	9.60%	
R206	Annia Galeria Faustina	-8	18	9.60%	
R128	Ti. Claudius Celsus	-8	20	10.70%	
R214	M. Valerius Homullus	-8	21	11.30%	
R88	Domitia Lucilla Minor	-7	22	11.90%	
R106	M. Annius Libo	-7	23	12.50%	
R69	Trajan	-7	24	13.00%	2nd Octile
R156	Peducaeus Lupulus	-6	25	13.60%	
R71	Rupilia Faustina	-5	26	14.20%	
R77	A_Villius_ Alexander	-5	27	14.70%	
R133	Villius Aug	-5	28	14.70%	
R123	Sergia Paulina	-5	29	15.90%	
R49	T. Flavius Ampliatus	-5	30	16.40%	
R160	Pontius Clodianus, A	-5	31	16.40%	
R122	Sabina Aug.	-5	32	17.60%	
R226	Maius servus	-5	33	18.10%	
R83	Caecilia Quinta	-5	34	18.70%	
R150	Myrinus	-5	35	19.30%	
R205	Annia Corneficia	-4	36	19.80%	
R93	Hadrian	-4	37	20.40%	
R87	Domitia Domitiani	-4	38	21.00%	
R124	Sex. Laelius Lelianus	-4	39	21.00%	
R99	L. C. Iuventus	-4	39	22.10%	
R41	Abascantus Aug lib	-4	39	22.70%	
R231	Marcus Aurelius	-4	39	23.20%	
R130	Ti. Tutinius Sentius Satrianus	-4	39	23.80%	
R91	Flavia Seia Isaurica	-4	39	24.40%	
R154	Oppius Iustus, Q	-4	45	25.00%	
R234	Brittidius Priscus	-3	46	25.50%	
R102	L. Man Theocritus	-3	47	26.10%	3rd Octile
R80	Arminius_Cestianus	-3	48	26.70%	
R95	Iulius Eutactus	-3	49	26.70%	
R108	M. Ulpius Ulpianus	-3	50	26.70%	
R110	Octavius Cestianus	-3	51	26.70%	
R115	Q. Cassius Caecilianus	-3	52	26.70%	
R117	Q. Pomp Pae	-3	53	26.70%	
R65	Cornelius Palina	-3	53	30.10%	
R66	L. Publilius Celsus	-3	55	30.10%	
R208	Ti. Cl. Iulianus	-3	56	31.20%	
R229	T. Travius Felix	-3	56	31.20%	
R104	L Turranius Pri	-3	56	32.30%	
R97	Iunius Iulianus	-3	56	32.90%	
R141	Claudius Secundinus, Ti	-3	56	33.50%	
R149	Mercurius Ti. Cl. Quinquatrailis	-3	56	33.50%	
R144	Gellius Prudens, L	-3	56	34.60%	
R148	Statius Marcius Lucifer	-3	56	34.60%	
R63	Arruntia Camilla Camilli	-3	56	35.70%	
R224	Aelius Felix	-3	56	36.30%	
R146	Statius Marcius Bassus	-2	56	36.90%	
R101	L. Iulius Ursus Valerius Flaccus	-2	56	37.50%	
R174	Petronius Mamertinus	-2	56	38.00%	4th Octile
R219	Calvius Crescens	-2	56	38.60%	
R131	Titia Quartilla	-2	56	39.20%	
R145	Iulius Fortunatus, C	-2	56	39.70%	
R134	Aelius Alex, P	-2	56	40.30%	
R142	Cominius Proculus, C	-2	56	40.30%	
R158	Pompe Felix	-1	56	41.40%	
R195	Annia Fundandia Faustina	-1	56	42.00%	
R147	Marcius Fyrmus	-1	56	42.60%	
R212	Q. Pomponius Mussa	-1	56	43.10%	
R220	M. Ummidius Quadratus	-1	56	43.70%	
R136	Allius Rufus, L	-1	56	44.30%	
R198	Arria Caesennia Paulina	-1	56	44.30%	
R138	Avienus Halys	-1	56	45.40%	
R58	Plotia Isaurica	-1	56	46.00%	
R103	L. Turranius Gal	-1	56	46.50%	
R193	Valeria Polla	0	56	47.10%	
R52	Domitia Cn. f. Lucilla	0	56	47.70%	
R227	Sex. Pompeius Aeli	0	86	48.20%	
R228	Sex. Publicius Consors	0	87	48.20%	
R98	Iunius Sulpicianus	0	88	49.40%	
R96	Iulius Stephanus	0	89	50.00%	
R175	Petronius Septimianus	0	90	50.50%	5th
R165	Valerius Priscus, M	0	91	51.10%	
R216	L. Plautius Aquilinus	0	91	51.70%	

R42	Allienus Proclus	0	93	52.20%	
R43	Aviedus Quietus	0	94	52.20%	
R44	L. Bellicus Sollers	0	95	52.20%	
R45	C. Biculeius Priscus	0	96	52.20%	
R47	Claudius Iullus	0	97	52.20%	
R50	Antonia Malliola	0	98	52.20%	
R51	Atilia Quintilla	0	98	52.20%	
R53	M. Herennius Pollio	0	100	52.20%	
R55	P. Marcius Crispus	0	101	52.20%	
R56	L. Memmius Rufus	0	102	52.20%	
R62	A. Vicirius Martialis	0	103	52.20%	
R64	Q. Anticuleius Paetus	0	104	52.20%	
R70	L Cornelius Priscus (later)	0	105	52.20%	
R90	Flavia Operata	0	105	52.20%	
R92	Flavius Posidonius	0	107	52.20%	
R105	L Volusius Nepos	0	108	52.20%	
R111	P. Cassius Secundus	0	109	52.20%	
R118	Q. Pompeius Mameius	0	110	52.20%	
R119	Q. Pompeius Mammeianus	0	111	52.20%	
R120	Q Sc P P	0	112	52.20%	
R194	Sabina Sabinilla	0	113	52.20%	
R199	M. Caelius Iulianus	0	114	52.20%	
R200	Cuspius Rufinus	0	115	52.20%	
R202	Pomponia Bassila	0	115	52.20%	
R209	Cusinia Gratilla	0	117	52.20%	
R213	C. Prastina Pacatus Messalinus	0	117	52.20%	
R215	C. Iulius Dioscorus	0	119	52.20%	
R217	Cornelius Atticus	0	120	52.20%	
R221	Q. Canusius Praenestinus	0	121	52.20%	
R222	C. Statius Capito	0	121	52.20%	
R223	C. Aelius Asclepiades	0	123	69.30%	6th Octile
R68	Curiatus Cosanus	0	123	69.80%	
R135	Alfius Amandus, Sex	0	125	69.80%	
R109	Memmia Macrina	0	125	71.00%	
R196	Vitrasius Pollio	0	125	71.50%	
R163	Rutilius Doretus, D	0	125	72.10%	
R67	M. Titius Marcellus	1	125	72.70%	
R201	Iulia Saturnina	1	125	73.20%	
R86	Cornelia Manliola	1	131	73.80%	
R61	Servilius Capito	1	132	74.40%	
R72	Sergia Paullina	2	133	75.00%	
R54	Iunius Rufus	2	134	75.50%	7th Octile
R59	Regina Capitolina	2	135	75.50%	
R161	Procilia Phila	2	136	75.50%	
R171	Iulia Albana	2	137	75.50%	
R203	M. Pontius Sabinus	2	138	75.50%	
R204	Larcia Sabina	2	139	75.50%	
R107	M. Annius Verus	2	139	78.90%	
R78	Agathyrsus_Aug_l	2	141	79.50%	
R94	Iulia Procula	3	142	80.10%	
R132	Trebicia Tertulla	3	143	80.60%	
R48	Cornelius Severus	3	144	81.20%	
R211	Matidia Aug. f	4	145	81.80%	
R112	Plaetorius Nepos	4	146	82.30%	
R162	Rausius Pamphilus, T	4	147	82.90%	
R100	Ceionius Commodus, L. (Lucius Verus)	4	147	83.50%	
R159	Pomponius Ianuarius, Q	4	149	84.00%	
R143	Domitius Rufinus	5	150	84.60%	
R137	Aristius Thallus, A	6	150	85.20%	
R75	Vismatius Felix	6	152	85.70%	
R210	Q. Marcius Hermogenes	7	153	86.30%	
R116	Q. Marcius Hermogene	7	154	86.90%	
R140	Claudius Fortunatus	8	155	87.50%	8th Octile
R125	Stertinia Bassula	9	156	88.00%	
R155	P P( ) B( )	10	157	88.60%	
R89	Flavia Procula	10	158	89.20%	
R46	Calpurnia Secunda	10	159	89.70%	
R197	M. Aemelius Proculus	10	159	89.70%	
R152	Nunnidius Restitutus, C	12	161	90.90%	
R225	C. Calpetanus Pannychus	15	162	91.40%	
R169	C. Bruttius Praesens	17	163	92.00%	
R170	Caetennia Chione	17	164	92.00%	
R57	Pedania Quintilla	17	165	93.10%	
R218	Terentius Iulianus	17	166	93.70%	
R76	Vismatius Successus	18	167	94.30%	
R121	Q. Servilius Pudens	23	168	94.80%	
R151	Nunnidius Fortunatus, C	24	169	95.40%	
R233	Flavius Probus	25	170	96.00%	
R232	Flavius Corinthus	26	171	96.50%	
R114	Q. Asinius Marcellus	32	172	97.10%	
R74	Sex. Vismatius Neritus	32	173	97.70%	
R230	Vibia Procilla	33	173	98.20%	
R60	M. Rutilius Lupus	44	175	98.80%	
R192	M. Flavius Aper	44	176	99.40%	
R207	Asinia Quadratilla	53	177	100.00%	

Dependents-based power (positive attenuation of Bonacich Power Index)

Network #	Person	Power Index	Ranking	Percent	Octile
R94	Iulia Procula	38	1	100.00%	
R99	L C Iuventus	36	2	99.40%	
R151	Nunnidius Fortunatus	34	3	98.80%	
R58	Plotia Isaurica	32	4	98.20%	
R210	Q. Marcius Hermogenes	29	5	97.70%	
R127	T. Stat. Max. Sev. Had	27	6	97.10%	
R157	Pettius Proculus	21	7	96.50%	
R116	Q. Marcius Hermogene	18	8	96.00%	
R205	Annia Corneficia Faustina	17	9	95.40%	
R73	Plotina	16	10	94.80%	
R169	C. Bruttius Praesens	16	11	93.70%	
R170	Caetennia Chione	16	11	93.70%	
R224	Aelius Felix	15	13	93.10%	
R91	Flavia Seia Isaurica	13	14	92.60%	
R168	Ulpius Anicetianus, M	12	15	92.00%	
R49	T. Flavius Ampliatus	11	16	91.40%	

R134	Aelius Alex, P	11	17	90.90%	
R232	Flavius Corinthus	11	18	90.30%	
R175	Petronius Septimianus	11	19	89.70%	
R230	Vibia Procilla	10	20	89.20%	
R132	Trebicia Tertulla	9	21	88.60%	
R48	Cornelius Severus	8	22	88.00%	
R155	P P( ) B( )	8	23	87.50%	1st Octile
R159	Pomponius Ianuarius, Q	8	24	86.90%	
R93	Hadrian	8	25	86.30%	
R106	M. Annius Libo	7	26	85.70%	
R212	Q. Pomponius Mussa	7	27	85.20%	
R207	Asinia Quadratilla	7	28	84.60%	
R163	Rutilius Doretus, D	6	29	84.00%	
R219	Calvius Crescens	6	30	83.50%	
R206	Annia Galeria Faustina	6	31	82.90%	
R76	Vismatius Successus	6	32	82.30%	
R103	L. Turranius Gal	6	33	81.80%	
R79	Antoninus Pius	6	34	81.20%	
R150	Myrinus	5	35	80.60%	
R227	Sex. Pompeius Aeli	5	36	79.50%	
R228	Sex. Publicius Consors	5	36	79.50%	
R166	Vibius Pudens	5	38	78.90%	
R96	Iulius Stephanus	5	39	78.40%	
R211	Matidia Aug. f	4	40	77.80%	
R63	Arruntia Camilla Camilli	4	41	77.20%	
R52	Domitia Cn. f. Lucilla	4	42	76.70%	
R86	Cornelia Manliola	3	43	75.50%	
R191	Arria Iulia Lupula	3	43	75.50%	
R60	M. Rutilius Lupus	3	45	75.00%	2nd Octile
R65	Cornelius Palina	3	46	73.80%	
R66	L. Publilius Celsus	3	46	73.80%	
R125	Stertinia Bassula	3	48	73.20%	
R167	Vismatius Fortunatus	3	49	72.70%	
R112	Plaetorius Nepos	3	50	72.10%	
R139	Caetennius Magnio	3	51	71.50%	
R88	Domitia Lucilla Minor	3	52	71.00%	
R196	Vitrasius Pollio	2	53	70.40%	
R131	Titia Quartilla	2	54	69.80%	
R128	Ti. Claudius Celsus	2	55	69.30%	
R223	C. Aelius Asclepiades	2	56	68.70%	
R98	Iunius Sulpicianus	2	57	68.10%	
R68	Curiatus Cosanus	2	58	67.60%	
R193	Valeria Polla	1	59	67.00%	
R57	Pedania Quintilla	1	60	66.40%	
R135	Alfius Amandus, Sex	1	61	65.90%	
R100	Ceionius Commodus, L. (Lucius Verus)	1	62	65.30%	
R90	Flavia Operata	1	63	63.00%	
R92	Flavius Posidonius	1	63	63.00%	
R118	Q. Pompeius Mameius	1	63	63.00%	
R119	Q. Pompeius Mammeianus	1	63	63.00%	
R109	Memmia Macrina	1	67	62.50%	3rd Octile
R54	Iunius Rufus	1	68	59.00%	
R59	Regina Capitolina	1	68	59.00%	
R161	Procilia Phila	1	68	59.00%	
R171	Iulia Albana	1	68	59.00%	
R203	M. Pontius Sabinus	1	68	59.00%	
R204	Larcia Sabina	1	68	59.00%	
R84	Coelius Fortunatus	0	74	57.90%	
R85	Coelius Philetus	0	74	57.90%	
R130	Ti. Tutinius Sentius Satrianus	0	76	57.30%	
R113	Q. Aburnius Caedicianus	0	77	56.80%	
R158	Pompe Felix	0	78	56.20%	
R77	A_Villius_ Alexander	0	79	55.10%	
R133	Villius Aug	0	79	55.10%	4th Octile
R42	Allienus Proclus	0	81	39.70%	
R43	Aviedus Quietus	0	81	39.70%	
R44	L. Bellicus Sollers	0	81	39.70%	
R45	C. Biculeius Priscus	0	81	39.70%	
R47	Claudius Iullus	0	81	39.70%	
R50	Antonia Malliola	0	81	39.70%	
R51	Atilia Quintilla	0	81	39.70%	
R53	M. Herennius Pollio	0	81	39.70%	
R55	P. Marcius Crispus	0	81	39.70%	
R56	L. Memmius Rufus	0	81	39.70%	
R62	A. Vicirius Martialis	0	81	39.70%	
R64	Q. Anticuleius Paetus	0	81	39.70%	
R70	L Cornelius Priscus (later)	0	81	39.70%	
R105	L Volusius Nepos	0	81	39.70%	
R111	P. Cassius Secundus	0	81	39.70%	
R120	Q Sc P P	0	81	39.70%	
R194	Sabina Sabinilla	0	81	39.70%	
R199	M. Caelius Iulianus	0	81	39.70%	
R200	Cuspius Rufinus	0	81	39.70%	
R202	Pomponia Bassila	0	81	39.70%	
R208	Ti. Cl. Iulianus	0	81	39.70%	
R209	Cusinia Gratilla	0	81	39.70%	
R213	C. Prastina Pacatus Messalinus	0	81	39.70%	
R215	C. Iulius Dioscorus	0	81	39.70%	
R217	Cornelius Atticus	0	81	39.70%	
R221	Q. Canusius Praenestinus	0	81	39.70%	
R222	C. Statius Capito	0	81	39.70%	
R137	Aristius Thallus, A	0	108	39.20%	
R216	L. Plautius Aquilinus	0	109	38.60%	
R74	Sex. Vismatius Neritus	0	110	38.00%	5th Octile
R141	Claudius Secundinus, Ti	0	111	36.90%	
R149	Mercurius Ti. Cl. Quinquatrailis	0	111	36.90%	
R229	T. Travius Felix	0	113	36.30%	
R82	C. Iulius Apollinaris	0	114	35.70%	
R226	Maius servus	-6	115	35.20%	
R143	Domitius Rufinus	-6	116	34.60%	
R121	Q. Servilius Pudens	-6	117	34.00%	
R67	M. Titius Marcellus	-6	118	33.50%	
R41	Abascantus Aug lib	-7	119	32.90%	

R234	Brittidius Priscus	-7	120	32.30%	
R225	C. Calpetanus Pannychus	-7	121	31.80%	
R201	Iulia Saturnina	-7	122	31.20%	
R97	Iunius Iulianus	-7	123	30.60%	
R142	Cominius Proculus, C	-8	124	30.10%	
R164	Servilius Firmus, P	-8	125	29.50%	
R80	Arminius_Cestianus	-8	126	28.90%	
R87	Domitia Domitiani	-8	127	28.40%	
R95	Iulius Eutactus	-8	128	27.80%	
R108	M. Ulpius Ulpianus	-8	129	27.20%	
R110	Octavius Cestianus	-9	130	26.70%	
R115	Q. Cassius Caecilianus	-9	131	26.10%	
R117	Q. Pomp Pae	-9	132	25.50%	
R124	Sex. Laelius Lelianus	-9	133	25.00%	6th Octile
R140	Claudius Fortunatus	-9	134	24.40%	
R214	M. Valerius Homullus	-10	135	23.80%	
R46	Calpurnia Secunda	-10	136	23.20%	
R197	M. Aemelius Proculus	-10	137	22.70%	
R126	T. Stat. Max	-10	138	22.10%	
R101	L. Iulius Ursus Valerius Flaccus	-10	139	21.50%	
R71	Rupilia Faustina	-11	140	21.00%	
R123	Sergia Paulina	-11	141	20.40%	
R233	Flavius Probus	-11	142	19.80%	
R195	Annia Fundandia Faustina	-11	143	19.30%	
R160	Pontius Clodianus, A	-11	144	18.70%	
R122	Sabina Aug.	-12	145	18.10%	
R147	Marcius Fyrmus	-12	146	17.60%	
R154	Oppius Iustus, Q	-12	147	17.00%	
R75	Vismatius Felix	-12	148	16.40%	
R218	Terentius Iulianus	-12	149	15.90%	
R72	Sergia Paullina	-12	150	15.30%	
R102	L. Man Theocritus	-13	151	14.70%	
R104	L Turranius Pri	-13	152	14.20%	
R136	Allius Rufus, L	-13	153	13.60%	
R198	Arria Caesennia Paulina	-13	154	13.00%	
R146	Statius Marcius Bassus	-13	155	12.50%	7th Octile
R69	Trajan	-14	156	11.90%	
R144	Gellius Prudens, L	-14	157	11.30%	
R148	Statius Marcius Lucifer	-14	158	10.70%	
R107	Annius Verus	-14	159	10.20%	
R78	Agathyrsus_Aug_l	-14	160	9.60%	
R153	Ocius Antiochus, P	-15	161	9.00%	
R138	Avienus Halys	-15	162	8.50%	
R220	M. Ummidius Quadratus	-15	163	7.90%	
R165	Valerius Priscus, M	-15	164	7.30%	
R61	Servilius Capito	-15	165	6.80%	
R145	Iulius Fortunatus, C	-16	166	6.20%	
R81	Arria_Fadilla	-16	167	5.60%	
R156	Peducaeus Lupulus	-16	168	5.10%	
R129	Ti. Claudius Maximus	-16	169	4.50%	
R192	M. Flavius Aper	-16	170	3.90%	
R162	Rausius Pamphilus, T	-16	171	3.40%	
R174	Petronius Mamertinus	-17	172	2.80%	
R231	Marcus Aurelius	-17	173	2.20%	
R89	Flavia Procula	-17	174	1.70%	
R114	Q. Asinius Marcellus	-17	175	1.10%	
R152	Nunnidius Restitutus, C	-17	176	.50%	
R83	Caecilia Quinta	-18	177	.00%	8th Octile

Severans

Centrality based power (negative attentuation of Bonacich Power Index)

Network #	Person	Power Index	Rank	Percent	Octile
R181	C. Fulvius Plautianus	-24	1	.00%	
R185	Septimius Severus	-10	2	6.20%	
R176	Petronius Mamertinus (grandson)	-4	3	12.50%	1st Octile
R187	C. Calpetanus Crescens	-2	3	18.70%	
R188	Fulvius Primitivus	-2	5	18.70%	
R189	L. Lanius Felicissimus	-2	5	18.70%	
R190	L. Numerius Iustus	-2	5	18.70%	2nd Octile
R172	Calpurnius Proculus	0	5	43.70%	
R177	Acilia Malliola	0	5	43.70%	
R178	Aelia Severa	0	5	43.70%	
R180	Flavius Titianus	0	11	43.70%	
R183	Mummia Vara	0	11	43.70%	
R186	T. Statilius Silianus	0	11	43.70%	3rd Octile
R182	Hortensius Paulinus	2	11	81.20%	
R184	Passenia Petronia	2	15	81.20%	7th Octile
R179	Aemilia Severa	112	16	93.70%	
R173	C. Iulius Flaccus Aelianus	114	17	100.00%	8th Octile

Dependents-based power (positive attenuation of Bonacich Power Index)

Network #	Person	Power Index	Ranking	Percent	Octile
R179	Aemilia Severa	12	1	100.00%	
R185	Septimius Severus	2	2	93.70%	
R182	Hortensius Paulinus	1.33	3	87.50%	1st Octile
R184	Passenia Petronia	0.67	4	81.20%	2nd Octile
R187	C. Calpetanus Crescens	0.33	5	56.20%	
R188	Fulvius Primitivus	0.33	5	56.20%	
R189	L. Lanius Felicissimus	0.33	5	56.20%	
R190	L. Numerius Iustus	0.33	5	56.20%	3rd Octile
R172	Calpurnius Proculus	0	9	6.20%	
R176	Petronius Mamertinus (grandson)	0	9	6.20%	
R177	Acilia Malliola	0	9	6.20%	
R178	Aelia Severa	0	9	6.20%	
R180	Flavius Titianus	0	9	6.20%	
R181	C. Fulvius Plautianus	0	9	6.20%	
R183	Mummia Vara	0	9	6.20%	
R186	T. Statilius Silianus	0	9	6.20%	
R173	C. Iulius Flaccus Aelianus	-10	17	.00%	8th Octile

Appendix E: Market Orientation Index

This data is derived from Filippi and Stanco's draft of their 2005 paper, 'Epigrafia e toponomastica della produzione laterizia nella Valle del Tevere: l'Umbria e la Sabina tra *Tuder* e *Crustumerium*; L'Etruria tra *Volsinii* e *Lucus Feroniae*', from *CIL*, and the SES collection. Future discoveries of stamped brick will alter the numbers presented here, so these figures should be considered reasonably accurate as of December 2002.

Julio-Claudian

Stamp type	# found in rural contexts	# found in urban (Rome) context	Total	index value
950	1	2	3	2
978	1	0	1	-10
1393	8	72	80	9
1460	7	2	9	-3.5
2196	1		1	-10
2263	2	1	3	-2
1254	1	1	2	1
1268	4	1	5	4
1269	3	5	8	1.6
1279	1	2	3	2
1325	2	2	4	1
1349/50.1	1		1	-10
1445a	1	2	3	2
1510b	5	7	12	1.4
1510c	1	1	2	1
1510d	2	3	5	1.5
1510dII	1	3	4	3
1511 vel S.395	2		2	-10
2179/80	1		1	-10
2316	1		1	-10
347	2	2	4	1
438	3	1	4	-3
666	5	1	6	-5
670a	1	1	2	1
670b	1	3	4	3
777	1	1	2	1
864	4	4	8	1
874	1	4	5	4
898/9	1		1	-10
Aristanius	7		7	-10
S. 83	3		3	-10
total	75	121	196	

Flavian

Stamp type	# found in rural contexts	# found in urban (Rome) context	Total	index value
1000	9	43	52	5
1002	1	4	5	4
1094b	1	4	5	4
1094d	3	5	8	2
1094e	1	16	17	16
1095	1	5	6	5
1096a	1	9	10	9
118a	10	22	32	2
1253c	1	4	5	4
1253f	1	4	5	4
1289	1	2	3	2
1346a	1	4	5	4
148	1	3	4	3
259	1	9	10	9
283	2	9	11	4.5
303b	1	3	4	3
304	1	2	3	2
308	1	0	1	-10
310	2	5	7	3
632	1	3	4	3
658b corr.	6	2	8	-3
659a	5	6	11	1
659c	7	8	15	1
660	1	1	2	1
661a	1	2	3	2
662a*	2	5	7	3
664c	1	3	4	3
981	1	3	4	3
990	1	8	9	8
992c	1	1	2	1
999a=c?	1	1	2	1
999g	1	9	10	9
1979	1	0	1	-10
S. 431	3	0	3	-10
S.269, 992	1	1	2	1
app 124d	1	0	1	-10
1096g	1	5	6	5
total	74	211	287	

Nerva - Hadrian

Stamp type	# found in rural contexts	# found in urban (Rome) context	Total	index value
1008	2	6	8	3
1014a	1	19	20	19
1015a	1	8	9	8

1020	1	3	4	3
1026a	1	8	9	8
1027	5	2	7	-2.5
1029 a	1	10	11	10
1029c	1	10	11	10
1033*	1	11	12	11
1036	1	9	10	9
1051	2	13	15	6.5
1053*	2	9	11	4.5
1057	1	6	7	6
1075a	1	46	47	46
1100	10	4	14	-2.5
1102	4	23	27	5.75
1103c	1	0	1	0
1106a	1	1	2	1
1110	1	5	6	5
1111	1	0	1	-10
1115	1	26	27	26
1116d*	1	3	4	3
1118a	3	4	7	1.33
1118b	3	2	5	-1.5
113b	2	2	4	1
1228a	1	2	3	2
124	1	3	4	3
130	1	3	4	3
1342	1	2	3	2
1347	1	16	17	16
1348a	1	20	21	20
1420	1	2	3	2
1423a	1	1	2	1
207	1	2	3	2
209	1	0	1	-10
210	1	1	2	1
226	2	9	11	4.5
245*	1	5	6	5
246	1	5	6	5
249=S. 55*	1	4	5	4
252	1	7	8	7
261a	1	10	11	10
263	1	3	4	3
267	1	9	10	9
268=S.59	1	0	1	-10
270a	1	6	7	6
273	1	0	1	-10
275	2	10	12	5
288	1	2	3	2
313	1	16	17	16
336	2	0	2	-10
337	2	1	3	-2
341	1	1	2	1
358.1	1	0	1	-10
359	2	1	3	-2
374	1	16	17	16
375	8	16	24	2
377b	1	13	14	13
416	5	1	6	-5
422	1	5	6	5
452*	1	1	2	1
453b*	2	1	3	-2
486a	2	6	8	3
518b	1	4	5	4
521	1	1	2	1
528	1	2	3	2
545 var. 2?	1	1	2	1
578a	2	9	11	4.5
585a	3	6	9	2
61	4	2	6	-2
635c	2	22	24	11
637	1	7	8	7
674 corr.*	2	16	18	8
708a	1	14	15	14
802	1	5	6	5
803	5	4	9	-1.25
804b	1	0	1	0
806	1	10	11	10
846	1	5	6	5
849	2	4	6	2
855	1	5	6	5
917	2	2	4	1
S.222	1	0	1	0
S. 276	2	0	2	0
S. 87	1	0	1	0
811f	2	16	18	8
532	1	1	2	1
947	1	0	1	-10
total	146	566	712	

Antoninus Pius – Commodus

Stamp type	# found in rural contexts	# found in urban (Rome) context	Total	index value
1019b	1	1	2	1
1050	1	2	3	2
1052 = S. 277*?	2	6	8	3
1078	1	5	6	5
1081	1	6	7	6
1088	1	2	3	2
1089*	1	3	4	3
1144	1	1	2	1
1145	2	12	14	6
1146a	2	5	7	2.5
1219*	1	2	3	2

131	1	0	1	-10
134	1	9	10	9
140	1	1	2	1
1422	1	3	4	3
1440a or b	3	2	5	-1.5
156	1	5	6	5
172	1	12	13	12
173	1	3	4	3
186	1	2	3	2
198	2	1	3	0.5
211	1	12	13	12
223a	7	3	10	-2.33
226	2	9	11	4.5
292	1	15	16	15
326	1	1	2	1
327?	1	4	5	4
368	2	8	10	4
468a	1	1	2	1
475	1	3	4	3
541a-b=VIII 22632.6=XI 1402	1	10	11	10
622=S.189	1	0	1	-10
652	1	2	3	2
653	2	6	8	3
685*	1	3	4	3
725	1	7	8	7
738	1	10	11	10
754a	1	6	7	6
758	1	2	3	2
765	1	8	9	8
860=850	1	19	20	19
861	8	11	19	1.38
862	6	13	19	2.17
731b	1	7	8	7
1533	1	1	2	1
total	72	244	316	3.39

Severans

Stamp type	# found in rural contexts	# found in urban (Romc) context	Total	index value
155	3	15	18	5
156?	1	5	6	5
157	6	14	20	2.33
158	1	13	14	13
159	1	14	15	14
162	2	8	10	4
163	3	67	70	22.33
176	1	4	5	4
178	1	17	18	17

189	4	1	5	-4
190	2	13	15	6.5
192	1	10	11	10
193	1	8	9	8
195	3	13	16	4.33
196	1	1	2	1
203	2	11	13	5.5
204	4	37	41	9.25
213 compl.	1	5	6	5
221a*	1	11	12	11
383*	1	6	7	6
404	1	39	40	39
405	2	11	13	5.5
408	10	92	102	9.2
414	1	3	4	3
427	5	6	11	1.2
430	10	7	17	-1.43
433	7	6	13	-1.17
435	1	3	4	3
526	6	30	36	5
624	3	14	17	4.67
625	11	25	36	2.27
626	2	28	30	14
629	1	1	2	1
686	2	4	6	2
687	2	9	11	4.5
689	1	3	4	3
759	1	12	13	12
762a	1	30	31	30
762b	2	9	11	4.5
764 compl.	1	19	20	19
980	1	1	2	1
Total	111	625	736	5.63

Late Antique

Stamp type	# found in rural contexts	# found in urban (Rome) context	Total	index value
1671	1	0	1	-10
1673	1	1	2	1
1709	1	1	2	1
1667	2	1	3	-2
1564	4	2	6	-2
1666	1	2	3	2
1675	3	2	5	-1.5
1674	1	2	3	2
1546	1	3	4	3
1726	4	4	8	1
1578a	1	7	8	7
1572	1	8	9	8

1552a	1	11	12	11
1574b	1	12	13	12
1581a	2	16	18	8
1669	9	23	32	2.56
1563a	1	26	27	26
1615a	1	29	30	29
1665	6	42	48	7
1569a	3	83	86	27.67
total	45	275	320	
Grand Total	523	2042	2567	

References and Bibliography

"Place Names in Rotherhithe" URL: http://www.bathspa.ac.uk/greenwood/lplaces.html [Aug. 28 2002]

"Santa Fe Institute Current Research Focus Areas" URL:http://www.santafe.edu/sfi/research/indexResearchAreas.html [Sept. 20th 2002]

"The History Of The Port Of London up to the Advent Of the Port Of London Authority" URL: http://www.portoflondon.co.uk/display_fixedpage.cfm?id+238&site=leisure [Aug. 28 2002]

Adam, J. P. (1994) *Roman Building: Materials and Techniques* Trans. A. Mathews. London: B.T. Batsford. Originally published 1984, *La construction romaine: materiaux et techniques* Paris: Grands manuels Pircard.

Ades, A. and E. Glaesar. (1995) "Trade And Circuses: Explaining Urban Giants". *Quarterly Economics Journal* **110**: 195-227.

Agostini, S. and W. Pellegrini. (1996). "Altre Risorse della Maiella: Testimonianze di archeologia industriale". A. R. Staffa (ed). In *La Presenza dell'Uomo sulla Maiella: Archeologia e paesaggio storico. Guida alla Sezione Archeologica del Museo Paolo Barrasso Centro Visitatori della Riserva naturale Valle dell'Orfento.* Caramanico Terme (PE), Ministero delle Risorse Agricole, Alimentari e Forestali, Corpo Forestale dello Stato PESCARA, Regione Abruzzo; Ministero per I Beni Culturali e Ambientali, Soprintendenza Archeologica dell'Abruzzo CHIETI, Parco Nazionale della Maiella.

Albert, R. and A.-L. Barabàsi. (2002). "Statistical mechanics of complex networks". Reviews of Modern Physics **74**: 47-98.

Albert, R., H. Jeong, and A.-L. Barabási. (1999) "Diameter of the World Wide Web", *Nature* **401**: 130-1.

Albert, R., H. Jeong, and A.-L. Barabási. (2000) "Attack and Error Tolerance of Complex Networks", *Nature* **406**: 378-81.

Albertazii, A., A. Failla, G. Filippi, G. Turci (1994). "Analisi archeometriche di laterizi bollati di età romana: un esempio dalla Sabina". Atti 10° Congresso A.N.M.S., Bologna. p. 347-372

Alexander, L. E. and Klug, H. P. (1948) "Basic Aspects of X-ray Absorption in Quantitative Diffraction Analysis of Powder Mixtures." *Analytical Chemistry.* **20:**886-894.

Amaral, L. A. Sacala, M. Barthélémy, and H. Stanley (2002). "Classes of Behaviour of Small-World Networks," URL: http://xxx.lanl.gov/cond-mat/0001458, January 31. [Aug. 14 2002]

Anderson, J. C. jr (1991). *Roman Brickstamps: The Thomas Ashby Collection.* London, Archaeological Monographs of the British School at Rome.

Anderson, J. C. jr (1997). *Roman Architecture and Society.* Baltimore: Johns Hopkins University Press.

Aubert, J. J. (1994). *Business managers in ancient rome. A social and economic study of institores, 200 BC- AD 250.* Columbia Studies in the Classical Tradition 21. New York: E.J. Brill.

Bak, P., and M. Paczuski (1991). "Self-organized criticality." *Scientific American* **264**(1): 46-53.

Baldi, G., E. Gliozzo, D. Manacorda (forthcoming). "Chemical analysis of Roman stamped bricks".

Baldi, G., M. Bertinetti, L. Camilli, et al (1999). *Epigrafia e archeometria della produzione laterizia bollata di Roma antica e suburbio.* XI Congresso internazionale di epigrafia greca e latina, Roma 18-24 settembre 1997, Roma, Quasar.

Barabàsi, A.-L. (2002). *Linked: The New Science of Networks.* Cambridge, MA, Perseus.

Barabási, A.-L. and R. Albert. (1999) "Emergence of Scaling in Random Networks". *Science* **268:** 509-512.

Battey, M. and P. Torrens. (2001). "Modeling Complexity: The Limits to Prediction". *Centre for Advanced Spatial Analysis Working Paper Series, #36.* URL: http://www.casa.ucl.ac.uk/working_papers.htm [Aug. 23rd 2002]

Bently, R. A. and. H. Maschner . (2001). "Stylistic Change as a Self-Organized Critical Phenomenon: An Archaeological Study in Complexity." *Journal of Archaeological Method and Theory* **8**(1): 35-66.

Betts, I., Ernest W. Black, and John Gower (1994). "A Corpus of Relief-Patterned Tiles in Roman Britain." *Journal of Roman Pottery Studies* 7.

Blake, M. (1959). *Roman Construction in Italy from Tiberius through the Flavians.* Washington DC: Carnegie Institution of Washington.

Bloch, H. (1947). *I bolli laterizi e la storia edilizia romana: contributi all'archeologia e alla storia di Roma.* Rome: Comune di Roma, Ripartizione antichità e belle arti.

Bloch, H. (1959). "The Serapeum of Ostia and the Brick-Stamp of 123. New Landmark in the History of Roman Architecture." *American Journal of Archaeology* **63**: 225-240.

Borden, I., Jane Rendell, Joe Kerr, Alicia Pivaro (2001). "Things, Flows, Filters, Tactics" In Ian Borden, Joe Kerr, Jane Rendell, Alicia Pivaroe (eds). *The Unknown City: Contesting Architecture and Social Space.* Cambridge Massachusetts: MIT Press.

Borgatti, S. P., M.G. Everett, and L.C. Freeman (1996). *UCINET IV Version 1.64.* Natick, MA, Analytic Technologies URL: http:www.analytic.com.

Bouchaud, J-P., and M. Mézard. (2000) "Wealth Condensation in a Simple Model of Economy" *Physica A* **282:** 536.

Brigham, T., Bruce Watson, Ian Tyers, Ryszard Bartkowiak (1996). "Current Archaeological Work at Regis House in the City of London (Part I)." *London Archeaologist* **8**(2): 31-38.

Broise, H. (2000). "Les Estampilles Anépigraphes sur *Bessales* de la Rome Impériale". In P. Bocheron, Henri Broise, and Yvon Thébert (eds). *La Brique Antique et Médiévale: Production et Commercialisation d'un Matériau.* Rome: École Française de Rome.

Bruun, C., ed. (1991). *The Water Supply of Ancient Rome: A Study of Roman Imperial Administration.* Commentationes Humnarum Litterarum **93**. Helsinki, Finland: Societas Scientiarum Fennica.

Bruun, C., ed. (2005). *Interpretare i bolli laterizi di Roma e della valle del tevere: produzione, storia economica e topografia. Atti dell convegno all'Ecole Française de Rome e all' Institutum Romanum Finlandiae, 31 marzo e 1 aprile 2000, organizzato da Christer Bruun e François Chausson* Acta Instituti Romani Finlandiae 32.

Buchanan, M. (2002). *Nexus: Small Worlds and the Groundbreaking Science of Networks.* New York, W. W. Norton.

Carettoni, Gianfilippo; Colini, Antonio; Cozza, Lucos; and Gatti, Guglielmo, eds. (1960). *La pianta marmorea di Roma antica. Forma urbis Romae.* Rome: Comune di Roma.

Carlisle, J. A. and L. Cheliak. (1984) "The Heyday of Valley Logging" In *Ottawa Valley Forestry: 1984 Forestry Capital of Canada.* Petawawa, Ontario: Ottawa Valley Forestry Capital Promotion Society.

Cilliers, P. (1998). *Complexity and Postmodernism: Understanding Complex Systems.* London: Routledge.

Coarelli, F. (2000). "Discussions, sous la presidence de J.-M. Pesez". In P. Boucheron, Henri Broise, and Yvon Thebert (eds). *La Brique Antique et Medievale.* Rome: École Française de Rome.

Coates-Stephens, R. and A. Parisi. (1999). "Indagine su un crollo delle Mura Aurelianae presso Porta Maggiore." *Analecta Romana Instituti Danici* **26**: 85-98.

Collingwood, R. and R. Wright. (1992) *The Roman Inscriptions of Britain Volume 2: Instrumentum Domesticum,* fascicule *4*. Frere, S. S. and R. S. O. Tomlin (eds.). Stroud, Glos.: Alan Sutton Publishing.

Cosentino, D., Maurizio Parotto, Antonio Praturlon, (Eds). (1993). *Lazio, 14 Itinerari.* Guide Geologiche Regionali. Rome: BE-MA editrice.

Craddock, P. (1995). *Early Metal Mining and Production.* Edinburgh: Edinburgh University Press.

Curtis, B. (2001). "That Place Where: Some thoughts on memory and the city". In Ian Borden, Joe Kerr, Jane Rendell, Alicia Pivaroe (eds) *The Unknown City: Contesting Architecture and Social Space.* Cambridge Massachusetts: MIT Press.

Dandeker, C. and T. Johnson (1989) "Patronage: Relation and System" In A. Wallace-Hadrill (ed) *Patronage in Ancient Society.* London: Routledge.

Darvill, T. and A. McWhirr. (1982). "Roman Brick Production and the Environment." In D. Miles, (ed) *The Romano-British Countryside: Studies in Rural Settlement and Economy.* BAR British Series **103**. Oxford: BAR Publishing.

Darvill, T. and A. McWhirr. (1984). "Brick and tile production in Roman Britain: models of economic organisation." *World Archaeology* **15**(3): 239-261.

Davies, J. K. (1998). "Ancient Economies: Models and Muddles". In H. Parkins and C. Smith (eds.) *Trade, Traders and the Ancient City*. London: Routledge.

De Kleijn, G. (2001) *The Water Supply of Ancient Rome. City Area, Water, and Population* Gieben: Amsterdam.

DeLaine, J. (1995). "The supply of building materials to the City of Rome". In N. Christie (ed). *Settlement and Economy in Italy 1500 BC – AD 1500. Papers of the Fifth Conference of Italian Archaeology*. Oxford: Oxbow Monograph. **41:** 555-562.

DeLaine, J. (1997) *The Baths of Caracalla: A Study in the Design, Construction, and Economics of Large-Scale Building Projects in Imperial Rome* Journal of Roman Archaeology Supplementary Series **25**. Portsmouth Rhode Island.

DeLaine, J. (2001). "Bricks and mortar: exploring the economics of building techniques at Rome and Ostia". In D. J. Mattingly and J. Salmon (eds). *Economies Beyond Agriculture in the Classical World*. London: Routledge.

DeLaine, J. (2002). "Building Activity in Ostia in the Second Century AD." *Acta Instituti Romanae Finlandiae*. 27.

De Neeve, P.W. (1984). *Colonus. Private Farm – Tenancy in Roman Italy during the Republic and the Early Principate*. Amsterdam: Gieben.

De Neeve, P.W. (1990) "A Roman landowner and his estates : Pliny the younger" *Athenaeum* **68**: 363-402.

Dressel, H. (1891). *Corpus Inscriptionum Latinarum XV.1*, Rome: Berolini.

Drinkwater, J. G. (2001). "The Gallo-Roman Woollen Industry and the Great Debate: The Igel Column Revisted". In D. Mattingly and J. Salmon (eds.) *Economies beyond agriculture in the classical world*. London: Routledge.

Drummond, A. (1989). "Early Roman Clientes". In A. Wallace-Hadrill (ed) *Patronage in Ancient Society*. London: Routledge.

Dyson, S. L. (1979). "New Methods and Models in the Study of Roman Town-Country Systems." *Ancient World* **2**: 91-95.

Dyson, S. L. (1992). *Community and Society in Roman Italy*. Baltimore: Johns Hopkins University Press.

Emirbayer, M. and J. Goodwin. (1994). "Network Analysis, Culture, and the Problem of Agency". *American Journal of Sociology* **99** (6): 1411-1454.

Erdkamp, P. (2001). "Beyond the Limits of the 'Consumer City'. A Model of the Urban and Rural Economy in the Roman World." *Historia* **2001**(3): 332-356.

Favro, D. G. (1996) *The Urban Image of Augustan Rome*. Cambridge: Cambridge University Press.

Filippi, G. (1992). "Scheda computerizzata per il materiale laterizio bollato". *Archeologia e Calcolatori* **3:** 219-252.

Filippi, G. and E. Stanco. (2005). 'Epigrafia e toponomastica della produzione laterizia nella Valle del Tevere: l'Umbria e la Sabina tra *Tuder* e *Crustumerium*; L'Etruria tra *Volsinii* e *Lucus Feroniae*' in C. Bruun, C. *Interpretare i bolli laterizi di Roma e della valle del tevere: produzione, storia economica e topografia. Atti dell convegno all'Ecole Française de Rome e all' Institutum Romanum Finlandiae, 31 marzo e 1 aprile 2000, organizzato da Christer Bruun e François Chausson* Acta Instituti Romani Finlandiae 32.

Finley, M. I. (1985). *The Ancient Economy*. 2nd edition. First published 1973. London: University of California Press.

Foxhall, L. (1990). 'The Dependent Tenant: Leasing and Labour in Italy and Greece'. *Journal of Roman Studies* **80**:97-114.

Freestone, I.C. (1997). "Analysis of the brick and tile". In Potter, T. W. and A. C. King (eds) *Excavations at the Mola di Monte Gelato. A Roman and Medieval Settlement in South Etruria*. London: Archaeological Monographs of the British School at Rome.

Gabler, R., R. Sayer, D. Wise et al. (1999). *Essentials of Physical Geography*. Orlando: Harcourt Brace.

Gaffney, V., H. Patterson, and P. Roberts (2001). "Forum Novum-Vescovio: Studying Urbanism in the Tiber Valley." *Journal of Roman Archaeology* **14**: 59-79.

Garnsey, P. and G. Woolf. (1989). "Patronage of the rural poor in the Roman world". In A. Wallace-Hadrill (ed) *Patronage in Ancient Society*. London: Routledge.

Graham, S. (2004) 'Three Brick Stamps from Castellum Amerinum' appendix to Paul Johnson, Simon Keay, and Martin Millett, 'Lesser urban sites in the Tiber valley: Baccanae, Forum Cassii and

Castellum Amerinum" *Papers of the British School at Rome* 72: 69-99.

Graham, S. (2005a) 'Of Lumberjacks and Brick Stamps: Working with the Tiber as Infrastructure'. In A. MacMahon and J. Price. (eds.) *Roman Urban Living*. Oxbow. pp106-124.

Graham, S. (2005b) 'Agent Based Modeling, Archaeology and Social Organisation: The Robustness of Rome' *The Archaeological Computing Newsletter* 63.

Graham, S. (2006) 'Who's in Charge? Studying Social Networks in the Roman Brick Industry in Central Italy', in Mattusch, C. and A. Donohue (eds.) ACTA *of the XVIth International Congress of Classical Archaeology*. Dave Brown Publishing.

Granovetter, M. (2002). "A Theoretical Agenda for Economic Sociology". In Mauro Guillen, Randall Collins, Paula England and Marshall Meyer, editors. *The New Economic Sociology: Developments in an Emerging Field*. New York: Russell Sage Foundation.

Greene, K. (1986). *The Archaeology of the Roman Economy*. London: B.T. Batsford.

Hanneman, R. and M. Riddle. 2005. Introduction to social network methods. Riverside, CA: University of California, Riverside, available online at url: http://faculty.ucr.edu/~hanneman/

Hansen, A. E. (1994). "Ancient Illiteracy". In A. K. Bowman and G. Woolf (eds.) *Literacy and Power in the Ancient World*. Cambridge: Cambridge University Press.

Harris, W. V. (Ed.) (1993) *The Inscribed Economy: Production and Distribution in the Roman Empire in the Light of Instrumentum Domesticum*. Journal of Roman Archaeology Supplementary Series **6**. Portsmouth Rhode Island.

Healy, J. F. (1978). *Mining and Metallurgy in the Greek and Roman World*. London: Thames and Hudson.

Helen, T. (1975). "Organization of Roman Brick Production in the First and Second Centuries AD- An Interpretation of Roman Brick Stamps." *Annales Academiae Scientiarum Fennicae- Dissertationes Humanarum Litterarum* **5**. Helsinki.

Herz, N. and E. Garrison. (1998). *Geological Methods for Archaeology*. Oxford: Oxford University Press.

Hodder, I. (1972). "Locational models and the study of Romano-British setlements". In D. Clark (ed) *Models in archaeology*. London: Methuen.

Hodder, I. (1974). "Some marketing models for Romano-British coarse pottery". *Britannia* **5**: 240-59.

Holland, L. and L. Holland (1950) "Down the Tiber on a Raft" *Archaeology* **3** (2): 87-94.

Hong, S., Jean-Pierre Candelone, Clair C. Patterson, Claude F. Boutron (1994). "Greenland Ice Evidence of Hemispheric Lead Pollution Two Millennia Ago by Greek and Roman Civilizations." *Science, New Series* **265** (5180) : 1841-1843.

Hooton, D.H. and Giorgetta, N.E. (1977). "Quantitative X-Ray Diffraction Analysis by a Direct Calculation Method". *X-Ray Spectrometry* **6**: 2-5.

Horden, P. and N. Purcell. (2000). *The Corrupting Sea: A Study of Mediterranean History*. Oxford: Blackwell.

Hughson, J. and C. Bond. (1987) *Hurling Down the Pine*. Historical Society of the Gatineau, Chelsea Quebec.

Janssen, M., M. Scheffer, and T. Kohler. (2002). "Sunk-cost Effects Made Ancient Societies Vulnerable to Collapse". URL: http://www.santafe.edu/sfi/publications/ Working-Papers/02-02-007.pdf [Jul. 17th 2002].

Kahane, A., Threipland, M. and J. B. Ward-Perkins (1968). "The Ager Veientanus, North and East of Veii". *Papers of the British School at Rome* **36**.

Kahane, A. and J. B. Ward-Perkins. (1972). "The Via Gabina." *Papers of the British School at Rome* **40**: 91-126.

Kahane, A. (1977). "Field Survey of an Area South and West of La Storta." *Papers of the British School at Rome* **45**: 138-90.

Kehoe, D. (1988). *The economics of agriculture on Roman imperial estates in North Africa.* Hypomnemata. Heft **89**. Göttingen: Vandenhoeck & Ruprecht.

Kehoe, D. (1997). *Investment, Profit, and Tenancy: The Jurists and the Roman Agrarian Economy*, Ann Arbor MI: University of Michigan Press.

Knoke, D. and J. Kuklinski. (1991). "Network Analysis: Basic Concepts". In G. Thompson, et al (eds). *Markets, Hierarchies and Networks: The Coordination of Social Life*. London: SAGE Publications in association with The Open University.

La Folette, L. (1999) "Thermae Decianae" in E. M. Steinby (ed) *Lexicon Topographicum Urbis Romae* V (T-Z). Rome: Qasar.

Laurence, R. (1997). "Writing the Roman Metropolis". In H. Parkins (ed*). Roman urbanism : beyond the consumer city*. London: Routledge.

Laurence, R. (1999). *The roads of Roman Italy : Mobility and Cultural Change*. London: Routledge.

Laurence, R. (2001a) "Roman Italy's Urban Revolution" In E. Lo Cascio and A. S. Marino (eds). *Modalit insediative e strutture agrarie nell'italia meridionale in et romana.* Bari: Edipuglia.

Laurence, R. (2001b). "The Creation of Geography: An Interpretation of Roman Britain" in C. Adams and R. Laurence (eds). *Travel and Geography in the Roman Empire*. London: Routledge.

Le Gall, J. (1953). *Le Tibre, fleuve de Rome dans l'antiquité*. Publications de l'Institut d'art et d'archéologie de l'Université de Paris 1. Paris: Presses universitairées de France.

Lewin, R. (1993). *Complexity: Life at the Edge of Chaos*. London: J.M. Dent.

Ling, R. (1990). "A Stranger in Town: Finding the Way in an Ancient City." *Greece and Rome* **37**(2): 204-214.

Lirb, Huib J. (1993). "Partners in agriculture : the pooling of resources in rural « societates » in Roman Italy". In H Sancisi-Weerdenburg, R. J. Van der Spek, H. C. Teitler and Herman T. Wallinga (eds). *De agricultura : in memoriam Pieter Willem De Neeve (1945-1990)*. Amsterdam: Gieben.

Lucas, C. (2001). *Self-Organizing Systems (SOS) FAQ: Frequently Asked Questions Version 2.6 August 2001*. USENET newsgroup comp.theory.self-org-sys. [Dec. 8th, 2001]

Lynch, K. (1960) *The Image of the City*. Cambridge MA.: M.I.T. Press & Harvard University Press.

Manacorda, D. (1993). "Appunti sulla bollatura in età romana". In W. V. Harris (ed) *The Inscribed Economy: Production and Distribution in the Roman Empire in the Light of Instrumentum Domesticum*. Journal of Roman Archaeology Supplementary Series **6**. Portsmouth Rhode Island.

Martin, A. (1999). "Ceramic Building Materials". In D. and S. Sorren (eds) *A Roman Villa and a Late Roman Infant Cemetery: Excavation at Poggio Gramignano, Lugnano in Teverina*. Rome: Erma di Bretschneider.

Massey, D., John Allen, Steve Pile, eds. (1999). *City Worlds: Understanding Cities 1*. London: Routledge.

Mattingly, D. and J. Salmon. (2001). *Economies beyond agriculture in the classical world.* London: Routledge.

McFadzean, D. and L. Tesfatsion. (1997). "An Agent-Based Computational Model for the Evolution of Trade Networks". In P. Angeline, R. Reynolds, J. McDonnell, and R. Eberhart (eds). *Evolutionary Programming VI* Proceedings of the Sixth International Conference on Evolutionary Programming. Berlin: Springer-Verlag.

McGrail, S. (1989). "The Shipment of Traded Goods and of Ballast in Antiquity." *Oxford Journal of Antiquity* **8**: 353-358.

Meiggs, R. (1982). *Trees and Timber in the Ancient Mediterranean World*. Oxford: Clarendon Press.

Millar, F. (1992). *The Emperor in the Roman World (31 BC - AD 337)*. 2nd edition. London: Duckworth.

Millett, M. (1982). "Town and Country: A Review of Some Material Evidence". In D. Miles (ed). *The Romano-British Countryside: Studies in Rural Settlement and Economy.* BAR British Series **103**. Oxford: BAR Publishing.

Millett, M. (1990) The Romanization of Britain : an essay in archaeological interpretation. New York: Cambridge University Press.

Mocchegiani Carpano, C. (1984) "Il Tevere, archeologia e commercio". *Bollettino di Numismatica* **2-3**: 21-83.

Mocchegiani Carpano, C. (2002) *Tredicesimo Itinerario: Il Tevere. RomArcheologica: Guida alle antichità della città eterna* Rome: Elio de Rosa.

Monacchi, D. (1999). "Bolli Laterizi". In D. and S. Sorren (eds) *A Roman Villa and a Late Roman Infant Cemetery: Excavation at Poggio Gramignano, Lugnano in Teverina*. Rome: Erma di Bretschneider.

Morley, N. (1996). *Metropolis and hinterland: The city of Rome and the Italian economy 200 B.C. - A.D. 200.* Cambridge: Cambridge University Press.

Nardi, G. (1980). *Le antichità di Orte : esame del territorio e dei materiali archeologici*. Rome: Consiglio nazionale delle ricerche, Centro di studio per l'archeologia etrusco-italica.

Ogilvie, R. M. (1965). "Eretum." *Papers of the British School at Rome* **33:**70-112.

Olcese, G. (1993). "Archeologia e archeometria dei laterizi bollati urbani: primi risultati e prospettive di ricerca". In W. V. Harris (ed) *The Inscribed Economy: Production and Distribution in the Roman Empire in the Light of Instrumentum Domesticum.* Journal of Roman Archaeology Supplementary Series **6**. Portsmouth Rhode Island.

Padgett, J. (2001). "Modeling Florentine Republicanism". URL: http://www.santafe.edu/sfi/publications/ Working-Papers/01-02-008.pdf [Aug. 5th 2002]

Padgett, J. and C. Ansell. (1993). "Robust Action and the Rise of the Medici, 1400-1434". *American Journal of Sociology* **98**.6: 1259-1319.

Page, S. (1998). "On the Emergence of Cities". URL: http://www.santafe.edu/sfi/publications/ Working-Papers/98-08-075.pdf [Sept. 4th 2002]

Palombi, D. (1999). "Thermae Aurelianae". In E.M. Steinby (ed). *Lexicon Topographicum Urbis Romae* V (T-Z). Rome: Quasar, Roma.

Parca, M. (2001). "Local languages and native cultures". In J. Bodel (ed). *Epigraphic Evidence: Ancient History from Inscriptions.* London, Routledge.

Parker, A.J. (1992). Ancient Shipwrecks of the Mediterranean and the Roman Provinces. BAR International Series **580**. Oxford: BAR Publishing.

Pastor-Satorras, R. and A. Vespignani. (2001). "Epidemic Spreading in Scale-Free Networks". *Physical Review Letters* **86**.14: 3200-3203.

Patterson, H. and M. Millett. (1998). "The Tiber Valley Project." *Papers of the British School at Rome* **66**: 1-20.

Patterson, H., R. Witcher, and H. di Giuseppe. (forthcoming) *Tiber Valley Project database CD-ROM, Version 1.*

Peacock, D. P. S. (1979). "An Ethnoarchaeological Approach to the Study of Roman Bricks and Tiles". In A. McWhirr (ed). *Roman Brick and Tile – Studies in Manufacture, Distribution and Use in the Western Empire*. BAR International Series, **68**. Oxford: BAR Publishing.

Peña, J. T. (1987). *Roman-period ceramic production in Etruria Tiberina: a geographical and compositional study*, Ph.D. Thesis, University of Michigan.

Peña, J. T. (1995). "The Organization of Pottery Production in Roman South Etruria". In P. Vincenzini (ed). *The Ceramics Cultural Heritage-Proceedings of the International Symposium The Ceramics Heritage of the 8th CIMTEC – World Ceramics Congress and Forum on New Materials Florence, Italy June 28 – July 2 1994*. Faenza Italy: Techna..

Pitts, F. (1978). "The Medieval River Trade Network of Russia Revisited". *Social Networks* **1**: 285-292.

Potter, T. W. (1979). *The Changing Landscape of Southern Etruria.* London: P. Elek.

Potter, T. W. and A. C. King, eds. (1997). *Excavations at the Mola di Monte Gelato. A Roman and Medieval Settlement in South Etruria.* London: Archaeological Monographs of the British School at Rome.

Powell, W. (1991). "Neither Market nor Hierarchy: Network Forms of Organization". In G Thompson (ed). *Markets, Hierarchies and Networks: The Coordination of Social Life.* London: SAGE Publications in association with The Open University.

Proietti, A. (1990). "Prodotti laterizi". In L. Sagui and L. Paroli (eds). *Archaeologia urbana a Roma: il progetto della Crypta Balbi.* Firenze.

Pucci, G. (2001). "Inscribed *instrumentum* and the ancient economy". In J. Bodel (ed). *Epigraphic Evidence: Ancient history from inscriptions.* London: Routledge.

Pujol, J. A. Flache, J. Delgado and R. Sangüesa (2005) 'How Can Social Networks Ever Become Complex? Modelling the Emergence of Complex Networks from Local Social Exchanges' *Journal of Artificial Societies and Social Simulation* 8.4 http://jasss.soc.surrey.ac.uk/8/4/12.html

Quilici, L. (1986). Il Tevere e l'aniene come vie d'acqua a monte di Roma in età imperiale. *Archeologia laziale*. **7** (2): 198-217.

Rappaport, R. (1984). *Pigs for the Ancestors: Ritual in the Ecology of a New Guinea People*, 2nd edition (first published 1968), New Haven: Yale University Press.

Rathbone, D. (1985) "Roman Farming". *Classical Review* **35** (2): 329-30.

Rathbone, D. (1991). *Economic Rationalism and Rural Society in Third-Century Egypt: The Heroninos*

Archive and the Appianus Estate. Cambridge: Cambridge University Press.

Raybould, M. E. (1999). A Study of Inscribed Material from Roman Britain. An Inquiry into Some Aspects of Literacy in Romano-British Society. BAR British Series **281.** Oxford: BAR Publishing.

Reid, R. M. (Ed.) (1990) *The Upper Ottawa Valley to 1855*, Ottawa: Carleton University Press.

Rickman, G. (1980). *The Corn Supply of Ancient Rome*. Oxford: Clarendon Press.

Rihll, T.E. and A. G. Wilson (1991). "Modelling settlement structures in ancient Greece : new approaches to the polis" In J. Rich and A. Wallace-Hadrill (eds). *City and country in the ancient world*. London : Routledge.

Rosenthal, N., M. Fingrutd, M. Ethier, R. Karant, and D. McDonald. (1985). "Social Movements and Network Analysis: A Case Study of Nineteenth-Century Women's Reform in New York State". *American Journal of Sociology* **90**: 635-48.

Rudling, D. (1986). "The Excavation of a Roman Tilery on Great Cansiron Farm, Hartfleld, Sussex". *Britannia* **17**: 191-230.

Sallares, R. (1991). *The Ecology of the Ancient Greek World* London: Duckworth.

Saller, R. (1989). "Patronage and friendship in early imperial Rome: drawing the distinction." In A. Wallace-Hadrill (ed). *Patronage in Ancient Society*. London, Routledge.

Sartorio, G. (1996). "Muri Aureliani". In E. M. Steinby (ed). *Lexicon Topigraphicum Urbis Romae* III (H-O). Rome: Qasar.

Setälä, P. (1977). "Private Domini in Roman Brick Stamps of the Empire: A Historical and Prosopographical Study of Landowners in the District of Rome." *Annales Academia Scientiarum Fennicae- Dissertationes Humanorum Litterarum* **10**. Helsinki.

Shennan, S. (1997). *Quantifying Archaeology*. Edinburgh: Edinburgh University Press.

Simpson, D. P. (1991). *Cassell's Latin Dictionary*, 5th edition. London: Cassell.

Steinby, E. M. (1977). "Lateres Signati Ostienses". *Acta instituti romani finlandiae* **7**.

Steinby, E. M. (1982). "I Senatori e l'Industria Laterizia Urbana." *Tituli* **4**: 227-237.

Steinby, E. M. (1987). "Indici complementari ai bolli doliari urbani (CIL. XV,1)". *Acta Instituti Romani Finlandiae* **11**.

Steinby, E. M. (1993). "L'Organizzazione produttiva dei laterizi: un modello interpretativo per l'instrumen in genere?" In W.V. Harris (ed). *The Inscribed Economy: Production and Distribution in the Roman Empire in the Light of Instrumentum Domesticum.* Journal of Roman Archaeology Supplementary Series **6**. Portsmouth Rhode Island.

Steinby, E.M. (1974) "La cronologia delle "figlinae" doliari urbane dalla fine dell'età repubblicana fino all'inizio dell III secolo" *Bullettino della Commissione Archeologica Comunale di Roma*, **84**: 7-132.

Steinby, E.M. (1978) "Zeiglestempel von Rom und Umgebung" *Reale Encyclopaedia* Suppl. **XV** cols. 1489-1531.

Steinby, E.M. (1981) "La diffusione dell'opus doliare urbano" In A. Giardina and. A. Schiabone (eds). *Merci, Mercati e Scambi nel Mediterraneo*, **2**.

Stiell, W. (1984) "Timber Marks" In *Ottawa Valley Forestry: 1984 Forestry Capital of Canada.* Petawawa Ontario: Ottawa Valley Forestry Capital Promotion Society.

Temin, P. (2001). "A Market Economy in the Early Roman Empire". *Journal of Roman Studies* **91**: 169-181.

Tesfatsion, L. "Agent-Based Computational Economics". URL: http://www.econ.iastate.edu/tesfatsi/ace.htm [Sept. 8th 2001]

Thébert, Y. (2000). "Transport à grande distance et magasinage de briques dans l'empire Romain: Quelques remarques sur les relations entre production et consomation". In P. Bocheron, Henri Broise, and Yvon Thébert (eds). *La Brique Antique et Médiévale: Production et Commercialisation d'un Matérlau.* Rome: École Française de Rome.

Theilheimer, Ish. (1984) "Timber on the Move: From River Drive to Overdrive" In *Ottawa Valley Forestry: 1984 Forestry Capital of Canada.* Petawawa Ontario: Ottawa Valley Forestry Capital Promotion Society.

Torrens, P. M. (2000). How Cellular Models of Urban Systems Work. 1: Theory. *Centre for Advanced Spatial Analysis Working Paper Series, #28*. URL:

http://www.casa.ucl.ac.uk/working_papers.htm [Jul. 27th 2001].

UNESCO. (1988). Compendium of Statistics on Illiteracy. Paris. 1988.

Velde, B. and I. Druc. (1999). *Archaeological Ceramic Materials, Origin and Utilization.* Boston: Springer.

Wallace-Hadrill, A. (1989) "Patronage in Roman society: from Republic to Empire" In A. Wallace-Hadrill (ed). *Patronage in Ancient Society*. London: Routledge.

Wallace-Hadrill, A. (1991). "Elites and trade in the Roman town". In J. Rich and A. Wallace-Hadrill (eds). *City and country in the ancient world.* London : Routledge.

Watts, D. (2000). "A Simple Model of Fads and Cascading Failures" URL: http://www.santafe.edu/sfi/publications/ Working-Papers/00-12-062.pdf [May 25th 2002]

Watts, D. J., and Strogatz, S. (1998). "Collective dynamics of 'small-world' networks". *Nature* **393**: 440-42.

Watts, D.J. (1999). *Small Worlds: The Dynamics of Networks between Order and Randomness.* Princeton Studies in Complexity. Princeton: Princeton University Press.

Weinreb, B. and. C. Hibbert, eds. (1995). *The London Encyclopaedia.* London: Macmillan.

Whittaker, C. R. (1995). "Do theories of the ancient city matter?". In T. Cornell and H. Lomas (eds). *Urban Society in Roman Italy*. London: University College London Press.

Whittaker, C.R. (1990) "The consumer city revisited: the *vicus* and the city". *Journal of Roman Archaeology* **3**: 110-117.

Wilensky, U., Mitchel Resnick (1998). "Thinking in Levels: A Dynamic Systems Approach to Making Sense of the World." *Journal of Science Education and Technology* **8**(1) 3-18.

Yeo, C.A. (1946). "Land and Sea Transportation in Imperial Italy". *Transactions of the American Philological Association* **77**:221-44.

Zaccaria, C. and C. Gomezel. (2000). "Aspetti della produzione e circolazione del laterizi nell'area adriatica settentrionale tra II secolo A.C. et II secolo D.C." In In P. Bocheron, Henri Broise, and Yvon Thébert (eds). *La Brique Antique et Médiévale: Production et Commercialisation d'un Matériau.* Rome: École Française de Rome.

Zanette, D. (2001). "Critical Behaviour of Propagation on small-World Networks". URL: http://xxx.lanl.gov/cond-mat/0105596, May 30. [Aug. 18 2002]

www.ingramcontent.com/pod-product-compliance
Lightning Source LLC
LaVergne TN
LVHW071213210926
848482LV00015B/800